Tyrosine Phosphorylation/ Dephosphorylation and Downstream Signalling

NATO ASI Series

Advanced Science Institutes Series

A series presenting the results of activities sponsored by the NATO Science Committee, which aims at the dissemination of advanced scientific and technological knowledge, with a view to strengthening links between scientific communities.

The Series is published by an international board of publishers in conjunction with the NATO Scientific Affairs Division

A Life Sciences B Physics	Plenum Publishing Corporation London and New York
C Mathematical and Physical Sciences D Behavioural and Social Sciences E Applied Sciences	Kluwer Academic Publishers Dordrecht, Boston and London
F Computer and Systems Sciences G Ecological Sciences H Cell Biology I Global Environmental Change	Springer-Verlag Berlin Heidelberg New York London Paris Tokyo Hong Kong Barcelona Budapest

NATO-PCO DATABASE

The electronic index to the NATO ASI Series provides full bibliographical references (with keywords and/or abstracts) to more than 30000 contributions from international scientists published in all sections of the NATO ASI Series. Access to the NATO-PCO DATABASE compiled by the NATO Publication Coordination Office is possible in two ways:

- via online FILE 128 (NATO-PCO DATABASE) hosted by ESRIN, Via Galileo Galilei, I-00044 Frascati, Italy.
- via CD-ROM "NATO Science & Technology Disk" with user-friendly retrieval software in English, French and German (© WTV GmbH and DATAWARE Technologies Inc. 1992).

The CD-ROM can be ordered through any member of the Board of Publishers or through NATO-PCO, Overijse, Belgium.

Series H: Cell Biology, Vol. 76

Tyrosine Phosphorylation/ Dephosphorylation and Downstream Signalling

Edited by

Ludwig M. G. Heilmeyer Jr.

Ruhr-Universität Bochum
Medizinische Fakultät
Institut für Physiologische Chemie
Abteilung für Biochemie Supramolekularer Systeme
D-44780 Bochum, Germany

Springer-Verlag
Berlin Heidelberg New York London Paris Tokyo
Hong Kong Barcelona Budapest
Published in cooperation with NATO Scientific Affairs Division

Proceedings of the NATO Advanced Study Institute on Tyrosine Phosphorylation/Dephosphorylation and Downstream Signalling, held at Acquafredda di Maratea, Italy, September 1–12, 1992

Library of Congress Cataloging-in-Publication Data
Tyrosinephosphorylation/dephosphorylation and downstream signalling / edited by Ludwig M. G. Heilmeyer Jr.
p. cm. -- (Nato ASI series. Series H, Cell biology : voll. 76)
"Proceedings of the NATO Advanced Study Institute of Tyrosine Phosphorylation/Dephosphorylationand Downstream Signalling, held at Acquafredda di Maratea, Italy, September 1-12, 1002" -- T. p. verso.
"Published in cooperation with NATO Scientific Affairs Division." Includes bibliographical references and index.
ISBN-13: 978-3-642-78249-7 e-ISBN-13: 978-3-642-78247-3
DOI: 10.1007/ 978-3-642-78247-3

1. Tyrosine--Congresses. 2. Phosphorylation--Congresses. 3. Protein-tyrosine kinase-Congresses. 4. Posttranslational modification--Congresses. I. Heilmeyer, L. M. G. (Ludwig M. G.), 1937- . II. NATO Advanced Study Institute on Tyrosine phosphorylation/Dephosphorylation and Downstream Signalling (1992 : Acquafredda di Maratea, Italy) III. Series.
QP562.TT9T97 1993
574.87'6042--dc20 93-15958

Softcover reprint of the hardcover 1st edition 1993

Typesetting: Camera ready by authors
31/3145 - 5 4 3 2 1 0 - Printed on acid-free paper

PREFACE

Protein phosphorylation/dephosphorylation has come to be recognized as the most prevalent mechanism by which nearly every cellular activity is regulated. The first regulated protein kinase was detected by Fischer and Krebs nearly 40 years ago; for this fundamental observation the Nobel Prize was awarded to these two extraordinary researchers in 1992. In cooperation with Prof. Fischer two NATO Advanced Study Institutes were already held on the topic of protein phosphorylation. The first one took place on the island of Spezai/Greece in September 1986 and was published in the NATO ASI Series "Signal Transduction and Protein Phosphorylation" (L.M.G. Heilmeyer, Jr. (ed.), Plenum Press New York, London 1987). The second one was held in La Londe les Maures/France and was published in the NATO ASI series "Cellular Regulation by Protein Phosphorylation" (L.M.G. Heilmeyer, Jr. (ed.), Springer Verlag Heidelberg, 1991). Meanwhile the area of protein phosphorylation has widened so much that only a limited aspect could be covered in the third NATO Advanced Study Institute on "Tyrosine Phosphorylation/Dephosphorylation and Downstream Signalling", which took place in Acquafredda di Maratea/Italy. In the previous meetings, serine/threonine phosphorylation was the main topic; tyrosine phosphorylation was more difficult to deal with since quantitatively serine/threonine phosphorylation greatly exceeds the level of tyrosine phosphorylation. However, it is now well recognized that this low amount of protein tyrosine phosphorylation regulates such diverse processes as nuclear division, protein synthesis, cell growth, proliferation, etc. It was therefore a challenge to review this area in a third NATO Advanced Study Institute and to visualize how these tyrosine phosphorylation signals are further transduced into the cell.

This volume covers the presentations that were made during this third NATO Advanced Study Institute. The primary structure and regulation of the various families of protein tyrosine kinases including the receptors for hormones and growth factors were reviewed emphasizing their relationship to different types of serine/threonine kinases. Many of these have oncogenic forms which were discussed in detail due to their importance in cell proliferation and transformation. Rapidly growing is the field of protein tyrosine phosphatases; their structure has been recently elucidated. Therefore a major part of the presentations concerned this topic.

A variety of tyrosine phosphorylated proteins are connected to the phosphatidylinositol signalling system thus allowing to influence many cellular reactions which opend the discussion on downstream signalling. Some serine/threonine protein kinases and protein phosphatases were also discussed in detail, especially S6 kinase as a means to regulate protein synthesis. The complexity of these interacting signalling pathways was covered in several lectures concerning the cell cycle control which summarized all parts of signal transduction so far discussed during the meeting.

The organizers would like to express their thanks to the coworkers for their help in the preparation of the manuscripts. We are especially grateful to Mrs. Humuza who organized these contributions in the camera ready form and to Mrs. Rosenbaum for her part in the practical organization.

February 1993 The organizers

Contents

I. Protein Tyrosine Kinases and Protein Tyrosine Phosphate Phosphatases

II. The Phosphatidylinositol Signalling System

III. Serine/Threonine Protein Kinases

IV. Protein Serine/Threonine Phosphate Phosphatases

V. S6-Phosphorylation

VI. Cell-Cycle-Control

VII. Control of Intracellular Ca^{2+}

I. Protein Tyrosine Kinases and Protein Tyrosine Phosphate Phosphatases

PROTEIN TYROSINE PHOSPHORYLATION-DEPHOSPHORYLATION AND CELL SIGNALING

Fischer, E.H., Zander, N.F., Lorenzen, J., Dadabay, C. and Cool, D.E.
Department of Biochemistry
University of Washington
Seattle, Washington 98195
USA

Tyrosine phosphorylation, discovered a little over ten years ago, has represented one of the most exciting developments in the field of cellular regulation. It has been well established that receptor or membrane-bound tyrosine kinases afford primary signals capable of affecting the activity of secondary enzymes downstream ultimately leading to cell growth, proliferation and differentiation. Furthermore, overexpression or constitutive activation by mutation of the receptor-linked or low molecular weight enzymes can result in oncogenic transformation (for review, see 1-5). However, for these processes to be fully regulated, there must be protein tyrosine phosphatases (PTPs) to catalyze the reverse reaction. This article addresses some structural characteristics of these enzymes and their involvement in cell cycle and transformation (for review, see 6-9). When this work was initiated, it was assumed that if transformation could be brought about by overexpression of tyrosine kinases, then overexpression of the phosphatases might reverse these reactions. As will be seen from this report, this is not necessarily true.

First described in the late '70's, a PTP was isolated in homogeneous form from human placenta. The enzyme was of low Mr (37 kDa) and totally specific for phosphotyrosyl residues. Its specific activity, like those of other low molecular weight PTPs that have been identified since then, was up to 3 orders of magnitude higher than those of the cellular or viral protein tyrosine kinases (PTKs), or the kinases linked to hormone receptors. Therefore, these enzymes would be expected to be tightly regulated to allow for those signals that are necessary for normal cell development.

The amino acid sequence of the placenta enzyme showed no structural relationship with any of the Ser/Thr protein phosphatases (10) but was homologous to the cytoplasmic domain of the leukocyte common antigen (11), also designated as CD45, T-200 or B-220. CD45 is present in all hematopoietic cells except mature erythrocytes; it has been implicated in modulating lymphocyte function, including cytotoxicity, proliferation and

NATO ASI Series, Vol. H 76
Tyrosine Phosphorylation/Dephosphorylation
and Downstream Signalling
Edited by L. M. G. Heilmeyer Jr.

differentiation but, depending on the systems employed and ligands used, proliferation can be either enhanced or inhibited (12).

Since then, many transmembrane forms have been identified (Fig. 1); with one exception, they all contain two conserved intracellular catalytic domains (6-9). Their extracellular segments suggest that they might be implicated in cell-cell or cell matrix interaction and metabolic regulation (6-9). One of these receptor-linked enzymes, RPTPα, is widely distributed and its CDNA has been isolated by a number of groups including that of J. Schlessinger with whom we have been collaborating (13). RPTPα was expressed using the baculovirus system, purified to homogeneity and some of its enzymatic properties have been characterized (14). Each of the cytoplasmic domains have been expressed in E. coli and their catalytic properties investigated (15).

A number of non-receptor intracellular PTPs have also been identified, starting with the 48 kDa TC-PTP cloned from a human T-cell cDNA library (16). They all contain a single, highly conserved catalytic domain but display a surprising diversity in their non-catalytic segments undoubtedly implicated in their regulation and localization (17) (Fig. 2). The T-cell and placenta enzymes (TC-PTP and PTP1B, respectively) have highly hydrophobic segments at their C-termini that might serve to anchor them to the ER or other subcellular structures (18,19). Recently, cDNAs for murine and human TC-PTP lacking the C-terminal hydrophobic stretch has been described (20). Some PTPs have segments homologous to cytoskeletal proteins such as band 4.1, ezrin and talin (21,22); others contain two SH2 (src-homology) domains (23-25) which might allow them to interact with phosphotyrosyl residues at sites of autophosphorylation of hormone or growth factor receptors (26). Tyrosine phosphatases are also found as the gene products (YOPs) of virulence plasmids from bacteria of the genus Yersinia (such as Y. pestis responsible for the bubonic plague) (27,28). One can surmise that information for pathogenicity must reside in the regulatory segments of these enzymes since overexpression of the catalytic domain alone would not be expected to have such drastic consequences.

PTPs in Signal Transduction

T-lymphocyte signaling was considerably enhanced when CD45 was cross-linked to the surface antigen CD4 (29) whose cytoplasmic domain is associated with the lymphocyte specific tyrosine kinase p56^{lck} (30), a src family kinase. The activity of these enzymes is repressed by tyrosine phosphorylation at the C-terminus (31,32). Thus, the cross-linking results leading to T-cell stimulation could be explained if CD45 catalyzed the dephosphorylation of tyrosine 505 of p56^{lck}, thereby causing its activation. This conclusion was greatly reinforced by data from Matt Thomas and Arthur Weiss' laboratories showing that a mutant of a human T-cell leukemia line lacking CD45 failed to signal through the T-cell receptor and displayed no phosphatidyl-inositol turnover in response to T-cell receptor stimulation (33,34). Particularly striking was the fact that in these cells lacking CD45, the overall level of tyrosine phosphorylation was greatly reduced (35). This finding, as well as others to be described below, indicates that, contrary to expectations, PTPs are capable of causing a positive as well as a negative response.

The cDNA of TC-PTP was cloned into various expression vectors (17,36). When expressed, the enzyme was associated with the particulate fraction of cells and was essentially inactive when assayed in vitro with certain artificial substrates unless subjected to limited proteolysis which removes a ca. 11 kDa fragment from the C-terminus of the molecule. Introduction of a premature stop codon into the cDNA to delete this carboxyl end gave a fully active, soluble enzyme species (C11). The C-terminal tail is entirely hydrophilic until one reaches the 19 last residues which are highly hydrophobic, and reminiscent of a transmembrane segment. The data indicate that the C-terminal segment plays an important role in determining the localization and regulation of TC-PTP (17).

PTPs in cell cycle progression

The morphology of BHK cells transfected with vector alone, wild-type TC.PTP or the truncated, deregulated C11 were remarkably different. With the truncated form, cells were cobblestone-shaped and many were multinucleate apparently due to a defect in cytokinesis. Remarkably, the syncytial nuclei enter mitosis asynchronously: often, sets of mitotic chromosomes co-exist with interphase nuclei (37). No such asynchrony was seen when multinucleation was induced by cytochalasin or in cells overexpressing the full-length PTP. It is not known at present what causes the failure in cytokinesis nor what disrupts the internuclear signaling mechanism. Possible points of control are at the G1/S and G2/M boundaries regulated in part by $p34^{cdc2}$ in association with various cyclins (38-42). Activation of the histone kinase by tyrosine dephosphorylation (43) induces mitosis, and this process may be required for its translocation to the centrosome in late G2 (44).

Although the full-length enzyme had no obvious effect on cell cycle or morphology, it prevented actin fiber disassembly by dihydrocytochalasin-B as observed in control BHK cells or cells overexpressing the truncated C11 enzyme. While the molecular basis for these effects have not been determined as yet, it is known that actin bundles are linked to focal adhesions through a number of cytoskeletal proteins. Several of these (vinculin, talin, ezrin, paxillin, tensin, etc.) are phosphorylated on tyrosyl residues, as is the ß subunit of the integrin receptor. Phosphorylation correlates with a reduced interaction with actin filaments and the extracellular matrix protein fibronectin (45-47). As of now, the impact of these reactions on focal adhesion organization and on cellular processes, including transformation, remains unclear.

PTPs in cell transformation

A study was undertaken to assess the involvement of PTPs in cell transformation and, in particular, the possible role that the C-terminal regulatory domain of the T-cell enzyme might play in this process. Overexpression of full-length PTP1B was shown by others to block transformation of NIH 3T3 cells infected with retroviruses containing the human neu oncogene (48) or v-src (19). By contrast, our studies indicate that a full-length T-cell PTP had no effect on (or, if anything, even increased) the tumorigenicity of BHK cells in nude mice. On the other hand, expression of the truncated/deregulated enzyme (C.11) inhibited growth on soft agar and suppressed tumor formation in nude mice.

Similar experiments were performed on another system in which the oncogene was known, namely, rat 2 cells transformed with v-fms (in collaboration with Larry Rohrschneider, Fred Hutchinson Cancer Research Center). Here again, overexpression of truncated C11.PTP, but not of wild-type TC.PTP, suppressed the transformed phenotype, as seen by: a) return of the cells from the spindle-shape to the non-transformed cobblestone morphology; b) loss of growth on soft agar; c) reduction if not total suppression of tumor formation in nude mice; and d) increased number of adhesion plaques and actin filaments as determined by immunocytochemistry. Reduced level of phosphorylation of the v-fms receptor was also observed. In a few instances, tumors did appear in nude mice even though the cells in culture contained high levels of C11 and did not grow on soft agar. However, analysis of tumor extracts indicated that while v-fms was present, C11 was no longer detectable.

The above data indicate that phosphatases cannot be viewed as simply providing the "off" switches in a "on-off" kinase/phosphatase system (49). Depending on the type of phosphatases involved and their location within the cell, PTPs can act either synergistically or antagonistically with the tyrosine kinases to elicit a particular physiological response. An understanding of tyrosine phosphatases and their specific involvement in particular steps of the signal pathway leading to oncogenic transformation may suggest therapeutic strategies by which these processes could be brought under control.

REFERENCES

1. Yarden, Y. & Ullrich, A. (1988) *Annu. Rev. Biochem.* **57**, 443-478.
2. Weinberg, R.A. (1990) *Trends Biochem.Sci.* **15**:199-202.
3. Bishop, J. M. (1991) *Cell* **64**, 235-248.
4. Cantley, L. C., Auger, K. R., Carpenter, C., Duckworth, B., Graziani, A., Kapeller, R. & Soltoff, S. (1991) *Cell* **64**, 281-302.
5. Hunter, T. (1991) *Cell* **64**, 249-270.
6. Fischer, E. H., Charbonneau, H. & Tonks, N. K. (1991) *Science* **253**, 401-406.
7. Shenolikar, S. & Nairn, A. (1990) in *Advances in Second Messenger and Phosphoprotein Research*, eds. Y.Nishizuka, M.Endo, & C.Tanaka, (Raven Press, New York, NY), pp. 1-121.
8. Alexander, D. R. (1990) *New Biol.* **2**, 1049-1062.
9. Saito, H. and Streuli, M. (1991) *Cell Growth and Differ.* 2:59-65.
10. Charbonneau, H., Tonks, N. K., Kumar, S., Diltz, C. D., Harrylock, M., Cool, D. E., Krebs, E. G., Fischer, E. H. & Walsh, K. A. (1989) *Proc. Natl. Acad. Sci. USA* **86**, 5252-5256.
11. Charbonneau, H., Tonks, N. K., Walsh, K. A. & Fischer, E. H. (1988) *Proc. Natl. Acad. Sci. USA* **85**, 7182-7186.
12. Thomas, M. L. (1989) *Annu. Rev. Immunol.* **7**, 339-369.
13. Kaplan, R., Morse, B., Huebner, K., Croce, C., Howk, R., Ravera, M., Ricca, G., Jaye, M. & Schelssinger, J. (1990) *Proc. Natl. Acad. Sci. USA* **87**, 7000-7004.
14. Daum G., Zander, N. F., Morse, B., Hurwitz, D., Schlessinger, J. & Fischer, E. H. (1991) *J. Biol. Chem.* **266**:1211-1215.
15. Wang, Y. & Pallen, C. J. (1991) *EMBO J.* **10**, 3231-3237.
16. Cool, D. E., Tonks, N. K., Charbonneau, H., Walsh, K. A., Fischer, E. H. & Krebs, E. G. (1989) *Proc. Natl. Acad. Sci. USA* **86**, 5257-5261.
17. Cool, D. E, Tonks, N. K., Charbonneau, H., Fischer, E. H. & Krebs, E. G. (1990) *Proc. Natl. Acad. Sci. USA* **87**, 7280-7284.
18. Frangloni, J. V., Beahm, P. H., Shifrin, V., Jost, C. A. & Neel, B. G. (1992) *Cell* **68**, 545-560.
19. Woodford-Thomas, T. A., Rhodes, J. D. & Dixon, J. E. (1992) *J. Cell Biol.* **117**, 401-414.
20. Mosinger, B.,Jr., Tillmann, U., Westphal, H. & Tremblay, M. L. (1992) *Proc. Natl. Acad. Sci. USA* **89**, 499-503.
21. Gu, M., York, J. D., Warshawsky, I. & Majerus, P. W. (1991) *Proc. Natl. Acad. Sci. USA* **88**, 5867-5871.
22. Yang, O. & Tonks, N. K. (1991) *Proc. Natl. Acad. Sci. USA* **88**, 5949-5953.
23. Shen, S. -H., Bastien, L., Posner, B. I. & Chretien, P. (1991) *Nature* **352**, 736-739.
24. Plutzky, J., Neel, B. G. & Rosenberg, R. D. (1992) *Proc. Natl. Acad. Sci. USA* **89**, 1123-1127.

25. Yi, T., Cleveland, J. L. & Ihle, J. N. (1992) *Mol. Cell Biol.* **12**, 836-946.
26. Koch, C. A., Anderson, D., Moran, M. F., Ellis, C. & Pawson, T. (1991) *Science* **252**, 668-674.
27. Guan, K. & Dixon, J. E. (1990) *Science* **249**, 553-556.
28. Bliska, J. B., Guan, K., Dixon, J. E. & Falkow, S. (1991) *Proc. Natl. Acad. Sci. USA* **88**, 1187-1191.
29. Ledbetter, J. A., Tonks, N. K., Fischer, E. H. & Clark, E. A. (1989) *Proc. Natl. Acad. Sci. USA* **85**, 8628-8632.
30. Veillette, A., Bookman, M. A., Horak, E. M. & Bolen, J. B. (1988) *Cell* **55**, 301-308.
31. Perlmutter, R. M., Marth, J. D., Ziegler, S. F., Garvin, A. M., Pawar, S., Cooke, M. P. & Abraham, K. M. (1989) *Biochim. Biophys. Acta* **948**, 245-262.
32. Cooper, J. A. & King, C. S. (1986) *Mol. Cell Biol.* **6**, 4467-4477.
33. Pingel, J. T. & Thomas, M. L. (1989) *Cell* **58**, 1055-1065.
34. Koretzky, G. A., Picus, J., Thomas, M. L. & Weiss, A. (1990) *Nature* **346**, 66-68.
35. Koretzky, G. A., Picus, J., Schultz, J. & Weiss, A. (1991) *Proc. Natl. Acad. Sci. USA* **88**, 2037-2041.
36. Zander, N. F., Lorenzen, J. A., Cool, D. E., Tonks, N. K., Daum, G., Krebs, E. G. & Fischer, E. H. (1991) *Biochemistry* **30**, 6964-6970.
37. Cool, D. E., Andreassen, P. R., Tonks, N. K., Krebs, E. G., Fischer, E. H. & Margolis, R. L. (1992) *Proc. Natl. Acad. Sci. USA* **89**, 5422-5426.
38. D'Urso, G., Marraccino, R. L., Marshak, D. R. & Roberts, J. M. (1990) *Science* **250**, 786-791.
39. Fang, F. & Newport, J. W. (1991) *Cell* **66**, 731-742.
40. Riabowol, K., Draetta, G., Brizuela, L., Vandre, D. & Beach, D. (1989) *Cell* **57**, 393-401.
41. Pines, J. & Hunter, T. (1990) *New Biol.* **2**, 389-401.
42. Th'ng, J. P., Wright, P. S., Hamaguchi, J., Lee, M. G., Norbury, C. J., Nurse, P. & Bradbury, E. M. (1990) *Cell* **63**, 313-324.
43. Gould, K. L. & Nurse, P. (1989) *Nature* **342**, 39-45.
44. Bailly, E., Doree, M., Nurse, P. & Bornens, M. (1989) *EMBO J.* **8**, 3985-3995.
45. Tapley, P., Horwitz, A., Buck, C., Duggan, K. & Rohrschneider, L. (1989) *Oncogene* **4**, 325-333.
46. Horvath, A. R., Elmore, M. A. & Kellie, S. (1991) *Oncogene* **5**, 1349-1357.
47. Otey, C., Pavalko, F. M. & Burridge, K. (1990) *J. Cell Biol.* **111**, 721-729.
48. Brown-Shimer, S., Johnson, K. A., Hill, D. E. & Bruskin, A. M. (1992) *Cancer Res.* **52**, 1234-1236.
49. Fischer, E. H. , Charbonneau, H. , Cool, D. E. & Tonks, N. K. (1992) in *Interactions Among Cell Signalling Systems*, eds. Sato, R. , Block, G. R. & Widdows, K. (John Wiley & Sons, Chichester, UK), pp. 132-144.
50. Chernoff, J., Schievella, A. R., Jost, C.A., Erickson, R. L., & Neal, B. G. (1990) *Proc. Natl. Acad. Sci. USA* **87**:2735-2739.

51. Brown-Shimer S., Johnson, K. A., Lawrence, J. B., Johnson, C., Bruskin, A., Green, N. R. & Hill, D. E. (1990) *Proc. Natl. Acad. Sci. USA* **87**:5148-5152.

52. Ottilie, S., Chernoff, J., Hannig, G., Hoffman, C. & Erikson, R. L. (1991) *Proc. Natl. Acad. Sci. USA* **88**:3455-3459

53. Guan, K., Deschenes, R.J., Qiu H. & Dixon, J.E. (1991) *J. Biol. Chem.* **266**:12964-12970.

54. Guan, K., Broyles, S. S. & Dixon, J.E. (1991) *Nature* **350**:359-362.

55. Russel, P. & Nurse, P. (1986) *Cell* **45**:145-153.

56. Gu, M., Warshawsky, I. and Majerus, P. W. (1992) *Proc. Natl. Acad. Sci. USA* **89**, 2980-2984.

REGULATION AND FUNCTIONS OF SRC-FAMILY KINASES: AN UPDATE

S. Kussick
J.A. Cooper
Fred Hutchinson Cancer Research Center
1124 Columbia Street
Seattle, Washington 98104
USA

Perhaps the most intensively studied tyrosine kinases are the products of the viral and cellular *src* genes. Mutations in v-*src* which abolish catalytic activity also prevent transformation, indicating that the tyrosine kinase activity of the viral oncoprotein p60$^{v\text{-}src}$ is required for its function (Martin, 1970; Sefton *et al.*, 1980; reviewed by Jove & Hanafusa, 1987). Indeed, cells transformed by v-*src* show about a ten-fold increase in the phosphorylation of cellular proteins on tyrosine, compared to normal controls (Hunter & Sefton, 1980; Sefton *et al.*, 1980; Jakobovits *et al.*, 1984). Presumably, the tyrosine phosphorylation of certain critical substrates initiates and maintains the transformed phenotype, since shifting cells transformed by temperature-sensitive mutants of v-*src* to the restrictive temperature causes reversion of transformed morphology within several hours (reviewed by Jove & Hanafusa, 1987).

Intrinsic tyrosine kinase activity is not the only requirement for v-*src* transformation. Membrane localization resulting from myristylation of p60$^{v\text{-}src}$ at Gly 2 is necessary for v-*src* transformation (Cross *et al.*, 1984; Schultz *et al.*, 1985). A Gly 2 to Ala mutation in p60$^{v\text{-}src}$ abolishes transformation without altering the catalytic activity of the molecule or its ability to phosphorylate many proteins in the cell (Kamps *et al.*, 1986), implying that p60$^{v\text{-}src}$ substrates which are critical for transformation reside at the membrane. In addition, certain mutations in a region in the N-terminal half of p60$^{v\text{-}src}$ known as the *src* homology 2, or SH2, domain alter the range of host cells which are sensitive to v-*src* transformation (reviewed in Parsons & Weber, 1989); it has been hypothesized that this region may help target p60$^{v\text{-}src}$ activity to critical substrates in cells.

With the cloning and sequencing of the c-*src* cellular homolog of v-*src*, it became possible to show that p60$^{c\text{-}src}$ is not oncogenic even when overexpressed ten-fold in cells (Shalloway *et al.*, 1984; Coussens *et al.*, 1985; Iba *et al.*, 1985). The oncogenicity of

NATO ASI Series, Vol. H 76
Tyrosine Phosphorylation/Dephosphorylation
and Downstream Signalling
Edited by L. M. G. Heilmeyer Jr.

p60$^{v\text{-}src}$ presumably resides in specific amino acid sequence differences between the two proteins. These differences include the following changes from c-*src* to v-*src* from the Schmidt-Ruppin strain of Rous sarcoma virus (Schwartz *et al.*, 1983; Takeya & Hanafusa, 1983; reviewed in Parsons & Weber, 1989; Cooper, 1990): 1) Arg 95 at the N-terminus of p60$^{c\text{-}src}$ changed to Trp in p60$^{v\text{-}src}$; 2) within the kinase domain, Thr 338 to Ile, Glu 378 to Gly, and Ile 441 to Phe; and 3) the replacement of the 19 C-terminal amino acids of p60$^{c\text{-}src}$ with an unrelated sequence of 12 amino acids in p60$^{v\text{-}src}$. These changes are each sufficient to render p60$^{c\text{-}src}$ partially or fully oncogenic, the most potent change being the substitution of the p60$^{v\text{-}src}$ C-terminus for that of p60$^{c\text{-}src}$. Indeed, the oncoprotein of another acutely transforming retrovirus, the S2-avian sarcoma virus (S2-ASV), is identical to p60$^{c\text{-}src}$ except for the replacement of the C-terminus (Shuntaro *et al.*, 1986).

Tyrosine Phosphorylation and p60$^{c\text{-}src}$ Regulation.

The C-terminus of p60$^{v\text{-}src}$ is one oncogenic feature of the molecule which is well understood. The proto-oncogenic p60$^{c\text{-}src}$ is normally phosphorylated in cells to high stoichiometry at Tyr 527 of the C-terminus (Cooper *et al.*, 1986). Either substitution of Phe for Tyr 527 (Cartwright *et al.*, 1987a; Kmiecik & Shalloway, 1987; Piwnica-Worms *et al.*, 1987), or C-terminal truncation of the molecule before Tyr 527 (Reynolds *et al.*, 1987), leads to roughly a ten-fold increase in p60$^{c\text{-}src}$ catalytic activity and enables the molecule to transform cells. Similarly, dephosphorylation of Tyr 527 in p60$^{c\text{-}src}$ immunoprecipitates leads to roughly a ten-fold stimulation of kinase activity *in vitro* (Cooper & King, 1986), and the dephosphorylation and activation of p60$^{c\text{-}src}$ complexed with polyoma virus middle T antigen (mT) plays a central role in mT transformation of fibroblasts (Bolen *et al.*, 1984). These observations therefore indicated that one way p60$^{v\text{-}src}$ acquired its transforming potential was by elimination of a negative-regulatory phosphorylation site during the transduction of c-*src* sequences by RSV.

It appears that p60$^{v\text{-}src}$ and mutationally activated p60$^{c\text{-}src}$ transform cells through very similar, if not identical, mechanisms. Cells transformed by either molecule show a similar pattern of tyrosine-phosphorylated proteins when analyzed by immunoblotting with antibodies to phosphotyrosine (Kamps & Sefton, 1988).

Membrane localization as a result of Gly 2 myristylation is required for activated p60^{c-src}, like p60^{v-src}, to transform cells. In addition, Tyr 416 in the catalytic domain, the principal *in vitro* autophosphorylation site of both molecules (Smart *et al.*, 1981; Patschinsky *et al.*, 1982), and the major *in vivo* site of tyrosine phosphorylation in p60^{v-src} and activated p60^{c-src} (reviewed by Cooper, 1990), is required for full transformation activity of both molecules (Snyder & Bishop, 1984; Kmiecik & Shalloway, 1987; Piwnica-Worms *et al.*, 1987). Indeed, phosphorylation of Tyr 416 is associated with a two- to three-fold elevation in *in vitro* kinase activity (Kmiecik *et al.*, 1988). Finally, both molecules require a lysine at position 295 in the catalytic domain - the putative ATP-binding site - for enzymatic activity and thus for transforming potential (Kamps *et al.*, 1984; Kamps & Sefton, 1986).

For the most part, the regulatory mechanisms characteristic of p60^{c-src} and p60^{v-src} apply to the other kinases related to *src* that have been investigated. For example, the transforming activities of p56^{lck} (Amrein & Sefton, 1988; Marth *et al.*, 1988) and p59^{hck} (Ziegler *et al.*, 1989) are stimulated by substitution of the Tyr 527 homologs. p56^{lck}, like p60^{src}, must be myristylated at Gly 2 and localized to membranes in order to transform fibroblasts (Abraham & Veillette, 1990). There are important differences, however, between the regulation of p60^{c-src} and the other *src* family kinases. Unlike p60^{c-src}, p56^{lck},and p59^{hck}, p59^{fyn} is not rendered transforming in fibroblasts by substitution at the Tyr 527 homolog (Cheng *et al.*, 1991), and it appears that additional regulatory mechanism(s) inhibit its kinase activity in cells. While the ability of v-*src* (Cross & Hanafusa, 1983; Snyder *et al.*, 1983) and activated c-*src* (Kmiecik & Shalloway, 1987; Piwnica-Worms *et al.*, 1987) to transform fibroblasts is only partially blunted in the absence of Tyr 416, C-terminally mutated p56^{lck} becomes unable to transform fibroblasts when its autophosphorylation site at Tyr 394 is mutated (Abraham & Veillette, 1990), suggesting that phosphorylation at this residue is more important to the activation of p56^{lck} than it is to p60^{src}.

Considerable effort has been made to identify the kinase or kinases which normally phosphorylate Tyr 527 in vertebrate cells. p60^{c-src} itself is capable of autophosphorylating Tyr 527 *in vitro* (Cooper & King, 1986; Cooper & Runge, 1987), although to much lower stoichiometry than the major autophosphorylation site, Tyr 416. When expressed in the budding yeast *Saccharomyces cerevisiae*, wild type p60^{c-src} is phosphorylated to approximately 20% stoichiometry at Tyr 527, in contrast to the nearly 100% phosphorylation believed to occur in fibroblasts (Jove *et al.*, 1987; Cooper

& MacAuley, 1988). A catalytically-inactive mutant of p60$^{c\text{-}src}$ is not detectably phosphorylated in *S. cerevisiae* (Jove *et al.*,1987; Cooper & MaAuley, 1988), suggesting that this yeast lacks endogenous tyrosine kinases capable of phosphorylating Tyr 527, and further suggesting that p60$^{c\text{-}src}$ is capable of autophosphorylating at Tyr 527 *in vivo*. This phosphorylation may be intermolecular, since kinase-inactive p60$^{c\text{-}src}$ co-expressed with catalytically active p60$^{c\text{-}src}$ mutated at Tyr 527 (to Phe) is phosphorylated as extensively at Tyr 527 as wild type p60$^{c\text{-}src}$ alone (Cooper & MacAuley, 1988). It has not been determined whether intramolecular Tyr 527 phosphorylation can occur, nor has it been ruled out that Tyr 527 phosphorylation in yeast is caused by an endogenous tyrosine kinase which is activated by p60$^{c\text{-}src}$.

The identification of the Tyr 527 kinase(s) in vertebrate cells has been considerably more complex than in yeast. Clearly, p60$^{c\text{-}src}$ catalytic activity is not required for high level Tyr 527 phosphorylation in fibroblasts; in a fibroblast cell line derived from mice homozygous for a null mutation in the c-*src* gene, introduced catalytically-inactive chicken p60$^{c\text{-}src}$ is phosphorylated to a similarly high level as in fibroblasts possessing a normal c-*src* gene (Thomas *et al.*, 1991). It is also clear that fibroblasts possess great reserves of Tyr 527-phosphorylating activity, since p60$^{c\text{-}src}$ overexpressed ten-fold in fibroblasts is fully phosphorylated at Tyr 527 and catalytically repressed (Cooper *et al.*, 1985; Iba *et al.*, 1985; Coussens *et al.*, 1985).

Recently, a tyrosine kinase with a singular specificity for the C-terminal tyrosine residues of p60$^{c\text{-}src}$ and the other mammalian *src* family kinases was purified from rat brain (Okada & Nakagawa, 1989). This protein is capable of phosphorylating Tyr 527 of catalytically-inactive p60$^{c\text{-}src}$ to high stoichiometry when the molecules are co-expressed in *S. cerevisiae* (Nada *et al.*, 1991). Cloning and sequencing of the gene encoding this protein, CSK (for c-*src* kinase), revealed a predicted protein tyrosine kinase with considerable similarity to the *src* family kinases with the following major exceptions (Nada *et al.*, 1991): 1) lack of an N-terminal myristylation signal, suggesting the protein is not membrane-associated; 2) lack of a tyrosine homologous to p60$^{c\text{-}src}$ Tyr 416; and 3) lack of the typical *src*-like carboxy terminus, with no Tyr 527 homolog. The absence of the two major tyrosine residues involved in p60$^{c\text{-}src}$ regulation suggests that CSK's activity may be controlled very differently than the activities of the other *src* family kinases in the cell.

Preliminary data suggest that CSK can cause repression of *src* family kinases *in vivo*. In a fibroblast line co-transformed by the oncoprotein p47$^{gag\text{-}crk}$ and overexpressed wild-type p60$^{c\text{-}src}$, transformation is suppressed when CSK is overexpressed in the cells (Sabe *et al.*, 1992b). CSK's action on p60$^{c\text{-}src}$ apparently does not involve stable association of the two proteins (Sabe *et al.*, 1992a). It has been reported that *fyn*-dependent transcription from the interleukin-2 (IL-2) promoter can be blunted by overexpression of CSK (Takeuchi *et al.*, 1992). It will be interesting to see if CSK overexpression can suppress transformation of cells caused by an overexpressed wild-type *src* family kinase. One such cell line is the murine T cell lymphoma line LSTRA, in which p56^{lck} is overexpressed approximately forty-fold relative to normal T cells as a result of a retoviral integration in the 5' flanking region of the *lck* gene (Marth *et al.*, 1985; Voronova, 1986). Suppression of LSTRA cell transformation by CSK would suggest an anti-oncogenic role for CSK.

At present, the extent to which CSK is responsible for C-terminal tyrosine phosphorylation in the various tissues of vertebrates is unknown. Current efforts to generate mice in which the CSK gene has been disrupted, or in which CSK is overexpressed, should contribute greatly to our understanding of CSK function.

In addition to the search for Tyr 527 kinases, a second facet of the studies of p60$^{c\text{-}src}$ catalytic repression has focussed on how the phosphorylated C-terminus of p60$^{c\text{-}src}$ serves to repress catalytic activity. In the absence of a crystal structure of p60$^{c\text{-}src}$ such studies have been indirect, but the general consensus is that the phosphorylated C-terminus serves as an allosteric inhibitor of kinase activity. Early evidence that the tail interacts with other parts of p60$^{c\text{-}src}$, particularly in the N-terminal half of the molecule, came from the following observations: 1) repressed and activated p60$^{c\text{-}src}$ show different patterns after partial proteolytic digestion, suggesting they have distinct conformations (MacAuley & Cooper, 1989); 2) binding of an antibody to the C-terminus of p60$^{c\text{-}src}$ relieves this repression and stimulates catalytic activity approximately ten-fold, perhaps by competing with another site on the molecule for binding to its "tail" (Cooper & King, 1986); 3) a proteolytic fragment of p60$^{c\text{-}src}$ containing only the catalytic domain and C-terminus is derepressed relative to full-length p60$^{c\text{-}src}$ and is no longer activated by antibody binding to the C-terminus, suggesting an inhibitory role for the N-terminus (MacAuley & Cooper, 1989); 4) various deletions and point mutations in the N-terminal half of p60$^{c\text{-}src}$ can lead to derepression of catalytic activity without markedly reducing the stoichiometry of Tyr 527 phosphorylation (Hirai & Varmus, 1990; Nemeth *et al.*, 1989; O'Brien *et al.*, 1990).

A number of the N-terminal mutations which lead to p60$^{c\text{-}src}$ activation localized to the SH2 domain of the molecule (Hirai & Varmus, 1990a; Nemeth *et al.*, 1989; O'Brien *et al.*, 1990), a domain which is conserved in all identified members of the *src*-family kinases. These observations, plus the growing body of data implicating SH2 domains in the binding of proteins to phosphorylated tyrosine residues (reviewed by Koch *et al.*, 1991), has led to the intriguing hypothesis that p60$^{c\text{-}src}$ expression results from an intramolecular interaction between the phosphorylated C-terminus and the SH2 domain. Several lines of evidence from *in vitro* biochemical studies (Roussel *et al.*, 1991) are consistent with such a model. First, a glutathione-S-transferase fusion protein containing the SH2-domain of p60$^{c\text{-}src}$ is able to bind to a phosphorylated synthetic C-terminal peptide, but not to a nonphosphorylated peptide. Second, this phosphorylated peptide is able to bind to the activated c-*src* mutant p60^{Phe527}, but not to wild type p60$^{c\text{-}src}$, implying that a binding site for the peptide is occupied in repressed p60$^{c\text{-}src}$, but not in derepressed p60$^{Phe\ 527}$. Finally, binding of the phosphorylated peptide to p60^{Phe527} reduces its *in vitro* kinase activity, suggesting that this interaction alters the molecule's conformation so as to repress activity. While these data are indeed consistent with the notion of intramolecular repression of catalytic activity, they do not rule out the possibility of *in vivo* repression by some other means, for example, as a result of an intermolecular interaction between the phosphorylated C-terminus of one molecule and the SH2 domain of another, or as a result of the tail interacting (intramolecularly) with the catalytic domain in a manner which cannot be mimicked by protein-peptide binding studies *in vitro*. Until the three-dimensional structure of p60$^{c\text{-}src}$ is determined, the precise structural basis of catalytic repression will remain unresolved.

Possible Functions of src Family Kinases in Cell Proliferation

While much is known about the biochemical characteristics of p60$^{c\text{-}src}$/p60$^{v\text{-}src}$ and related kinases, and a catalog of potential *in vivo* substrates has been assembled, it remains difficult to firmly ascribe a precise biological function to any of the *src* family kinases. The functions of the cellular *src* family kinases have been investigated by studying the properties of cells in which they are normally expressed, by studying factors which alter their activities in those cells, and by studying the characteristics of cells in which the wild-type kinases are overexpressed, or in which various mutants are expressed. With the recent disruptions of a number of *src* family kinases in mouse embryonic stem cells, and the generation of animals lacking both copies of the genes for these kinases, it has become possible to study cellular regulation in the complete absence of one or more of these kinases.

$p60^{c\text{-}src}$ is not essential for cell proliferation. Mice homozygous for c-*src* disruptions are viable (Soriano *et al.*, 1991), and fibroblasts from these animals have been used to generate immortal cell lines (Thomas *et al.*, 1991). These cells show no obvious growth differences compared with fibroblast cell lines from normal rodents, implying that either: 1) $p60^{c\text{-}src}$ does not normally function in the fibroblast mitogenic response; or 2) other molecules in the cell, such as additional members of the *src* family like $p62^{c\text{-}yes}$ and $p59^{fyn}$, subsume the essential *src* function when $p60^{c\text{-}src}$ is missing from the cells.

Although c-*src* is not absolutely necessary for cell proliferation it cannot be ruled out that *src* is necessary for *optimal* proliferation in some cells, since the c-*src* null mice are quite sick compared with normal litter mates (Soriano *et al.*, 1991). In addition, examination of situations in which $p60^{c\text{-}src}$ is activated constitutively or transiently suggest that it *can* function to initiate mitogenesis. For example, v-*src* or activated c-*src* can transform fibroblasts from a broad range of vertebrate organisms (reviewed in Jove & Hanafusa, 1987; Parsons & Weber, 1989; Cooper, 1990), and expression of ts mutants of v-*src* in cells leads to transformation/ proliferation which is rapidly halted following a shift to the non-permissive temperature. Polyoma virus middle T antigen (mT) transforms fibroblasts in culture (reviewed in Cantley, 1991) and endothelial cells in transgenic mice (Bautch *et al.*, 1987), at least in part, by causing the dephosphorylation of Tyr 527 and activation of $p60^{c\text{-}src}$ molecules complexed with mT (Courtneidge, 1985). Finally, a variety of growth factor-dependent hematopoietic cell lines can be made growth factor-independent by expression of v-*src* or activated c-*src* (Engelman & Rosenberg, 1990; Keller & Wagner, 1989; Ohta *et al.*, 1989; Watson *et al.*, 1987).

Other evidence from fibroblasts suggests that endogenous *src*-family kinases may function in the amplification of mitogenic signals transmitted through growth factor receptor tyrosine kinases. First, high level overexpression of c-*src* in fibroblasts increases their saturation density (Johnson *et al.*. 1985), perhaps indicating a role in the response to growth factors. Second, C3H10T-1/2 cells overexpressing $p60^{c\text{-}src}$ have two-fold enhanced DNA synthesis in response to EGF (Luttrell *et al.*, 1988), and about a three-fold increase in the tyrosine phosphorylation of a set of cell proteins following

EGF stimulation (Wilson & Parsons, 1990); this effect is due to phosphorylation and activation of the EGF receptor (Wasilenko *et al.*, 1991), suggesting that p60$^{c\text{-}src}$ can potentiate the EGF mitogenic response. Third, p60$^{c\text{-}src}$, p59^{fyn}, and p62^{yes} are transiently activated approximately two-fold after stimulation of fibroblasts with PDGF (Gould & Hunter, 1988; Kypta *et al.*, 1990). In addition, about 5 to 10 percent of fibroblast p60$^{c\text{-}src}$, p62$^{c\text{-}yes}$, and p59^{fyn} physically associates with PDGF receptors following PDGF stimulation (Kypta *et al.*, 1990), and it may be only the associated molecules which are activated. It is not clear which changes (e.g., phosphorylation of dephosphorylation) occur in the *src*-family kinases to account for this activation. Finally, recent evidence that the *src*-family kinases may play an important role in PDGF-stimulated mitogenesis comes from experiments in which microinjection into fibroblasts of a putative dominant negative, catalytically-inactive mutant of p59^{fyn} greatly reduced DNA synthesis following PDGF treatment (Twamley *et al.*, 1992).

A final set of data suggesting a role for p60$^{c\text{-}src}$ in the cell cycle come from studies of p60$^{c\text{-}src}$ at mitosis. In mitotic fibroblasts isolated either by nocodazole arrest or by shake-off from a monolayer, p60$^{c\text{-}src}$ is transiently activated approximately five-fold relative to interphase cells, and undergoes novel serine/threonine phosphorylation at three sites in the N-terminus of the molecule (Chackalaparampil, 1988). The activation of p60$^{c\text{-}src}$ is apparently due to a reduction in the stoichiometry of phosphorylation at Tyr 527 (Bagrodia *et al.*, 1991); at present, it is not clear whether Tyr 527 dephosphorylation is due to increased phosphatase activity directed against Tyr 527, or decreased activity of Tyr 527 kinases. While there is reasonably good evidence that the mammalian p34^{cdc2} homolog catalyzes the novel N-terminal phosphorylations of p60$^{c\text{-}src}$ at mitosis (Morgan *et al.*, 1989; Shenoy *et al.*, 1989), the functional significance of these phosphorylations is not clear: perhaps they affect the redistribution of p60$^{c\text{-}src}$ from centrosomes during G2 to diffuse positions in the cell at mitosis (David-Pfeuty *et al.*, 1990).

Several lines of evidence suggest that the *src*-related kinase p56^{lck} is important for the proliferation of developing T cells in the thymus. First, overexpression of either activated p56^{lck} mutated at the residue homologous to Tyr 527 of p60$^{c\text{-}src}$ in the thymus of transgenic mice leads to thymic tumorigenesis (Abraham *et al.*, 1991); overexpression of wild-type p56^{lck} has the same effect, but higher levels of expression are needed. Second, high level expression of a kinase-inactive mutant of p56^{lck} blocks the proliferation of immature thymocytes in a dose-dependent manner proportional to the number of transgene copies in the genome (Steven Levin, personal communication); this result may be due to a dominant negative effect of the kinase-inactive mutant on the function of endogenous p56*lck*, perhaps by competing for positive effectors or substrates. Finally, mice homozygous for a disrupted *lck* gene show pronounced thymic atrophy, a dramatic reduction in $CD4^+/CD8^+$ thymocytes, and very few peripheral T cells (Molina *et al.*, 1992).

A second context in which p56*lck* may directly or indirectly influence proliferation occurs when peripheral T cells are stimulated by the coordinated binding of the antigen/MHC complex to the T cell antigen receptor and the co-receptor molecule CD4. As a result of this engagement p56*lck*, which is non-covalently coupled via its N-terminus to the short cytoplasmic domain of CD4 (Veillette *et al.*, 1988; Shaw *et al.*, 1989), is transiently activated, perhaps as a result of the dephosphorylation of its regulatory C-terminal Tyr 505 (reviewed in Klausner, 1991). T cells then go on to produce interleukin-2 (IL-2) receptors, secrete IL-2, and proliferate via an autocrine/paracrine mechanism. p56*lck* may also function in this mechanism of proliferation, as the kinase is also activated in response to IL-2 binding to T cells (Horak *et al.*, 1991). Recent evidence suggests a physical association between p56*lck* and the β chain of the IL-2 receptor (Hatakeyama *et al.*, 1991). Expression of a constitutively active Tyr 505 to Phe mutant of p56*lck* in T cell lines results in hyperresponsiveness of those lines to antigenic stimulation (Abraham *et al.*, 1991), further suggesting that p56*lck* or a similar kinase normally functions in this process.

While the c-*src* and *lck* gene disruptions produce phenotypes in mice, almost no detectable abnormalities result from homozygous disruption of either c-*yes* or *fyn* (Stein *et al.*, 1992), other than a diminished calcium flux in *fyn*-minus T cells following antigenic stimulation. One explanation for these results is that there are certain functions of c-*src* and *lck* which cannot be complemented by other molecules in mice, while the individual functions of c-*yes* and *fyn* can largely be performed by other molecules. Since c-*src*, c-*yes*, and *fyn* are expressed in many of the same cell types (reviewed by Cooper, 1990; Eisenman & Bolen, 1990), it is likely that animals bearing disruptions in two or all three of these genes will show more profound phenotypic abnormalities and illuminate functions which are redundant among these kinases. Preliminary analysis of animals bearing disruptions in two of these genes suggests this is the case (Stein *et al.*, 1992). Functional redundancy between the *hck* and *fgr* genes in myeloid cells may explain why preliminary analysis of mice with individual disruptions in these genes revealed no striking abnormalities (Lowell *et al.*, 1992); again, one might predict considerable abnormalities in myeloid cells from animals bearing the double disruption, assuming that such animals are viable.

While *fyn* is not essential for thymocyte proliferation, since thymic development in *fyn*-null mice appears to be normal, the role of *fyn* in the proliferation of peripheral T cells following antigenic stimulation appears more complex. A potential role for *fyn* in this process is suggested by the following observations: 1) transgenic mice overexpressing *fyn* in thymocytes have T cells which are hyperresponsive to antigen (Cooke *et al.*, 1991); 2) expression of enzymatically-inactive $p60^{fyn}$ in transgenic mice inhibits TCR-mediated signalling (Cooke *et al.*, 1991); and 3) there is diminished calcium flux in *fyn*-minus T cells following antigenic stimulation (Stein *et al.*, 1992). However, the fact that T cells from the *fyn*-disrupted mice do, in fact, secrete IL-2 and proliferate following antigenic stimulation suggests that: 1) other T cell molecules, perhaps the *src* family kinases $p62^{yes}$ and $p56^{lck}$, can assume the function of $p60^{fyn}$, and 2) that wild-type levels of calcium flux are not required for adequate T cell function. Again, studies of mice in which multiple *src* family members have been disrupted may help resolve the issue of functional redundancy.

Possible Functions of *src* Family Kinases: Post-Mitotic Cells

Perhaps the most compelling data concerning the function of $p60^{c\text{-}src}$ and related kinases suggest that they play a variety of roles in the differentiation and mature functioning of cells. A variety of early, circumstantial data indicated that these kinases might not only function in cell proliferation, because they were expressed at high levels in post-mitotic cells. While c-*src* is expressed in almost every vertebrate cell type examined, its level of expression is particularly high in neurons (Cotton & Brugge, 1983) and blood platelets (Golden *et al.*, 1986), both of which are post-mitotic. Moreover, expression of c-*src* increases during the terminal differentiation of monocytes (Barnekow & Gessler, 1986) and neurons (Cartwright *et al.*, 1987b). c-*yes* and *fyn* are also widely expressed, but c-*yes* is most highly expressed in differentiated mast cells, T cells, and natural killer (NK) cells, and *fyn* is most highly expressed in T cells, B cells, NK cells and brain (reviewed in Eiseman & Bolen, 1990). A *yes*-related kinase gene (*yrk*) was recently cloned and, by Northern analysis, is most highly expressed in cerebellum and spleen (Sudol *et al.*. 1992).

Five additional members of the *src* family - *lyn*, *hck*, *lck*, c-*fgr* and *blk* - almost exclusively expressed in hematopoietic cells, and tend to be most highly expressed in differentiated cells (reviewed in Eiseman & Bolen, 1990). For example, the *lyn* gene is most highly expressed in macrophages, monocytes and B cells, and expression is higher in mature B cells than in pre-B cells. Increased *hck* expression in myeloid cells parallels increased abundance of mature cell-surface markers in these cells (Ziegler *et al.*, 1988). As mentioned earlier $p56^{lck}$, the most abundant *src* family kinase in mature T cells, probably functions in the response of these cells to antigen, in addition to its role in early thymocyte proliferation and development. The c-*fgr* gene, like *hck*, is most highly expressed in differentiated myeloid cells, while the *blk* appears to be something of a B lymphocyte equivalent of *lck*, in that it is present in B cell precursors as well as mature B cells.

In addition to the expression data, a variety of experiments have suggested that $p60^{c\text{-}src}$ plays a direct role in specific differentiation processes. In the PC12 pheochromocytoma line, transfection with v-*src* does not lead to transformation but rather causes the cells to extend neurites and express markers of neuronal differentiation (Alema *et al.*, 1985). In addition, the stimulation of PC12 cell differentiation by nerve growth factor (NGF) or fibroblast growth factor (FGF) is blocked by pre-injecting the anti-$p60^{c\text{-}src}$ monoclonal antibody 327 (Kremer *et al.*, 1991), suggesting that the endogenous $p60^{c\text{-}src}$ in PC12 cells plays a normal role in neuronal differentiation. Over half of the $p60^{c\text{-}src}$ in PC12 and other neuronal cell types has a six amino acid insert at position 117 not seen in $p60^{c\text{-}src}$ in non-neuronal cell types (Brugge *et al.*, 1987; Levy *et al.*, 1987; Martinez, 1987), generating a molecule referred to as $p60^{c\text{-}src(+)}$. $p60^{c\text{-}src(+)}$ appears to have considerably higher specific activity than $p60^{c\text{-}src}$ (Brugge *et al.*, 1985), and it may be that $p60^{c\text{-}src(+)}$ is specifically linked to a differentiation function. In addition to *src* effects on neuronal differentiation, it has been shown that polyoma middle T antigen, which is believed to act through *src* family kinases to transform fibroblasts, can trigger the differentiation of ML1 myeloid cells (Gee *et al.*, 1986) and the P19 cell line (Boulter & Wagner, 1988). Thus, the variation in the effects of v-*src* and mT expression from transformation to differentiation suggests that whatever role *src* (and especially activated *src*) might play in differentiation is very complex, and clearly varies with the type of cell in which *src* and mT are acting.

Additional evidence of a role for c-*src* in a differentiated function comes from studies of *src*-null mice (Soriano *et al.*, 1991). In addition to being small and poorly viable compared to their heterozygous or wild-type littermates, homozygotes lacking $p60^{c\text{-}src}$ show the histological and anatomical signs of osteopetrosis in which osteoclast cells are present but dysfunctional. While it is known that c-*src* is highly expressed in the monocytic cells from which osteoclasts derive (Barnekow & Gessler, 1986), the role of $p60^{c\text{-}src}$ in osteoclast function remains to be elucidated. Loss of the CSF-1 gene product in homozygous *op* mice also leads to osteopetrosis (Wiktor-Jedrzejcak *et al.*, 1990; Yoshida *et al.*, 1990), raising the possibility of a functional link between CSF-1 and $p60^{c\text{-}src}$, perhaps mediated by the CSF-1 receptor tyrosine kinase (the c-*fms* gene product). It recently was reported (Twamley *et al*, 1992) that an activated subpopulation of $p60^{c\text{-}src}$ in macrophages binds to the CSF-1 receptor following CSF-1 treatment of these monocytic cells, although it remains to be seen if loss of this association is related to the osteopetrotic phenotype in the *src*-null mice.

Models for Src's Involvement in Signalling Pathways

By piecing together data from a variety of studies in different cell types, it is possible to fit the *src* family kinases into tentative signalling pathways. However, because the immediate upstream effectors/physiological activators and substrates of the *src* family kinases remain unknown, the exact position of these kinases in such pathways cannot yet be defined.

A major theme emerging from recent studies of signal transduction in organisms ranging from man to yeast is that similar molecules appear to beinvolved whether the signal ultimately leads to proliferation or differentiation. An excellent example of this is the divergent responses of PC12 cells to EGF and NGF (reviewed by Chao, 1992). Both factors induce a similar constellation of early events following PC12 cell stimulation: increased tyrosine phosphorylation of a similar profile of cell proteins (including phospholipase C (PLC) γ and mitogen activated protein (MAP) kinase), membrane ruffling, immediate early gene induction, and activation of $p21^{ras}$, the MAP kinases, and the ribosomal S6 (RSK) kinases. Yet, paradoxically, prolonged NGF treatment leads to neuronal differentiation of these cells, while EGF stimulation results in mitogenesis.

As in the EGF/NGF stimulation of PC12 cells, similar molecules have been implicated in the processes by which v-*src* transforms one cell type, such as fibroblasts, and causes the differentiation of another, such as PC12. v-*src* transformation of fibroblasts appear to proceed through p21^{ras}, since injection of neutralizing antibodies to p21^{ras} into these cells (Smith *et al.*, 1986) or overexpression of dominant interfering *ras* alleles (Feig & Cooper, 1988), blocks both of these v-*src* effects. Additional evidence that *ras* acts downstream of *src* in fibroblasts comes from experiments showing that overexpression of the *ras* GTPase activating protein (GAP) suppresses v-*src* transformation (DeClue *et al.*, 1991; Nori *et al.*, 1991). Evidence for the same *src-ras* pathway order in PC12 cell differentiation comes from the ability of neutralizing p21^{ras} antibodies to block v-*src* induced differentiation, and failure of anti-p60$^{c\text{-}src}$ neutralizing antibodies to block the differentiation of PC12 cells stimulated by activated p21^{ras} (Kremer *et al.*, 1991). In both fibroblast mitogenesis and PC12 cell differentiation, activation of p21^{ras} leads to activation of common downstream proteins such as the MAP and RSK/S6 kinases (reviewed by Cantley, 1991; Chao, 1992), and it is not currently clear how the cell ultimately decides whether to grow or differentiate. Presumably, the expression of different genes is induced by the mitogenic and non-mitogenic stimuli. The identification of such genes, perhaps by subtractive cloning approaches, may provide clues to how a cell chooses to divide or differentiate.

The proto-oncoprotein c-*raf*, a serine/threonine kinase, may also function downstream of p60$^{v\text{-}src}$ in certain signalling pathways. It has been shown that a kinase-inactive form of c-*raf*, presumably acting as a dominant negative, can block v-*src* induced activation of the Egf-1 promoter (Qureshi *et al.*, 1991). Additional evidence has indicated that c-*raf* acts downstream of c-*ras*, suggesting a potential *src-ras-raf* signal transduction pathway (Lowy et al., 1991; Heidecker et al., 1992).

While *src* can be placed downstream of the NGF and FGF receptors in PC12 cells (since anti-p60$^{c\text{-}src}$ antibodies block NGF and FGF-induced neuronal differentiation) and upstream of p21^{ras}, it is not clear if cellular *src* normally helps in the transmission of mitogenic signals from growth factor receptors to p21^{ras}. A fraction of the p60$^{c\text{-}src}$ in fibroblasts appears to bind to *ras*-GAP (Brott *et al.*, 1991), and GAP is phosphorylated on tyrosine in *src*-transformed cells (Ellis *et al.*, 1990), but it is not yet known if *src* protein influences the level of (activated) *ras*-GTP in a cell by affecting GAP function. Now that the other major determinant of cellular *ras*-GTP levels - the guanine nucleotide exchange factors - are being identified in mammalian cells (Shou &

Feig, 1992), insights into the mechanisms by which p60$^{c\text{-}src}$ influences the activation of p21^{ras} should be possible.

Although the data are currently preliminary, it is intriguing to speculate that the *src* family kinases, especially p60$^{c\text{-}src}$, are involved in a second signal transduction pathway leading from integrins at the cell surface. Since p125^{FAK} is activated both by v-*src* transformation and by integrin binding, it is possible that p60$^{c\text{-}src}$ (or related kinases) function between the integrins and p125^{FAK} in the generation of tyrosine phosphorylation events following integrin binding to the extracellular matrix. p60$^{c\text{-}src}$ might thus influence cell shape changes in response to the external environment. Along these lines, it is not unreasonable to think that p60$^{c\text{-}src}$ in nerve growth cones might also influence cell shape change during neuronal growth and migration. Specific testing of these pathways with such reagents as anti-p60$^{c\text{-}src}$ antibodies or kinase-inactive (dominant negative) *src* family kinases, or in cells from mice bearing disruptions in *src* family genes, should shed light on the validity of these models.

REFERENCES

Abraham, K.M., Levin, S.D., Marth, J.D., Forbush, K.A. & Perlmutter, R.M. (1991). Thymic tumorigenesis induced by overexpression of p56^{lck}. Proc. Natl. Acad. Sci. USA, 88, 3977-3981.

Abraham, N. & Veillette, A. (1990). Activataion of p56^{lck} through mutation of a regulatory carboxy-terminal tyrosine requires intact sites of autophosphorylation and myristylation. Mol. Cell. Biol., 10, 5197-5206.

Alema, S., Casalbore, P., Agostini, E. & Tato, F. (1985). Differentiation of PC12 phaeochromocytoma cells induced by v-*src* oncogene. Nature, 316, 557-559.

Amrein, K.E. & Sefton, B.M. (1988). Mutation of a site of tyrosine phosphorylation in the lymphocyte-specific tyrosine protein kinase, p56^{lck}, reveals its oncogenic potential in fibroblasts. Proc. Natl. Acad. Sci. USA, 85, 4247-4251.

Bagrodia, S., Chackalaparampil, I., Kmiecik, T.E. & Shalloway, D. (1991). Altered tyrosine 527 phosphorylation and mitotic activation of p60$^{c\text{-}src}$. Nature, 349, 172-175.

Barnekow, A. & Gessler, M. (1986). Activation of the pp60$^{c\text{-}src}$ kinase during differentiation of monomyelocytic cells *in vitro*. EMBO J., 5, 701-705.

Bautch, V.L., Toda, S., Hassell, J.A. & Hanahan, D. (1987). Endothelial cell tumors develop in transgenic mice carrying polyoma virus middle T oncogene. Cell, 51, 529-538.

Bolen, J.B., Thiele, C.J., Israel, M., Yonemoto, W., Lipsich, L.A. & Brugge, J.S. (1984). Enhancement of cellular *src* gene product associated tyrosyl kinase activity following polyoma virus infection and transformation. Cell, 38, 767-777.

Boulter, C.A. & Wagner, E.F. (1988). The effects of v-*src* expression on the differentiation of embryonal carcinoma cells. Oncogene, 2, 207-214.

Brott, B.K., Decker, S., Shafer, J., Gibbs, J.B. & Jove, R. (1991). GTPase-activating protein interactions with the viral and cellular Src kinases. Proc. Natl. Acad. Sci. USA, 88, 755-759.

Brugge, J.S., Cotton, P.C., Queral, A.E., Barrett, J.N., Nonner, D. & Keane, R.W. (1985). Neurones express high levels of structurally modified activated form of pp60$^{c\text{-}src}$. Nature, 316, 554-557.

Brugge, J., Cotton, P., Lustig, A., Yonemoto, W., Lipsich, L., Coussens, P., Barrett, J.N., Nonner, D. & Keane, R.W. (1987). Characterization of the altered form of the c-*src* gene product in neuronal cells. Genes Dev., 1, 287-296.

Cantley, L.C., Auger, K.R., Carpenter, C., *et al.* (1991). Oncogenes and signal transduction. Cell, 64, 281-302.

Cartwright, C.A., Eckhart, W., Simon, S. & Kaplan, P.L. (1987a). Cell transformation by pp60$^{c\text{-}src}$ mutated in the carboxy-terminal regulatory domain. Cell, 49, 83-91.

Cartwright, C.A., Simantov, R., Kaplan, P.L., Hunter, T. & Eckhart, W. (1987b). Alterations in pp60$^{c\text{-}src}$ accompany differentiation of neurons from rat embryo striatum. Mol. Cell. Biol., 7, 1830-1840.

Chakalaparampil, I. & Shalloway, D. (1988). Altered phosphorylation and activation of pp60$^{c\text{-}src}$ during fibroblast mitosis. Cell, 52, 801-810.

Chao, M.V. (1992). Growth factor signalling: Where is the specificity? Cell, 68, 995-997.

Cheng, S.H., Espino, P.C., Marshall, J., Harvey, R., Merrill, J. & Smith, A.E. (1991). Structural elements that regulate pp59^{fyn} catalytic activity, transforming potential, and ability to associate with polyome-virus middle-T antigen. J. Virol., 65, 170-179.

Cooke, M.P., Abraham, R.M., Forbush, K.A. & Perlmutter, R.M. (1991). Regulation of T cell receptor signalling by a *src* family protein-tyrosine kinase (p59^{fyn}). Cell, 65, 281-291.

Cooper, J.A. (1990). The *src*-family of protein tyrosine kinases. In Kemp B (ed.), Peptides and Protein Phosphorylation, Boca Raton FL, CRC Press, pp 85-113.

Cooper, J.A., Gould, K.L., Cartwright, C.A. & Hunter, T. (1986). Tyr^{527} is phosphorylated in pp60$^{c\text{-}src}$: implications for regulation. Science, 231, 1431-1434.

Cooper, J.A., Hunter, T. & Shalloway, D. (1985). Protein-tyrosine kinase activity of pp60$^{c\text{-}src}$ is restricted in intact cells. *In* Cancer Cells, Vol. 3, Feramisco, J., Ozanne, B. & Stiles, C., Eds., Cold Spring Harbor Laboratory, Cold Spring Harbor, NY, 321-328.

Cooper, J.A. & King, C.S. (1986). Dephosphorylation or antibody binding to the carboxy terminus stimulates pp60$^{c\text{-}src}$. Mol. Cell. Biol., 6, 4467-4477.

Cooper, J.A. & MacAuley, A. (1988). Potential positive and negative autoregulation of p60$^{c\text{-}src}$ by intermolecular autophosphorylation. Proc. Natl. Acad. Sci. USA, 85, 4232-4236.

Cooper, J.A. & Runge, K. (1987). Avian pp60$^{c\text{-}src}$ is more active when expressed in yeast than in vertebrate fibroblasts. Oncogene Res., 1, 297-310.

Cotton, P.C. & Brugge, J.S. (1983). Neural tissues express high levels of the cellular *src* gene product pp60$^{c\text{-}src}$. Mol. Cell. Biol., 3, 1157-1162.

Courtneidge, S.A. (1985). Activation of the pp60$^{c\text{-}src}$ kinase by middle T antigen binding or by dephosphorylation. EMBO J., 4, 1471-1477.

Coussens, P.M., Cooper, J.A., Hunter, T. & Shalloway, D. (1985). Restriction of the *in vitro* and *in vivo* tyrosine protein kinase activities of pp60$^{c\text{-}src}$ relative to pp60$^{v\text{-}src}$. Mol. Cell. Biol., 5, 2753-2763.

Cross, F.R., Garber, E.A., Pellman, D. & Hanafusa, H. (1984). A short sequence in the p60^{src} N terminus is rquired for p60$^{c\text{-}src}$ myristylation and membrane association and for cell transformation. Mol. Cell. Biol., 4, 1834-1842.

Cross, F.R. & Hanafusa, H. (1983). Local mutagenesis of Rous sarcoma virus: the major sites of tyrosine and serine phosphorylation of p60^{src} are dispensable for transformation. Cell, 34, 597-607.

David-Pfeuty, T. & Nouvian-Dooghe, Y. (1990). Immunolocalization of the cellular *src* protein in interphase and mitotic NIH c-*src* overexpresser cells. J. Cell Biol., 111, 3097-3116.

DeClue, J.E., Zhang, K., Redford, P., Vass, W.C. & Lowy, D.R. (1991). Suppression of *src* transformation by overexpression of full-length GTPase-activating protein (GAP) or of the GAP C terminus. Mol. Cell. Biol., 11, 2819-2825.

Eiseman, E. & Bolen, J.B. (1990). *src*-related tyrosine protein kinases as signaling components in hematopoietic cells. Cancer Cells, 2, 303-310.

Ellis, C., Moran, M., McCormick, F. & Pawson, T. (1990). Phosphorylation of GAP and GAP-associated proteins by transforming and mutagenic tyrosine kinases. Nature, 343, 377-381.

Engelman, A. & Rosenberg, N. (1990). *bcr/abl* and *src* but not *myc* and *ras* replace v-*abl* in lymphoid transformation. Mol. Cell. Biol., 10, 4365-4369.

Feig, L.A. & Cooper, G.M. (1988). Inhibition of N1H3T3 cell proliferation by a mutant ras protein with preferential affinity for GDP. Mol. Cell. Biol., 8, 3235-3243.

Gee, C.E., Griffin, J., Sastre, L., Miller, L.J., Springer, T.A., Piwnica-Worms, H. & Roberts, T.M. (1986). Differentiation of myeloid cells is accompanied by increased levels of pp60$^{c\text{-}src}$ protein and kinase activity. Proc. Natl. Acad. Sci. USA, 83, 5131-5135.

Golden, A., Nemeth, S.P. & Brugge, J.S. (1986). Blood platelets express high levels of the pp60$^{c\text{-}src}$-specific tyrosine kinase activity. Proc. Natl. Acad. Sci. USA, 83, 852-856.

Gould, K.L. & Hunter, T. (1988). Platelet-derived growth factor induces multisite phosphorylation of pp60$^{c\text{-}src}$ and increases its protein-tyrosine kinase activity. Mol. Cell. Biol., 8, 3345-3356.

Hatakeyama, M., Kono, T., Kobayashi, N., Kawahara, A., Levin, S.D., Perlmutter, R.M. & Taniguchi, T. (1991). Interaction of the IL-2 receptor with the *src*-family kinase p56^{lck}: Identification of novel intermolecular association. Science, 252, 1523-1528.

Heidecker, G., Kolch, W., Morrison, D.K. & Rapp, U.R. (1992). The role of Raf-1 phosphorylation in signal transduction. Adv. Cancer Res., 58, 53-73.

Hirai, H. & Varmus, H.E. (1990). Site-directed mutagenesis of the SH2- and SH3-coding domains of c-*src* produce varied phenotypes, including oncogenic activation of p60$^{c\text{-}src}$. Mol. Cell. Biol., 10, 1307-1318.

Horak, I.D., Gress, R.E., Lucas, P.J., Horak, E.M., Waldmann, T.A. & Bolen, J.B. (1991). T-lymphocyte interleukin 2-dependent tyrosine protein kinase signal transduction involves the activation of p56^{lck}. Proc. Natl. Acad. Sci. USA, 88, 1996-2000.

Hunter, T. & Sefton, B.M. (1980). Transforming gene product of Rous sarcoma virus phosphorylates tyrosine. Proc. Natl. Acad. Sci. USA, 77, 1311-1315.

Iba, H., Cross, F.R., Garber, E.A. & Hanafusa, H. (1985). Low level of cellular protein phosphorylation by nontransforming overproduced pp60$^{c\text{-}src}$. Mol. Cell. Biol., 5, 1058-1066.

Jakobovits, E.B., Majors, J.E. & Varmus, H.E. (1984). Hormonal regulation of the Rous sarcoma virus *src* gene via a heterologous promoter defines a threshold dose for cellular transformation. Cell, 38, 757-765.

Johnson, P.J., Coussens, P.M., Danko, A.V. & Shalloway, D. (1985). Overexpressed pp60$^{c\text{-}src}$ can induce focus formation without complete transformation of N1H3T3 cells. Mol. Cell. Biol., 5, 1073-1083.

Jove, R. & Hanafusa, H. (1987). Cell transformation by the viral *src* oncogene. Ann. Rev. Cell. Biol., 3, 31-56.

Jove, R., Kornbluth, S. & Hanafusa, H. (1987). Enzymatically inactive p60$^{c\text{-}src}$ mutant with altered ATP-binding site is fully phosphorylated in its carboxy-terminal region. Cell, 50, 937-943.

Kamps, M.P., Buss, J.E. & Sefton, B.M. (1986). Rous sarcoma virus transforming protein lacking myristic acid phosphorylates known polypeptide substrates without inducing transformation. Cell, 45, 105-112.

Kamps M.P. & Sefton B.M. (1986). Neither arginine nor histidine can carry out the function of lysine-295 in the ATP-binding site of p60^{src}. Mol. Cell. Biol., 6, 751-757.

Kamps, M.P. & Sefton, B.M. (1988). Most of the substrates of oncogenic viral tyrosine protein kinases can be phosphorylated by cellular tyrosine protein kinases in normal cells. Oncogene Research, 3, 105-115.

Kamps M.P., Taylor S.S. & Sefton B.M. (1984). Direct evidence that oncogenic tyrosine kinases and cyclic AMP-dependent protein kinase have homologous ATP-binding sites. Nature, 310, 589-91.

Keller, G. & Wagner, E.F. (1989). Expression of v-*src* induces a mycloproliferative disease in bone-marrow-reconstituted mice. Genes and Devel., 3, 827-37.

Klausner, R.D. & Samelson, L.E. (1991). T cell antigen receptor activation pathways: The tyrosine kinase connection. Cell, 64, 875-878.

Kmiecik, T.E., Johnson, P.J. & Shalloway, D. (1988). Regulation by the autophosphorylation site in overexpressed pp60$^{c\text{-}src}$. Mol. Cell. Biol., 8, 4541-4546.

Kmiecik, T.E. & Shalloway, D. (1987). Activation and suppression of pp60$^{c\text{-}src}$ transforming ability by mutation of its primary sites of tyrosine phosphorylation. Cell, 49, 65-83.

Koch, C.A., Anderson, D., Moran, M.F., Ellis, C. & Pawson, T. (1991). SH2 and SH3 domains: Elements that control interactions of cytoplasmic signalling proteins. Science, 252, 668-674.

Kremer, N.E., D'Arcangelo, G., Thomas, S.M., DeMarco, M., Brugge, J.S. & Halegoua, S. (1991). Signal transduction by nerve growth factor and fibroblast growth factor in PC12 cells requires a sequence of *src* and *ras* actions. J. Cell Biol., 115, 809-819.

Kypta, R., Goldberg, Y., Ulug, E. & Courtneidge, S.A. (1990). Association between the PDGF receptor and members of the *src* family of tyrosine kinases. Cell, 62, 481-492.

Levy, J.B., Dorai, T., Wang, L-H. & Brugge, J.S. (1987). The structurally distinct form of pp60$^{c\text{-}src}$ detected in neuronal cells is encoded by a unique c-*src* mRNA. Mol. Cell. Biol., 7, 4142-4145.

Lowell, C., Soriano, P. & Varmus, H.E. (1992). Targeted disruption of *hck* and c-*fgr* in mice (abstract #154). 8th Annual Meeting on Oncogenes, Frederick, Maryland.

Lowy, D.R., Zhang, K., DeClue, J.E. & Willumsen, B.M. (1991). Regulation of p21^{ras} activity. Trends Genet, 7, 346-351.

Luttrell, D.K., Luttrell, L.M. & Parsons, S.J. (1988). Augmented mitogenic responsiveness to epidermal growth factor in murine fibroblasts that over-express pp60$^{c\text{-}src}$. Mol. Cell. Biol., 8, 497-501.

MacAuley, A. & Cooper, J.A. (1989). Structural differences between repressed and derepressed forms of p60$^{c\text{-}src}$. Mol. Cell. Biol., 9, 2648-2656.

Marth, J.D., Cooper, J.A., King, C.S., Ziegler, S.F., Tinker, D.A., Overell, R.W., Krebs, E.G. & Perlmutter, R.M. (1988). Neoplastic transformation induced by an activated lymphocyte-specific protein-tyrosine kinase. Mol. Cell. Biol., 8, 540-550.

Marth, J.D., Peet, R., Krebs, E.G. & Perlmutter, R.M. (1985). A lymphocyte-specific protein-tyrosine kinase is rearranged and overexpressed in the murine T cell lymphoma LSTRA. Cell, 43, 393-404.

Martin, G.S. (1970). Rous sarcoma virus: a function required for the maintenance of the transformed state. Nature (London), 227, 1021-1023.

Martinez, R., Mathey-Prevot, B., Bernards, A. & Baltimore, D. (1987). cDNA cloning of the neuronal form of pp60$^{c\text{-}src}$. Science, 237, 411-415.

Molina, T.J., Kishihara, K., Siderovski, D.P. *et al.* (1992). Profound block in thymocyte development in mice lacking p56^{lck}. Nature (London), 357, 161-164.

Morgan, D.O., Kaplan, J.M., Bishop, J.M. & Varmus, H.E. (1989). Mitosis-specific phosphorylation of p60$^{c\text{-}src}$ by p34^{cdc2}-associated protein kinase. Cell, 57, 775-786.

Nada, S., Okada, M., MacAuley, A., Cooper, J.A. & Nakagawa, H. (1991). Cloning of a complementary DNA for a protein tyrosine kinase that specifically phosphorylates a negative regulatory site of p60$^{c\text{-}src}$. Nature, 351, 69-72.

Nemeth, S.P., Fox, L.G., DeMarco, M. & Brugge, J.S. (1989). Deletions within the amino-terminal half of the c-*src* gene product that alters the functional activity of the protein. Mol. Cell. Biol., 9, 1109-1119.

Nori, M., Vogel, U.S., Gibbs, J.B. & Weber, M.J. (1991). Inhibition of v-*src*-induced transformation by a GTPase-activating protein. Mol. Cell. Biol., 11, 2812-2818.

O'Brien, M.C., Fukui, Y. & Hanafusa, H. (1990). Activation of the proto-oncogene p60$^{c\text{-}src}$ by point mutations in the SH2 domain. Mol. Cell. Biol., 10, 2855-2862.

Ohta M, Anklesaria P., & Wheaton M. (1989). Retroviral *src* gene expression in continuous marrow culture increases the self-renewal capacity of multilineage hematopoietic stem cells. Leukemia, 3, 206-226.

Okada, M. & Nakagawa, H. (1989). A protein tyrosine kinase involved in pp60$^{c\text{-}src}$ function. J. Biol. Chem., 264, 20886-20893.

Parsons, J.T. & Weber, M.J. (1989). Genetics of *src*: structure and functional organization of a protein tyrosine kinase. Curr. Top. Microbiol. Immunol., 147:79-127.

Patschinsky, T., Hunter, T., Esch, F.S., Cooper, J.A. & Sefton, B.M. (1982). Analysis of the sequence of amino acids surrounding sites of tyrosine phosphorylation. Proc. Natl. Acad. Sci. USA, 79, 973-977.

Piwnica-Worms, H., Saunders, K.B., Roberts, T.M., Smith, A.E. & Cheng, S.H. (1987). Tyrosine phosphorylation regulates the biochemical and biological properties of pp60$^{c\text{-}src}$. Cell, 49, 75-82.

Qureshi, S.A., Rim, M., Bruder, J., Kolch, W., Rapp, U., Sukhatme, V.P. & Foster, D.A. (1991). An inhibitory mutant of c-Raf-1 blocks v-Src-induced activation of the Egr-1 promoter. J. Biol. Chem., 266, 20594-20597.

Reynolds, A.B., Vila, J., Lansing, T.J., Potts, W.M., Weber, M.J. & Parsons, J.T. (1987). Activation of the oncogenic potential of the avian cellular *src* protein by specific structural alteration of the carboxy terminus. EMBO J., 6, 2359-2364.

Roussel, R.R., Brodeur, S.R., Shalloway, D., & Laudano, A.P. (1991). Selective binding of activated pp60$^{c\text{-}src}$ by an immobilized synthetic peptide modeled on the carboxyl terminus of pp60$^{c\text{-}src}$. Proc. Natl. Acad. Sci. USA, 88, 10696-10700.

Sabe, H., Knudsen, B., Okada, M., Nada, S., Nakagawa, H. & Hanafusa, H. (1992a). Molecular cloning and expression of chicken C-terminal *src* kinase: lack of stable association with c-*src* protein. Proc. Natl. Acad. Sci. USA, 89, 2190-2194.

Sabe, H., Okada, M., Knudsen, B., Nada, S., Nakagawa, H. & Hanafusa, H. (1992b). CSK induced reversion of rat 3Y1 cells transformed by coexpression of c-*src* and v-*crk*. J. Cell. Biochem., Supplement 16B, H345.

Schultz AM, Henderson LE, Oroszlan, *et al.* (1985). Amino terminal myristylation of the protein kinase $p60^{src}$, a retroviral tranforming protein. Science, 227, 427-429.

Schwartz, D.E., Tizard, R. & Gilbert, W. (1983). Nucleotide sequence of Rous sarcoma virus. Cell, 32, 853-869.

Sefton, B.M., Hunter, T., Beemon, K. & Eckhart, W. (1980). Evidence that the phosphorylation of tyrosine is essential for cellular transformation by Rous sarcoma virus. Cell, 20, 807-816.

Shalloway, D., Coussens, P.M. & Yaciuk, P. (1984). Overexpression of the c-*src* protein does not induce transformation of NIH 3T3 cells. Proc. Natl. Acad. Sci. U.S.A., 81, 7071-7075.

Shaw, A.S., Amrein, K.E., Hammond, C., Stern, D.F., Sefton, B.M. & Rose, J.K. (1989). The *lck* tyrosine protein kinase interacts with the cytoplasmic tail of the CD4 glycoprotein through its unique amino-terminal domain. Cell, 59, 627-636.

Shenoy, S., Choi, J-K., Bagrodia, S., Copeland, T.D., Maller, J.L. & Shalloway, D. (1989). Purified maturation promoting factor phosphorylates $pp60^{c\text{-}src}$ at the sites phosphorylated during fibroblast mitosis. Cell, 57, 763-774.

Shou, C. & Feig, L.A. (1992). Cloning and characterization of a guanine-nucleotide releasing factor for mammalian $p21^{ras}$ (abstract #290). 8th Annual Meeting on Oncogenes, Frederick, Maryland.

Shuntaro, I., Yamamoto, T. & Toyoshima, K. (1986). Modification of carboxyl-terminal region is the cause of activation of the src gene in avian sarcoma virus. Jpn. J. Cancer Res., 77, 611.

Smart, J.E., Opperman, H., Czernilofsky, A.P., Purchio, A.F., Erikson, R.L. & Bishop, J.M. (1981). Characterization of sites for tyrosine phosphorylation in the transforming protein of Rous sarcoma virus ($pp60^{v\text{-}src}$) and its normal cellular homolog. Proc. Natl. Acad. Sci. USA, 78, 6013-6017.

Smith, M.R., De Gudicibus, S.J. & Stacey, D.W. (1986). Requirement for C-ras proteins during viral oncogene transformation. Nature, 320, 540-543.

Snyder, M.A. & Bishop, J.M. (1984). A mutation at the major phosphotyrosine in $pp60^{v\text{-}src}$ alters oncogenic potential. J. Virol., 136, 375-386.

Snyder, M.A., Bishop, J.M., Colby, W.W. & Levinson, A.D. (1983). Phosphorylation of tyrosine-416 is not required for the transforming properties and kinase activity of $pp60^{v\text{-}src}$. Cell, 32, 891-901.

Soriano, P., Montgomery, C., Geske, R. & Bradley, A. (1991). Targeted disruption of the c-*src* proto-oncogene leads to osteopetrosis in mice. Cell, 64, 693-702.

Stein, P.L., Lee, H.-M., Rich, S. & Soriano, P. (1992). Targeted disruption of the murine *fyn* gene leads to impaired thymocyte activation (abstract #152). 8th Annual meeting on Oncogenes, Frederick, Maryland.

Sudol, M., Martins-Green, M. Greulich, H., Sukegawa, J. & Yamamoto, T. (1992). Yes-related kinase, Yrk, is expressed in neural and hematopoietic tissues (abstract #36). 8th Annual Meeting on Oncogenes, Frederick, Maryland.

Takeuchi, M., Nada, S., Fusaki, N., Kuramochi, S., Semba, K., Okada, M. & Yamamoto, T. (1992). CSK downregulates fyn activity which is important in T cell activation (abstract #153). 8th Annual Meeting on Oncogenes, Frederick, Maryland.

Takeya, T. & Hanafusa, H. (1983). Structure and sequence of the cellular gene homologous to the RSV *src* gene and the mechanism for generating the transforming virus. Cell, 32, 881-890.

Thomas, J.E., Soriano, P. & Brugge, J.S. (1991). Phosphorylation of c-*Src* on tyrosine 527 by another protein tyrosine kinase. Science, 254, 568-571.

Twamley, G.M., Kypta, R.M., Hall, B. & Courtneidge, S.A. (1992). The involvement of src family tyrosine kinases in the response to growth factors (abstract #289). 8th Annual Meeting on Oncogenes, Frederick, Maryland.

Veillette, A., Bookman, M.A., Horak, E.M. & Bolen, J.B. (1988). The CD4 and CD8 T cell surface antigens are associated with the internal membrane protein-tyrosine kinase $p56^{lck}$. Cell, 55, 301-308.

Voronova, A.F. & Sefton, B.M. (1986). Expression of a new tyrosine protein kinase is stimulated by retrovirus promoter insertion. Nature (London), 319, 682-685.

Wasilenko, W.J., Payne, D.M., Fitzgerald, D.L. & Weber, M.J. (1991). Phosphorylation and activation of epidermal growth factor receptors in cells transformed by the *src* oncogene. Mol. Cell. Biol., 11, 309-321.

Watson JD, Eszes M, Overall R, *et al.* (1987). Effect of infection with murine recombinant retroviruses containing the v-*src* oncogene on interleukin 2 and interleukin 3-dependent growth states. J. Immunol., 139, 123-9.

Wilson, L.K. & Parsons, S.J. (1990). Enhanced EGF mitogenic response is associated with enhanced tyrosine phosphorylation of specific cellular proteins in fibroblasts overexpressing c-*src*. Oncogene, 5, 1471-1480.

Wiktor-Jedrzejczak, W., Bartocci, A., Ferrante, A.W., Jr., Ahmed-Ansari, A., Sell, K.W., Pollard, J.W. & Stanley, E.R. (1990). Total absence of colony-stimulating factor 1 in the macrophage-deficient osteopetrotic (op/op) mouse. Proc. Natl. Acad. Sci. USA, 87, 4828-4832.

Yoshida, H., Hayashi, S.-I., Kunisada, T., Ogawa, M., Nishikawa, S. Okamura, H., Sudo, T., Shultz, L.D. & Nishikawa, S.-I. (1990). The murine mutation osteopetrosis is in the coding region of the macrophage colony stimulating factor gene. Nature, 345, 442-444.

Ziegler, S.F., Levin, S.D., & Perlmutter, R.M. (1989). Transformation of fibroblasts by an activated form of $p59^{hck}$. Mol. Cell. Biol., 9, 2724-2727.

Ziegler, S.F., Wilson, C.B. & Perlmutter, R.M. (1988). Augmented expression of a myeloid-specific protein-tyrosine kinase gene (*hck*) after macrophage activation. J. Exp. Med., 168, 1801-1810.

MITOGEN-ACTIVATED SIGNALLING PATHWAYS

George Thomas
Friedrich Miescher Institute
P.O. Box 2543
4002 Basel, Switzerland

Historical Perspective

The purpose of this lecture was to introduce the topic of signalling pathways and some of the components which make up these pathways. Since both of these subjects will be covered in greater detail by individual speakers during the meeting, this lecture should be viewed as a preview. In brief, it will begin with a discussion of two types of receptors, one which contains an inherent tyrosine kinase activity (Yarden Y and Ullrich A, 1988; Ullrich A and Schlessinger J, 1990; Cantley LC et al., 1991) and a second set of receptors which are characterized by having seven transmembrane domains (Dohlmann HG et al., 1991; Battey JF et al., 1991; Vu T-KH et al., 1991; Rasmussen UB et al., 1991). I will largely restrict the tyrosine kinase receptor part of the lecture to the epidermal growth factor (EGF)-receptor (Ullrich A and Schlessinger J, 1990; Carpenter G and Cohen S, 1990), and the seven transmembrane receptor part to the bombesin receptor (Battey JF et al., 1991). Next, I will discuss the mechanisms which couple these receptors to signalling pathways (Cantley LC et al., 1991) and conclude by presenting a newly described family of mitogen-activated Ser/Thr kinases which propagate the mitogenic signal through phosphorylation of specific target proteins (Kozma SC and Thomas G, 1992).

When speaking of signalling pathways I like to begin with an observation made by Tomkins and co-workers more than twenty years ago. They noted that the activation of cell growth in the whole animal and in tissue culture led to the stimulation of a number of seemingly unrelated metabolic events. However, because these events were expressed in a coordinate fashion

NATO ASI Series, Vol. H 76
Tyrosine Phosphorylation/Dephosphorylation
and Downstream Signalling
Edited by L. M. G. Heilmeyer Jr.

they were referred to collectively as the pleiotypic response (Hershko A et al., 1971). The model they proposed to explain their hypothesis was that growth factors (at that time serum) acting on the resting cell surface, stimulated the production of some common pleiotypic mediator which in turn led to the activation of a number of key metabolic events associated with cell growth. These events included the activation of protein synthesis and RNA synthesis as well as the uptake of glucose and other metabolites from the media. This model had A far-reaching effect on the field: it tended to divide it in half. There were those who set out to purify the factors in serum which lead to activation of cell growth with the long-range goal of using these factors as a probe to search for the corresponding unknown receptor. The other group began with one of the obligatory steps in mitogenesis and attempted to describe the molecular events involved in the activation of that event with the ultimate objective of working back to the receptor.

Epidermal Growth Factor Receptor

Employing the first approach described above has led to the identification of a number of growth factors and their respective receptors (Yarden Y and Ullrich A, 1988). Most of these receptors contain a tyrosine kinase activity which is elevated when the appropriate ligand binds to its receptor (Yarden Y and Ullrich A, 1988; Ullrich A and Schlessinger J, 1990; Cantley LC et al., 1991). All of these receptor tyrosine kinases have several features in common; a large extracellular domain involved in ligand binding, a single hydrophobic transmembrane domain involved in anchoring the receptor in the membrane and an intracellular domain containing the highly conserved catalytic motif of kinases (Figure). Based on sequence similarity, it has been possible to divide these receptors into sub-classes (Ullrich A and Schlessinger J, 1990). The EGF receptor belongs to sub-class 1 whose members are unique in having two cysteine-rich repeat sequences. Other members of this class include neu receptor or human receptor 2 (HER2), as well as HER3, or erbB-3.

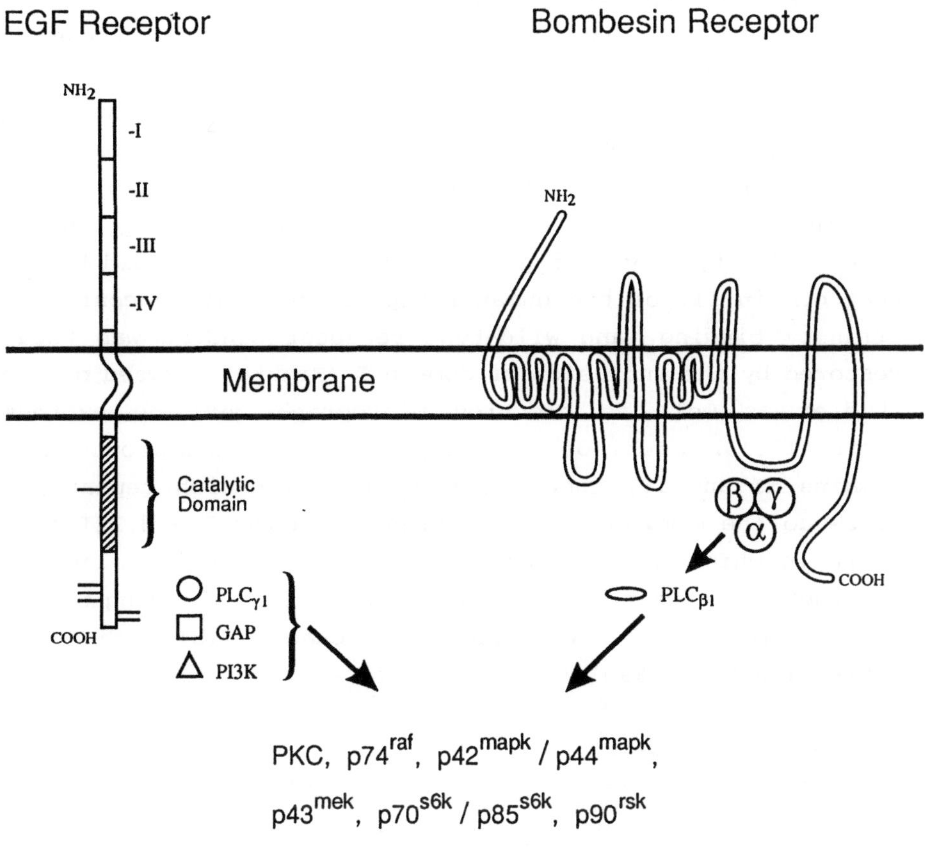

EGF and bombesin receptor signalling pathways. The figure is not meant to be structurally correct or to contain all the known components of the mitogenic signalling pathway. The abbreviations are given in the text, except (-), which indicates autophosphorylation sites.

Ligand Binding

The first step in ligand activation of the receptor is binding, followed by its rapid oligomerization, which results in enhanced tyrosine kinase activity of the receptor and substrate phosphorylation (Ullrich A and Schlessinger J, 1990 and references therein). Though the functional importance of the cysteine-rich motifs is unknown, some progress has been made in understanding the ligand binding domain through the use of chimeric EGF receptors derived from chicken and human (Lax I et al., 1989). This approach was based on the observation that even though the chicken receptor is highly homologous to the human receptor it binds EGF with approximately 250-fold lower affinity. By making chimeric receptors it could be shown that domain III of the human receptor could alone confer high affinity binding, and wild-type receptor binding was totally restored by the inclusion of domain I (Figure). Based on these studies as well as electron micrographs of the expressed extracellular domain of the receptor, it has been proposed that domains II and IV, containing the cysteine-rich repeats, lie close to the membrane. In contrast, domains I and III form a cleft in which EGF can bind, where ligand binding alters the interaction between the neighboring subunits, allowing transfer of an allosteric conformational change (Ullrich A and Schlessinger J, 1990).

Transmembrane Domain

Initially there was a great deal of excitement over the finding that a point mutation in the region of the HER2 receptor conferred the potential to induce transformation (Stern DF et al., 1988 and references therein). This finding led to the concept that this region of the protein was critical in signalling. However, similar mutations in other receptors did not result in increased transformation potential. Furthermore, in numerous chimeric receptor constructs the origin of the transmembrane domain had no effect on heterologous ligand signalling (Riedel H et al., 1989). Thus the main function of this domain appears to be to anchor it to the membrane. It should also be noted that the short juxtamembrane sequence, which links the transmembrane domain with the catalytic domain of the receptor, also appears to play only a minor role in signalling. Initially it was thought that this domain may be involved in a phenomenon termed transmodulation, or the ability of one type of receptor to affect the activity of a second receptor of another type (Ullrich A and Schlessinger J, 1990 and references therein). However, a number of recent studies have placed this hypothesis in question.

Tyrosine Kinase Domain

The kinase domain is the most highly conserved motif among the tyrosine kinase receptors. Mutation of Lys 721, which lies just downstream of the GXGXXG conserved motif in subdomain 1 of all kinases, renders the EGF receptor kinase inactive (Yarden Y and Ullrich A, 1988). This mutation, however, has no effect on the ability of the receptor to be expressed or targeted to the cell surface. Though the receptor can still bind ligand with the same affinity, it is unable to induce early mitogenic responses such as Ca^{2+} influx, Na^+/H^+ exchange, S6 phosphorylation and DNA synthesis (Honegger AM et al, 1987). Surprisingly, the mutation also prevents targeting of the receptor to lysosomes following ligand binding. Instead, the receptors are recycled to the membrane (Honegger AM et al., 1987).

Carboxy Tail

Within the carboxy tail domain reside most of the autophosphorylation sites (Carpenter G and Cohen S, 1990). Though the carboxy tail sequences are among the most divergent in tyrosine kinase receptors, these autophosphorylation residues are highly conserved within each subclass. Changing these residues individually from Tyr $\rightarrow$ Phe had little effect on the activity of the receptor in triggering ligand activated mitogenesis. However, simultaneous mutation of all the sites increased the sensitivity of the receptor for the ligand in terms of driving the growth response (Honegger A et al., 1988). This led to the speculation that the autophosphorylation domain may compete for the tyrosine kinase active site of the receptor. Thus, mutation of these residues would release this inhibition, as does phosphorylation, allowing the kinase to phosphorylate key targets (Ullrich A and Schlessinger J, 1990). Consistent with this view, deletions of this domain increased the oncogenic capacity of cells without inducing an oncogenic lesion (Khazaie K et al, 1988).

Substrate Targets

The search for direct targets of the receptor has been a frustrating task. However, over the past few years a number of targets have been identified on the basis of their ability to first interact with the activated receptor, followed by their phosphorylation on tyrosine residues and their enzymatic activation (Cantley LC et al., 1991). These targets include phospholipase $C_{\gamma1}$ ($PLC_{\gamma1}$), ras-GTPase-activating protein (GAP), and the phosphotidylinositol 3-kinase (PI3K) (see Figure). Each of these molecules in their active state could serve as an agent which couples the activated receptor to an intracellular signalling pathway (Figure). For example, the activation of $PLC_{\gamma1}$ leads to the hydrolysis of phosphatidylinositol 4,5-bisphosphate (PIP_2) to diacylglycerol (DG) and inositol 1,4,5-triphosphate (IP_3). DG leads to the direct activation of most members of the protein kinase C family. This response is augmented in many cases by the release of intracellular stores of Ca^{2+} which is triggered by the binding of IP_3 to its receptor (Nishizuka Y, 1988). In each case these targets have been shown to directly interact with their respective receptors by co-immunoprecipitation employing specific antibodies (Cantley LC et al., 1991). Since this interaction required the activated receptor, it led to the hypothesis that autophosphorylation of the receptor released the carboxyl tail from the active site as well as putative inhibitory proteins from the receptor (Ullrich A and Schlessinger J, 1990). In the activated state the receptor was then free to interact directly with its respective target proteins, leading to the concept of signal transfer particles, activated receptors complexed with their targeted substrates, which in turn coupled the receptor to specific intracellular signalling pathways (Ullrich A and Schlessinger J, 1990).

SH2 Domains

All the proteins listed above are distinct within their catalytic regions; however, largely through the observations of Pawson and collaborators, it was noted that they contain two highly conserved src homology regions, 2 and 3 (SH2 and SH3), respectively). The SH2 domains were first noted in the oncogenes v-fps and v-src (Sadowski I et al., 1986). Furthermore, mutants with alterations in the SH2 domain were found to be temperature-sensitive, defective, or host-dependent in their transforming capacity (Mayer BJ et al., 1988 and references therein). These data suggested that SH2 domains might be involved in regulating protein-protein interactions. This argument was largely supported by the observation that the $P47^{gag-crk}$ oncogene, a protein made up almost exclusively of SH2 - SH3 domains and lacking any apparent catalytic activity, can induce transformation by directly interacting with tyrosine phosphorylated proteins (Mayer BJ et al., 1988). These observations led to the finding that SH2 domains directly interact with tyrosine phosphorylated residues. Indeed, the crystal structure of the SH2 domain of v-src complexed with two phosphotyrosyl peptides from the EGF- and platelet-derived growth factor receptor has recently been resolved (Waksman G et al., 1992). Thus, the function of receptor activation and autophosphorylation on tyrosine residues does not release inhibitors but creates high affinity binding sites for $PLC_{\gamma 1}$, GAP and PI3K. This is followed by phosphorylation and activation of the target protein. In contrast to SH2 domains,

SH3 domains appear to be involved in targeting protein interactions at the level of the cytoskeleton rather than the membrane.

Seven Transmembrane Domain Receptors

In contrast to tyrosine kinase growth factor receptors, seven transmembrane domain receptors do not contain an inherent kinase activity (Honegger A et al., 1988). These receptors have only recently been implicated in mitogenesis through the finding that the receptors for two potent mitogens, bombesin and thrombin, belong to this family of receptors (Battey JF et al., 1991; Vu T-KH et al., 1991; Rasmussen UB et al., 1991). Interest in these receptors has also sprung from the observation that bombesin-like peptides have been implicated in the pathogenesis of some human small-cell lung carcinomas (Cuttitta F et al, 1985). The receptors are coupled to intracellular responses through direct interaction of heterotrimeric GTPases (Figure) or G proteins (Bourne HR et al., 1990). The G protein complex is made up of α, β and γ subunits. This family of GTPases is distinct from that of small G proteins whose most notable member is the protooncogene ras (Bourne HR et al., 1990). Large G proteins go through the same series of reactions: ligand-induced binding of GTP through the γ subunit, release of the βγ subunits as a complex, interaction with the effector molecule, hydrolysis of GTP, reformation of the heterotrimeric complex, exchange of GTP for GDP promoted by exchange factors, and repetition of the cycle. Thus, binding and hydrolysis of GTP drive the transition between the active GTP-bound state, and the inactive GDP-bound state. The rate of hydrolysis is apparently dictated by the interaction of the G protein with its effector or target molecule (Bourne HR and Stryer L, 1992). Thus, the target molecule not only propagates the signal but is also involved in its desensitization or downregulation (Bourne HR and Stryer L, 1992). The longest known examples of this type of reaction are the initiation and elongation steps of protein synthesis. Here eIF-2 and eEF-1 serve as the G proteins and the ribosome is the effector molecule and also the agent promoting or catalysis of GTP (Bourne HR and Stryer L, 1992 and references therein).

Structure of Gγ Proteins

To date relevant G proteins involved in mitogenesis have not been identified, though at least one potential candidate is emerging. Identification of this molecule was largely facilitated through studies of Simon and co-workers, who, through the use of the polymerase chain rection, have identified over 100 new G proteins (Simon MI et al., 1991). Furthermore, based on sequence homology they have been able to divide these G proteins into four families termed G_s, G_i, G_q and G_{12} (Simon MI et al., 1991). All G proteins appear to be largely made up of the same structural motifs even though the sequences between individual families are quite divergent. In brief, the amino terminus of the G_α protein is thought to be involved in βγ interaction and thus responsible for determining the specificity of this interaction. The carboxyl half of the protein appears to be involved in effector interaction and the carboxyl terminus with interacting with the receptor. Indeed, ADP-ribosylation by pertussis toxin of many G proteins at the carboxy terminus abrogates receptor interaction (Simon MI et al., 1991; Kaziro Y et al., 1991). As mentioned above, recent identification of new G proteins through molecular cloning approaches has facilitated the identification of a G protein which appears to couple the receptor with intracellular mitogenic signalling (see Figure). The protein, which was partially purified, has sequence identity with G_q family (Berstein G et al., 1992). Furthermore, recent studies show that unlike the tyrosine kinase receptors, this G_q-like protein interacts and activates only phospholipase$_{\beta 1}$. However, the end result is the same: the hydrolysis of PIP_2 generating DG and IP_3 leading to the activation of protein kinase C. The number of seven transmembrane domain receptors involved in mitogenesis remains unknown as does the number of G proteins which couple these receptors to effector molecules such as $PLC_{\beta 1}$. This will obviously be a field of intensive study over the next years.

Mitogen-activated Ser/Thr Kinases

Though growth factors trigger mitogenesis through ligand-activated tyrosine kinase receptors or through seven transmembrane domain receptors coupled to G proteins, both pathways apparently converge on a family of newly described mitogen-activated Ser/Thr kinases (Figure and Kozma SC and Thomas G, 1992). It is thought that these kinases then propagate the mitogenic response through the phosphorylation of target enzymes involved in transcription of specific gene products as well as the upregulation of protein synthesis. In many cases these kinases themselves are regulated by phosphorylation and dephosphorylation. This family now includes well-known members such as the cyclic AMP-dependent protein kinase (PKA) and the protein kinase C (PKC) family of enzymes. The latter group falls into two broad categories according to their dependence on Ca^{2+} for activation. The Ca^{2+}-dependent group includes PKC subtypes α, β_{I}, β_{II} and γ and the Ca^{2+}-independent group includes PKC subtypes δ, ε, ε', ζ and η. In contrast to these enzymes which are regulated by second messengers, another group of kinases is regulated by phosphorylation. These include the protooncogene raf or $p74^{raf}$, the mitogen-activated kinases or $p42^{mapk}$, $p44^{mapk}$ and $p54^{mapk}$, and the ribosomal S6 kinases, which can be divided into two families, the $p90^{rsk}$ and $p70^{s6k}/p85^{s6k}$. The $p90^{rsk}$ family is apparently made up of a multigene family and contains the unique feature of having two Ser/Thr kinase domains (Jones SW et al., 1988). In contrast, the $p70^{s6k}/p85^{s6k}$ family is apparently derived from a single gene giving rise to two protein products by differential splicing (Reinhard C et al., 1992). Recently there has been a great deal of excitement over the discovery of the $p42^{mapk}/p44^{mapk}$ kinase ($p43^{mek}$) (Ahn NG et al., 1991; Crews CM et al., 1992) which itself is activated directly by $p74^{raf}$ or an as yet unidentified $p42^{mapk}/p42^{mapk}$

kinase kinase (Kyriakis JM et al., 1992; Dent P et al., 1992; Howe LR et al., 1992). In turn, it has been known for some time that the $p90^{rsk}$ family of kinases are phosphorylated and partially reactivated by $p42^{mapk}/p44^{mapk}$ (Sturgill TW et al., 1988). Indeed, these two enzymes appear to have affinity for one another *in vivo* as well as *in vitro* (Scimeca JC et al., 1992). These latter findings are beginning to give us a glimpse of a complicated network of regulatory protein kinases, a number of which must still be identified. During the following lectures many of these topics will be presented in greater detail, which in some cases may allow us to put together some rough models of different mitogenic pathways.

Acknowledgements

I am greatly indebted to Drs. R. Pearson and M. Stewart for their rapid proof-reading of this manuscript. In addition I would like to thank C. Wiedmer for editing suggestions and and for typing the manuscript.

References

Ahn NG, Seger R, Bralien RL, Diltz CD, Tonks NK, Krebs EG (1991) Multiple components in an epidermal growth factor-stimulated protein kinase cascade. J Biol Chem 266:4220-4227

Battey JF, Way JM, Corjay MH, Shapira H, Kusano K, Harkins R, Wu JM, Slattery T, Mann E, Feldman RI (1991) Molecular cloning of the bombesin/gastrin-releasing peptide receptor from Swiss 3T3 cells. Proc Natl Acad Sci USA 88:395-399

Berstein G, Blank JL, Jhon D-Y, Exton JH, Rhee SG, Ross EM (1992) Phospholipase C-β1 is a GTPase-activating protein for $G_{q/11}$, its physiologic regulator. Cell 70:411-418

Bourne HR, Sanders DA, McCormick F (1990) The GTPase superfamily: a conserved switch for diverse cell functions. Nature 348:125-132

Bourne HR and Stryer L (1992) The target sets the tempo. Nature 358:541-543

Cantley LC, Auger KR, Carpenter C, Duckworth B, Graziani A, Kapeller R, Soltoff S (1991) Oncogenes and signal transduction. Cell 64:281-302

Carpenter G and Cohen S (1990) Epidermal growth factor. J Biol Chem 265:7709-7712

Crews CM, Alessandrini A, Erikson RL (1992) The primary structure of MEK, a protein kinase that phosphorylates the *ERK* gene product. Science 258:478-480

Cuttitta F, Desmond NC, Mulshine J, Moody TW, Fedorko J, Fischler A, Minna JD (1985) Bombesin-like peptides can function as autocrine growth factors in human small-cell lung cancer. Nature 316:823-826

Dent P, Haser W, Haystead TAJ, Vincent LA, Roberts TM, Sturgill TW (1992) Activation of mitogen-activated protein kinase kinase by v-*raf* in NIH 3T3 cells and *in vitro*. Science 257:1404-1407

Dohlmann HG, Thorner J,, Caron MG, Lefkowitz RJ (1991) Model systems for the study of seven-transmembrane-segment receptors. Annu Rev Biochem 60:653-688

Hershko A, Mamont P, Shields R, Tomkins GM (1971) Pleiotypic response. Nature New Biol 232:206-211

Honegger AM, Dull TJ, Felder S, Van Obberghen E, Bellot F, Szapary D, Schmidt A, Ullrich A, Schlessinger J (1987) Point mutation at the ATP binding site of EGF receptor abolishes protein-tyrosine kinase activity and alters cellular routing. Cell 51:199-209

Honegger A, Dull TJ, Szapary D, Komoriya A, Kris R, Ullrich A, Schlessinger J (1988) Kinetic parameters of the protein tyrosine kinase activity of EGF-receptor mutants with individually altered autophosphorylation sites. EMBO J 7:3053-3060

Howe LR, Leevers SJ, Gómez N, Nakielny S, Cohen P, Marshall CJ (1992) Activation of the MAP kinase pathway by the protein kinase raf. Cell 71:335-342

Jones SW, Erikson E, Blenis J, Maller JL, Erikson RL (1988) A *Xenopus* ribosomal S6 kinase has two apparent kinase domains that are each similar to distinct protein kinases. Proc Natl Acad Sci USA 85:3377-3381

Kaziro Y, Itoh H, Kozasa T, Masato N, Satoh T (1991) Structure and function of signal-transducing GTP-binding proteins. Annu Rev Biochem 60:349-400

Khazaie K, Dull TJ, Graf T, Schlessinger J, Ullrich A, Beug H, Vennström B (1988) Truncation of the human EGF receptor leads to differential transforming potentials in primary avian fibroblasts and erythroblasts. EMBO J 7:3061-3071

Kozma SC and Thomas G (1992) Serine/threonine kinases in the propagation of the early mitogenic response. Rev Physiol Biochem Pharmacol 119:123-155 Springer-Verlag

Kyriakis JM, App H, Zhang X-F, Banerjee P, Brautigan DL, Rapp UR, Avruch J (1992) Raf-1 activates MAP kinase-kinase. Nature 358:417-421

Lax I, Bello F, Howk R, Ullrich A, Givol D, Schlessinger J (1989) Functional analysis of the ligand binding site of EGF-receptor utilizing chimeric human/chicken receptor molecules. EMBO J 8:421-427

Mayer BJ, Michinari H, Hanafusa H (1988) A novel viral oncogene with structural similarity to phospholipase C. Nature 332:272-275

Nishizuka Y (1988) The molecular heterogeneity of protein kinase C and its implications for cellular regulation. Nature 334:661-665

Rasmussen UB, Vouret-Craviari V, Jallat S, Schlessinger Y, Pagès G, Pavirani A, Lecocq J-P, Pouysségur J, Van Obberghen-Schilling E (1991) cDNA cloning and expression of hamster alpha-thrombin receptor coupled to Ca^{2+} mobilization. FEBS Lett 288:123-128

Reinhard C, Thomas G, Kozma SC (1992) A single gene encodes two isoforms of the p70 S6 kinase: Activation upon mitogenic stimulation. Proc Natl Acad Sci USA 89:4052-4056

Riedel H, Dull TJ, Honegger AM, Schlessinger J and Ullrich A (1989) Cytoplasmic domains determine signal specificity, cellular routing characteristics and influence ligand binding of epidermal growth factor and insulin receptors. EMBO J 8:2943-2954

Sadowski I, Stone JC, Pawson T (1986) A noncatalytic domain conserved among cytoplasmic protein-tyrosine kinases modifies the kinase function and transforming activity of Fujinami Sarcoma virus $P130^{gag-fps}$. Mol Cell Biol 6:4396-4408

Scimeca J-C, Nguyen TT, Filloux C, Van Obberghen E (1992) Nerve growth factor-induced phosphorylation cascade in PC12 pheochromocytoma cells. J Biol Chem 267:17369-17374

Simon MI, Strathmann MP, Gautam N (1991) Diversity of G proteins in signal transduction. Science 252:802-808

Stern DF, Kamps MP, Lao H (1988) Oncogenic activation of $p185^{neu}$ stimulates tyrosine phosphorylation in vivo. Mol Cell Biol 8:3969-3973

Sturgill TW, Ray LB, Erikson E, Maller JL (1988) Insulin-stimulated MAP-2 kinase phosphorylates and activates ribosomal protein S6 kinase II. Nature 334:715-718

Ullrich A and Schlessinger J (1990) Signal transduction by receptors with tyrosine kinase activity. Cell 61:203-212

Vu T-KH, Hung DT, Wheaton VI, Coughlin SR (1991) Molecular cloning of a functional thrombin receptor reveals a novel proteolytic mechanism of receptor activation. Cell 64:1057-1068

Waksman G, Kominos D, Robertson SC, Pant N, Baltimore D, Birge RB, Cowburn D, Hanafusa H, Mayer BJ, Overduin M, Resh MD, Rios CB, Silverman L, Kuriyan J (1992) Crystal structure of the phosphotyrosine recognition domain SH2 of v-*src* complexed with tyrosine-phosphorylated peptides. Nature 358:646-653

Yarden Y and Ullrich A (1988) Growth factor receptor tyrosine kinases. Annu Rev Biochem 57:443-4781

STUDIES OF THE VERY EARLY RESPONSES OF A RECEPTOR TYROSINE KINASE TO GROWTH FACTOR BINDING AND THEIR APPLICATION TO THE PURIFICATION AND IDENTIFICATION OF PROTEINS THAT ARE TYROSINE PHOSPHORYLATED IN THE GROWTH FACTOR RESPONSE

E. Richard Stanley, Yee-Guide Yeung, Karen L. Berg & Fiona J. Pixley
Department of Developmental Biology and Cancer
Albert Einstein College of Medicine
1300 Morris Park Avenue
Bronx, NY 10461
USA

Abstract:

Growth factors whose effects are mediated via receptor tyrosine kinases induce changes in these receptors and other membrane and cytoplasmic proteins within 30 sec of addition. This paper describes kinetic analyses of these changes using the colony stimulating factor-1 (CSF-1)/CSF-1 receptor system and indicates how these studies have been useful in developing methods for the purification and identification of proteins that are rapidly phosphorylated in tyrosine in response to growth factor addition.

Introduction:

Growth factors whose effects are mediated via receptor tyrosine kinases induce changes in these receptors within 30 sec of addition to target cells. Changes that take place in the receptor itself can involve ligand-induced dimerization, activation of the tyrosine kinase, tyrosine phosphorylation, kinase inactivation, serine phosphorylation, ubiquitination and internalization. During the first 30 seconds of stimulation, many other, mostly cytoplasmic, proteins also exhibit increased phosphorylation in tyrosine. These proteins may be directly phosphorylated

NATO ASI Series, Vol. H 76
Tyrosine Phosphorylation/Dephosphorylation and Downstream Signalling
Edited by L. M. G. Heilmeyer Jr.

by the activated receptor, indirectly by non-receptor tyrosine kinases that are activated as part of a signal transduction process, or their tyrosine phosphorylation may increase due to growth factor-induced inhibition of a protein tyrosine phosphatase. In the case of some receptors, identification of several of these proteins has been possible due to their association with the activated, tyrosine phosphorylated receptor via their SH2 domains. However the identities of the majority are still unknown.

The CSF-1/CSF-1 receptor (CSF-1R) system is one system being used to investigate ligand-induced changes in the receptor with the aim of eventually determining how they might cause altered behavior or activity of the receptor. This work is reviewed here together with the approaches that we have used for the purification, identification and characterization of the proteins that are rapidly tyrosine phosphorylated in response to CSF-1.

CSF-1 and the CSF-1 Receptor:

CSF-1 is a homodimeric growth factor that regulates the survival, proliferation and differentiation of mononuclear phagocytes, the differentiation of osteoclast progenitors and the function of several cell types in the female reproductive tract during pregnancy (reviewed in Stanley, 1990). Some target populations are regulated by circulating CSF-1, while others are regulated by locally presented growth factor (Wiktor-Jedrzejczak et al., 1991). CSF-1 can be secreted as a glycoprotein or a proteoglycan (Price et al., 1992), or alternatively be expressed as a membrane-spanning glycoprotein (Rettenmier et al., 1987) on the cell surface of producing cells. This latter form of growth factor and perhaps also the proteoglycan species, may be involved in local regulation.

The CSF-1R possesses a single class of high affinity binding sites (K_d=0.1 nM) for its ligand (Guilbert and Stanley, 1986). The receptor was purified, shown to possess protein tyrosine kinase activity (Yeung et al., 1987) and then shown to share identity with the c-*fms* proto-oncogene product (Sherr et al., 1985). The c-*fms* cDNA sequence revealed that the receptor belongs to a subclass of tyrosine kinase receptors, possessing an extracellular ligand-binding domain with a cysteine spacing characteristic of the immunoglobulin gene superfamily, a transmembrane domain and an intracellular protein tyrosine kinase domain interrupted by an kinase insert of 72 hydrophilic amino acids (Coussens et al., 1986).

The response of target macrophages to CSF-1 is pleiotropic including both early morphological changes (within seconds) (Boocock et al., 1989) and later events such as an increase in the protein synthetic rate (within minutes) (Tushinski and Stanley, 1983) and the subsequent entry into S phase (within hours) (Tushinski and Stanley, 1985). CSF-1, in the presence or absence of bovine serum, is required for almost all of the lag-phase preceding the entry of quiescent cells into S phase (Tushinski and Stanley, 1985). CSF-1 is required for macrophage survival and effects of this growth factor on survival and proliferation can be separated in different ways (Tushinski et al., 1982; Pollard et al., 1991).

Analysis of the Structure-Function Relationships of the CSF-1R by Site-Directed Mutagenesis:

Considerable information concerning the structure-function relationships of the CSF-1 receptor have been obtained from studies utilizing site-directed mutagenesis. This approach has been used with several kinds of cell lines, predominantly with mouse NIH 3T3 cells expressing the human CSF-1R and rat fibroblast cell lines expressing the mouse or feline CSF-1R. Fibroblasts produce CSF-1, but high concentrations of mouse or rat CSF-1 are unable to bind the human CSF-1R and only very high concentrations of mouse CSF-1 bind the feline CSF-1R. Thus cell transformation in these systems is dependent on the addition of the cognate CSF-1 for the CSF-1R expressed by the cells.

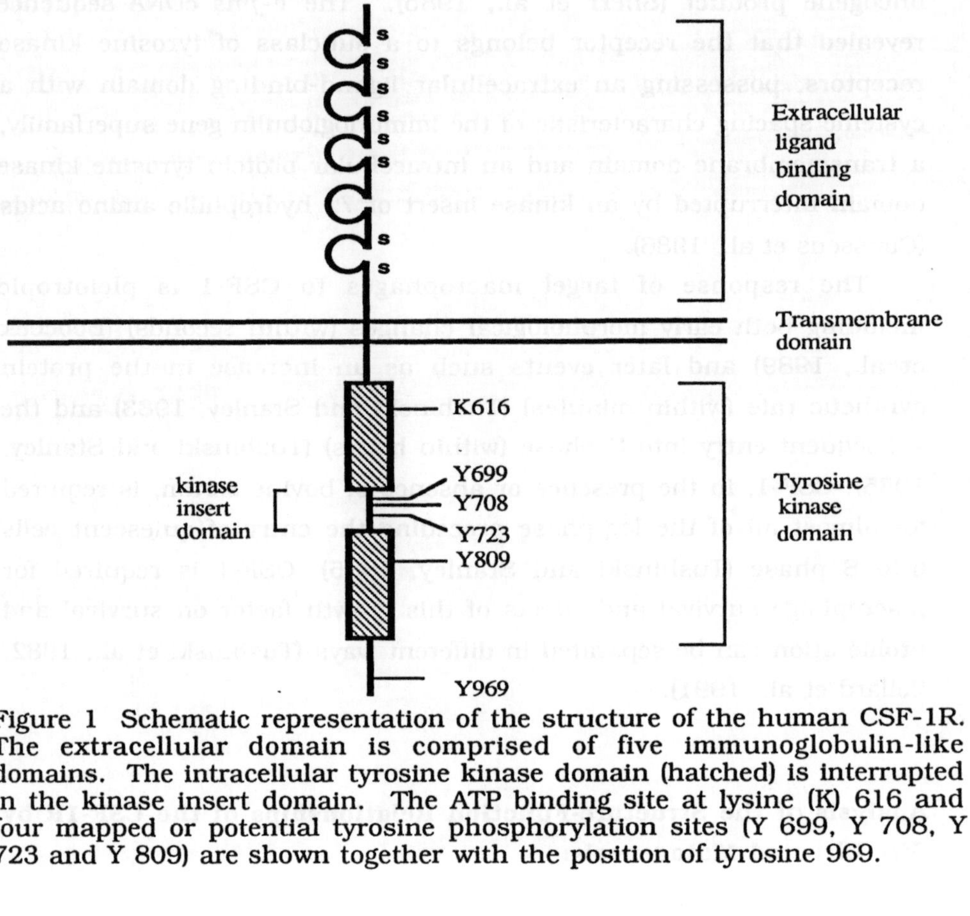

Figure 1 Schematic representation of the structure of the human CSF-1R. The extracellular domain is comprised of five immunoglobulin-like domains. The intracellular tyrosine kinase domain (hatched) is interrupted in the kinase insert domain. The ATP binding site at lysine (K) 616 and four mapped or potential tyrosine phosphorylation sites (Y 699, Y 708, Y 723 and Y 809) are shown together with the position of tyrosine 969.

Mutations in the extracellular domain of the CSF-1R, in some cases in the context of a C terminal truncation (Woolford et al., 1988), constitutively activate the kinase, resulting in ligand-independent transformation, consistent with their mimicry of ligand-binding (Woolford et al., 1988; Roussel et al., 1988). Kinase negative receptor mutants in which the lysine involved in ATP binding is mutated are unable to signal exhibit CSF-1-dependent internalization, but fail to exhibit CSF-1 dependent degradation (Downing et al., 1989b; Carlberg et al., 1991). One kinase negative mutant has been used to demonstrate that ligand-induced tyrosine phosphorylation of the CSF-1R can occur through an intermolecular reaction (Ohtsuka et al., 1990).

The system utilizing the human CSF-1R expressed in NIH 3T3 cells is the most extensively studied. In this system, the ligand-induced tyrosine phosphorylation sites Tyr 708, Tyr 723 (in the kinase insert domain) appear to be important for the association of phosphatidyl-inositol-3 kinase (Downing et al., 1991), but elimination of the entire kinase insert domain did not affect CSF-1-induced cell proliferation (Reedijk et al., 1990; Shurtleff et al., 1990). Replacement of Tyr 809 with phenylalanine substantially reduces the ligand-induced mitogenic response and elevation in *myc* mRNA without affecting *fos* and *junB* mRNA induction, only slightly lowering the protein tyrosine kinase activity of the receptor and protein tyrosine phosphorylation (Roussel et al., 1990, 1991). Although other interpretations are possible, this result raises the possibility that a bifurcation of signal transduction pathways occurs at the level of the receptor itself. Alteration of Tyr 969 to phenylalanine causes enhanced signalling by the receptor (Roussel et al., 1987). As Tyr 969 is apparently not a tyrosine phosphorylation site, this result suggests that the tyrosine is important at that position for ligand-induced conformational changes that negatively regulate signalling.

While considerable information has been learned from these site-directed mutagenesis studies, it is important to realize that these are carried out in heterologous systems in which the receptor of one species is expressed in an inappropriate lineage of another species.

Kinetic Analysis of Ligand-Induced Changes in the CSF-1R:

Because we wished to use a somatic cell genetic approach to the analysis of CSF-1 signal transduction all of these studies were carried out in the CSF-1-dependent mouse macrophage cell line, BAC1.2F5 (Morgan et al., 1987). Cells of this line possess approximately 120,000 CSF-1R per cell, proliferate well in response to CSF-1 and less well in response to granulocyte-macrophage colony stimulating factor (GM-CSF) (Morgan et al., 1987; Morgan and Stanley, 1984). Like mature macrophages, BAC1.2F5 cells exhibit rapid morphological changes that are apparent within 60 sec of addition of CSF-1.

For our studies of the kinetics of the ligand-induced changes in the CSF-1R, several techniques were critical:

1. The development of methods for the quantitative recovery of cell-surface receptors from unstimulated and stimulated cells by reacting the cells with anti-CSF-1R antibody prior to cell lysis and processing (Li and Stanley, 1991; Baccarini et al., 1991b).
2. Prevention of protein dephosphorylation after cell lysis by the elimination of EDTA, an activator of protein tyrosine phosphatases and the inclusion of Zn^{++}, an inhibitor of some protein tyrosine phosphatases, in the buffers usually used for cell lysis and immunoprecipitation (Baccarini et al., 1991b, Li et al., 1991).
3. Dual-labelling of cells with [^{3}H]-leucine and $^{32}P_i$ in order to obtain the relative specific activity of phosphate incorporated into the CSF-1R tyrosine and serine residues.
4. The use of cross-linking immediately after cell lysis (highly efficient), in conjunction with cross-linking prior to lysis (not very efficient), to establish the multimeric state of the receptor (Li and Stanley, 1991).
5. The use of low concentration gradient (3-10% acrylamide) SDS-PAGE in order to observe the higher molecular mass dimeric forms of the CSF-1R (Li and Stanley, 1991).
6. By studying these reactions at 4°C, the time course of the events of the first 30 sec are expanded to 120 min, allowing clear resolution of the order of events and the isolation of intermediates (with the exception of ligand dissociation and receptor internalization, the ligand-induced receptor events which take place within 30 sec of ligand addition at 37°C also take place at 4°C) (Sengupta et al., 1988).
7. The stimulation of cells with very high concentrations of

CSF-1 so that receptor sites are saturated within 1 min at 4°C and ligand binding does not limit the kinetic analysis of subsequent events (Sengupta et al., 1988; Li and Stanley, 1991).

The application of these and other techniques to the analysis of the ligand-induced changes in the CSF-1R has provided the following information. Addition of CSF-1 to cells at 4°C initially induces (by 2-5 min) the appearance of non-covalent 380-kDa CSF-1 tyrosine phosphorylated receptor homodimers that are associated with activation of the CSF-1 receptor kinase. The tyrosine phosphorylation of a specific group of primarily cytosolic proteins is then observed, becoming maximal by 45 min. However, prior to this point and as the rate of appearance of the tyrosine phosphorylated proteins slows, extracellular covalent linkage of the 380-kDa homodimers occurs via disulfide bonds. This covalent linking of the receptor subunits leads to the formation, by 90 min, of a 450-kDa heterodimer comprising a 165-kDa and a modified 215-kDa or 250-kDa form of the receptor (either covalently or non-covalently associated) (Li and Stanley, 1991). Experiments in this and other systems indicate that this modification is polyubiquitination (Mori et al., 1992), presumably required for degradation of the cytoplasmic domain of the CSF-1R. Associated with the formation of the heterodimeric species, there is a further increase in CSF-1R tyrosine and serine phosphorylation (Li and Stanlcy, 1991; Baccarini et al., 1991b).

In experiments at 37°C or on temperature shift to 37°C, the heterodimeric 450-kDa species are selectively internalized. Consistent with this observation and the correlation of active signalling with the presence of the non-covalently associated 380-kDa homodimers, pretreatment of the cells with the alkylating agent iodoacetic acid prevents formation of both the disulfide-bonded 380-kDa homodimer and the 450-kDa heterodimer, inhibits internalization and causes an approximately 8-fold increase in the cellular protein tyrosine phosphorylation (Li and Stanley, 1991).

Thus it appears that ligand-induced non-covalent dimerization activates the CSF-1 receptor kinase, whereas the covalent dimerization and subsequent modifications lead to kinase inactivation, phosphotyrosine dephosphorylation and internalization of the receptor-ligand complex. The kinetically later modification of polyubiquitination appears to be preparatory to intracellular CSF-1R degradation. This kind of kinetic analysis provides a solid base from which to look for causal relationships between alterations in the receptor structure and its function. Studies involving the determination of the sequence of phosphorylations at the tyrosine and serine phosphorylation sites and the effects of site-directed mutagenesis of these sites and the cysteines involved in the disulfide bonding can now be carried out in a more rational way.

An example of the usefulness of this kind of kinetic analysis was demonstrated in our analysis of the involvement of the cytoplasmic serine/threonine kinase, RAF-1, which was claimed to be tyrosine phosphorylated in the response of cells to platelet derived growth factor (PDGF) (Morrison et al., 1988; Morrison et al., 1989). Detailed analysis of the phosphorylation of RAF-1 in response to CSF-1 and platelet derived growth factor PDGF (Baccarini et al., 1990) or epidermal growth factor (EGF) (Baccarini et al., 1991a) revealed that RAF-1 is phosphorylated on serine and perhaps threonine, but not tyrosine. In contrast to the proteins that were rapidly phosphorylated at 37°C, which were also tyrosine phosphorylated at 4°C, RAF-1 was not phosphorylated when cells were stimulated with CSF-1 at 4°C. Furthermore, at 37°C, protein tyrosine phosphorylation was essentially complete by 30 sec of ligand addition, whereas RAF-1 serine phosphorylation did not commence until 1.5 min and was not maximal until at least 15 min (Baccarini et al., 1990). In the case of EGF, this increase in RAF-1 serine phosphorylation continued, even following dissociation of EGF from its receptor and substantial dephosphorylation of the EGF receptor (Baccarini et al., 1991a). These studies provided evidence that the activated tyrosine kinase activities of these receptors enhance serine phosphorylation of RAF-1 and the activation of the RAF-1 kinase via an intermediary molecule(s).

Purification and Identification of the Tyrosine Phosphorylated Proteins:

Analysis of the kinetics of appearance of the anti-phosphotyrosine (α-PY)-reactive proteins, carried out at 4°C, indicated that their order of appearance was the 165-kDa (CSF-1 receptor) protein by 3min, 115- and 99-kDa proteins by 10 min, the 86-, 75-, 57- and 36-kDa proteins by 30 min and the modified 250-kDa form of the CSF-1 receptor by 90 min (Li et al., 1991). With the exception of the receptor, most of the proteins are cytoplasmic (Sengupta et al., 1988). In contrast to the EGF and PDGF receptors for which phospholipase C-gamma has been shown to be substrate, phospholipase C-gamma is not a CSF-1 receptor substrate (Downing et al., 1989a) and none of the major intracellular CSF-1 receptor substrates, apart from the receptor itself, have yet been identified. However, CSF-1 has been shown to activate phosphatidyl-inositol kinase and to associate with this enzyme following stimulation (Varticovski et al., 1989). By analogy with the effects of PDGF and EGF stimulation of their target cells (Otsu et al., 1991; Escobedo et al., 1991; Skolnik et al., 1991), it is likely that the 86-kDa protein that is present in the α-PY-reactive fraction of CSF-1 stimulated cells is the subunit of this enzyme.

In contrast to the EGF and PDGF receptors, we have found that few of the tyrosine phosphorylated proteins appear to stably associate with the CSF-1 receptor, indicating that receptor-based purification or cloning were not generally applicable methods for their identification. We have therefore adopted the approach of direct identification by purification and microsequencing. As these proteins collectively represent less than 0.1% of the total cellular protein, in adopting the purification approach, we have focused on optimizing their generation in response to CSF-1 and their recovery during purification. Earlier studies, some of which have been outlined above, provided important information relative to the optimization of their generation and recovery. This information may be summarized as follows:

1. Stimulation of cells at 4°C increased the total yield of tyrosine phosphorylated proteins and slowed the rate of

their appearance (Sengupta et al., 1988). Use of 4°C stimulation therefore, not only increased the amount of starting phosphotyrosyl protein, but also meant that larger numbers of plates could be easily handled due to the longer stimulation time (2 h at 4°C versus 30 sec at 37°C).

2. The majority of the tyrosine phosphorylated proteins are recovered in the cytoplasmic fraction (Sengupta et al., 1988). Purification from this fraction, rather that whole cell lysates, simplified the procedures required.
3. Elimination of EDTA from lysis and fractionation buffers, coupled with the inclusion of 5 μM Zn^{++} prevented substantial tyrosine dephosphorylation encountered after cell lysis (Li et al., 1991). Use of these buffer conditions in the initial stages of fractionation was extremely important.
4. Pretreatment of the cells with iodoacetic acid increased the rate and extent of CSF-1 stimulated protein tyrosine phosphorylation (Li and Stanley, 1991). Use of IAA pretreated cells greatly increased the initial yield of tyrosine phosphorylated proteins.
5. Preliminary experiments utilizing SDS-PAGE-based purification steps for the proteins and their peptides (after fragmentation) were unsatisfactory because higher amounts were required to obtain amino acid sequence, compared with material obtained from HPLC fractionation.

An outline of the purification procedure is shown in Fig. 2. PY-containing proteins from the cytosol of 2,000 subconfluent 100 mm dish cultures of BAC1.2F5 cells and from 2,000 subconfluent cultures of CSF-1-stimulated BAC1.2F5 cells are separately processed as outlined in detail elsewhere (Yeung et al., 1992). Following size-exclusion chromatography under denaturing conditions, selected pool fractions corresponding to different molecular weight ranges are separated by reverse phase high pressure liquid chromatography (RP-HPLC) on a C_4 column. Protein peaks of interest are then subjected to tryptic digestion and the tryptic peptides separated by RP-HPLC on a C_8 column. Individual tryptic peptide peaks are then sequenced using a microsequencer.

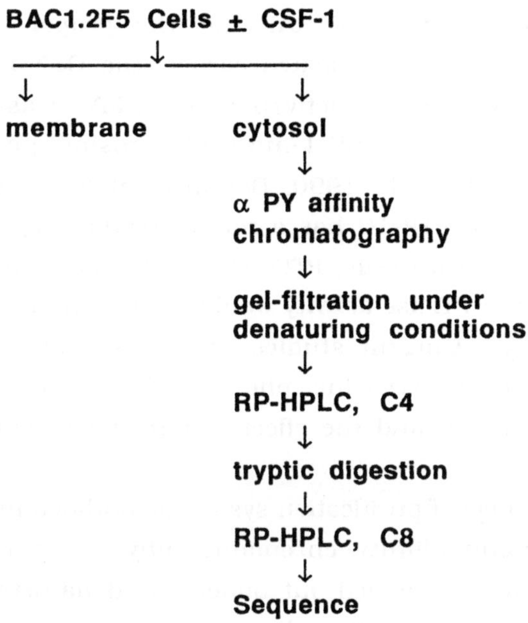

Fig.2 Scheme for the purification and sequencing of cytosolic α-PY reactive proteins.

The yields for proteins of 70-kDa or less are excellent (~3 μg) and sequencing was successful at the 5 picomole level. For proteins with molecular masses above 70-kDa, significant losses are encountered in the HPLC step, necessitating an alternative approach.

The first tyrosine phosphorylated protein to be identified in this fashion was protein tyrosine phosphatase-1C (PTP-1C) (Yeung et al., 1992). Its tyrosine phosphorylation was confirmed by anti-phosphotyrosine Western blotting of the purified protein. PTP-1C mRNA is expressed in hemopoietic cells and in some epithelial cells, but not in fibroblasts (Plutzky et al., 1992; Matthews et al., 1992; Yi et al., 1992; Yi et al., 1991). Interestingly, in a recent study, Perkins et al. (1992) have shown that the *Drosophila* gene, *corkscrew* (*csw*) encodes a PTPase that shows 52% amino acid sequence homology with human PTP-1C. They have provided evidence that *csw* acts downstream of the receptor tyrosine kinase *torso*, and is required in concert with the cytoplasmic serine/threonine kinase, D-RAF, for the proper determination of terminal cell fates. As indicated earlier, the CSF-1R-induced RAF-1 serine phosphorylation and activation of the RAF-1 associated kinase is temporally later than CSF-1-induced tyrosine phosphorylation of PTP-1C (Baccarini et al., 1990; Baccarini et al., 1991a). If tyrosine phosphorylation of PTP-1C has a role in regulating its PTPase activity and if, by analogy with *csw* , PTP-1C is in the same pathway as RAF-1, then regulation of PTPase activity would be upstream of RAF-1 activation in the pathway. Current studies are focused on the association of PTP-1C with other tyrosine phosphorylated proteins, its role in mitogenic signalling, and the effect of tyrosine phosphorylation on its catalytic activity.

One advantage of purification system as outlined in Fig. 2 is that the first step, immunoaffinity chromatography on anti-phosphotyrosine antibody columns is carried out under non-denaturing conditions and the subsequent size-exclusion chromatography step can also be carried out under non-denaturing conditions. When fractions from a non-denaturing size-exclusion column are examined by SDS-PAGE, it is clear that many of the anti-PY reactive proteins are associated with other proteins in complexes or multi-subunit proteins. Proteins co-fractionating with identified proteins (putative subunits or complex components) may be identified by Western blotting fractions for the identified protein or by silver staining and chemical crosslinking experiments with relatively small amounts of material. Once identified in the large-scale denaturing purification, they can be targeted for sequencing.

Biological Significance of Early Events in Growth Factor Signal Transduction:

A major problem in the analysis of early events in signal transduction, is to determine which subsequent events they regulate. The approaches taken to this problem generally involve genetically or otherwise abrogating or stimulating a particular response by altering the expression or behavior of a component involved in the early event. Such approaches as microinjection or scrape-loading of antibodies to proteins or the proteins themselves, the use of anti-sense RNA and dominant negative mutant constructs require that the identity of the component be known. These approaches have been covered by other lecturers at this meeting. Another important approach has been the analysis of cells expressing mutant receptors that defectively trigger some early events but not others. The failure of such mutant receptors to trigger certain long-term responses can provide a correlation between long-term responses and early events.

In addition to these approaches we have utilized two other methods. One of these has been to identify common early events associated with long-term responses in different systems. This approach was applied to the tyrosine phosphorylated proteins in an attempt to identify those whose tyrosine phosphorylation is important for the proliferative response. Among the proteins that are phosphorylated in tyrosine in response to CSF-1, a 57-kDa protein is also phosphorylated in tyrosine in fibroblasts responding to PDGF or in fibroblasts transformed by v-*fms*. Interestingly, this protein is the only protein that is reproducibly phosphorylated in tyrosine in the proliferative response of macrophages to GM-CSF(Li et al., 1991). Mouse fibroblasts expressing the human CSF-1 receptor have been shown to exhibit human CSF-1 dependent proliferation in semisolid culture medium (Roussel and Sherr, 1989). Thus tyrosine phosphorylation of the 57-kDa protein may represent an initial step in a common post-receptor pathway for the regulation of cell proliferation.

In a second approach to identifying those initial events which are important for the regulation of cell proliferation, we have isolated approximately 20 independently arising, mutagen-induced, factor-independent mutant cell lines from BAC1.2F5 cells (Pollard et al., 1991; Dello Sbarba et al., 1991). Among these cell lines, excluding the three which produce CSF-1, there are three lines which possess altered patterns of cytoplasmic protein tyrosine phosphorylation (Dello Sbarba et al., 1991), consistent with an important role for protein tyrosine phosphorylation in the immediate post-receptor pathway regulating the entry of macrophages into S-phase. It is anticipated that these mutant cell lines will be useful in the further analysis of CSF-1 signal transduction.

Acknowledgements

This work was supported by National Institutes of Health Grant CA 26504, Albert Einstein Core Cancer Grant P30-CA 1330, and a grant from the Lucille P. Markey Charitable Trust.

Literature References

Baccarini M, Sabatini DM, App H, Rapp UR, Stanley ER (1990) Colony stimulating factor-1 (CSF-1) stimulates temperature dependent phosphorylation and activation of the RAF-1 proto-oncogene product. EMBO J 9:3649-3657

Baccarini M, Gill GN, Stanley ER (1991a) Epidermal growth factor stimulates phosphorylation of RAF-1 independently of receptor autophosphorylation and internalization. J Biol Chem 266:10941-10945

Baccarini M, Li W, Dello Sbarba P, Stanley ER (1991b) Increased phosphorylation of the colony stimulating factor-1 receptor following transmembrane signaling. Receptor 1:243-259

Boocock CA, Jones GE, Stanley ER, Pollard JW (1989) Colony-stimulating factor-1 induces rapid behavioural responses in the mouse macrophage cell line, BAC1.2F5. J Cell Sci 93:447-456

Carlberg K, Tapley P, Haystead C, Rohrschneider L (1991) The role of kinase activity and the kinase insert region in ligand-induced internalization and degradation of the c-*fms* protein. EMBO J 10:877-883

Coussens L, Van Beveren C, Smith D, et al. (1986) Structural alteration of viral homologue of receptor proto-oncogene fms at carboxy-terminus. Nature 32:277-280

Dello Sbarba P, Pollard JW, Stanley ER (1991) Alterations in CSF-1 receptor expression and protein tyrosine phosphorylation in autonomous mutants of a CSF-1 dependent macrophage cell line. Growth Factors 5:75-85

Downing JR, Margolis BL, Zilberstein A, et al. (1989a) Phospholipase C-gamma, a substrate for PDGF receptor kinase, is not phosphorylated on tyrosine during the mitogenic response to CSF-1. EMBO J 8:3345-3350

Downing JR, Roussel MF, Sherr CJ (1989b) Ligand and protein kinase C downmodulate the colony-stimulating factor 1 receptor by independent mechanisms. Mol Cell Biol 9:2890-2896

Downing JR, Shurtleff SA, Sherr CJ (1991) Peptide antisera to human colony-stimulating factor 1 receptor detect ligand-induced conformational changes and a binding site for phosphatidylinositol 3-kinase. Mol Cell Biol 11:2489-2495

Escobedo JA, Navankasattusas S, Kavanaugh WM, Milfay D, Fried VA, Williams LT (1991) cDNA cloning of a novel 85 kd protein that has SH2 domains and regulates binding of PI3-kinase to the PDGF ß-receptor. Cell 65:72-82

Guilbert LJ, Stanley ER (1986) The interaction of ^{125}I-colony stimulating factor-1 with bone marrow-derived macrophages. J Biol Chem 261:4024-4032

Li W, Stanley ER (1991) Role of dimerization and modification of the CSF-1 receptor in its activation and internalization during the CSF-1 response. EMBO J 10:277-288

Li W, Yeung YG, Stanley ER (1991) Tyrosine phosphorylation of a common 57-kDa protein in growth factor-stimulated and -transformed cells. J Biol Chem 266:6808-6814

Matthews, RJ, Browne DB, Flores, E, Thomas, ML (1992) Characterization of hematopoietic intracellular protein tyrosine phosphatases: Description of a phosphatase containing an SH2 domain and another enriched in proline-, glutamic acid-, serine-, and threonine-rich sequence. Mol Cell Biol 12:2396-2405

Morgan CJ, Stanley ER (1984) Chemical crosslinking of the mononuclear phagocyte specific growth factor CSF-1 to its receptor at the cell surface. Biochem Biophys Res Commun 119:35-41

Morgan CJ, Pollard JW, Stanley ER (1987) Isolation and characterization of a cloned growth factor dependent macrophage cell line, BAC12F5. J Cell Physiol 130:420-427

Mori, S, Heldin, CH, Claesson,-Welsh, L (1992) Ligand-induced polyubiquitiination of the platelet-derived growth factor beta-receptor. J Biol Chem 267:6429-6434

Morrison DK, Kaplan DR, Rapp U, Roberts TM (1988) Signal transduction from membrane to cytoplasm: Growth factors and membrane-bound oncogene products increase Raf-1 phosphorylation and associated protein kinase activity. Proc Natl Acad Sci USA 85:8855-8859

Morrison DK, Kaplan DR, Escobedo JA, Rapp UR, Roberts TM, Williams LT (1989) Direct activation of the serine/threonine kinase activity of Raf-1 through tyrosine phosphorylation by the PDGF ß- receptor. Cell 58:649-657

Ohtsuka M, Roussel MF, Sherr CJ, Downing JR (1990) Ligand-induced phosphorylation of the colony-stimulating factor 1 receptor can occur through an intermolecular reaction that triggers receptor down modulation. Mol Cell Biol 10:1664-1671

Otsu M, Hiles I, Gout I, et al. (1991) Characterization of two 85 kd proteins that associate with receptor tyrosine kinases, middle-T/pp60$^{c\text{-}src}$ complexes, and PI3-kinase. Cell 65:91-104

Perkins LA, Larsen, I, Perrimon, N (1992) *corkscrew* encodes a putative protein tyrosine phosphatase that functions to transduce the terminal signal from the receptor tyrosine kinase *torso*. Cell 70:225-236.

Plutzky J, Neel BG, Rosenberg RD (1992) Isolation of a src homology 2-containing tyrosine phosphatase. Proc Natl Acad Sci USA 89:1123-1127

Pollard JW, Morgan CJ, Dello Sbarba P, Cheers C, Stanley ER (1991) Independently arising macrophage mutants dissociate growth factor-regulated survival and proliferation. Proc Natl Acad Sci USA 88:1474-1478

Price LKH, Choi HU, Rosenberg L, Stanley ER (1992) The predominant form of secreted colony stimulating factor-1 is a proteoglycan. J Biol Chem 267:2190-2199

Reedijk M, Liu X, Pawson T (1990) Interactions of phosphatidylinositol kinase, GTPase-activating protein (GAP), and GAP-associated proteins with the colony-stimulating factor 1 receptor. Mol Cell Biol 10:5601-5608

Rettenmier CW, Roussel MF, Ashmun RA, Ralph P, Price K, Sherr CJ (1987) Synthesis of membrane-bound colony-stimulating factor 1 (CSF-1) and downmodulation of CSF-1 receptors in NIH 3T3 cells transformed by cotransfection of the human CSF-1 and c-fms (CSF-1 receptor) genes. Mol Cell Biol 7:2378-2387

Roussel MF, Dull TJ, Rettenmier CW, Ralph P, Ullrich A, Sherr CJ (1987) Transforming potential of the c-fms proto-oncogene (CSF- 1 receptor). Nature 325:549-552

Roussel MF, Downing JR, Rettenmier CW, Sherr CJ (1988) A point mutation in the extracellular domain of the human CSF-1 receptor (c-fms proto-oncogene product) activates its transforming potential. Cell 55:979-988

Roussel MF, Sherr CJ (1989) Mouse NIH 3T3 cells expressing human colony-stimulating factor 1 (CSF-1) receptors overgrow in serum-free medium containing human CSF-1 as their only growth factor. Proc Natl Acad Sci USA 86:7924-7927

Roussel MF, Shurtleff SA, Downing JR, Sherr CJ (1990) A point mutation at tyrosine-809 in the human colony-stimulating factor 1 receptor impairs mitogenesis without abrogating tyrosine kinase activity, association with phosphatidylinositol 3-kinase, or induction of c-*fos* and *junB* genes. Proc Natl Acad Sci USA 87:6738-6742

Roussel, M.F., Cleveland, J.L., Shurtleff, S.A. and Sherr, C.J. (1991) *Myc* rescue of a mutant CSF-1 receptor impaired in mitogenic signalling. Nature 353:361-363.

Sengupta A, Liu W-K, Yeung Y-G, Yeung DC-Y, Frackelton AR, Stanley ER (1988) Identification and subcellular localization of proteins that are rapidly phosphorylated in tyrosine in response to colony stimulating factor 1. Proc Natl Acad Sci USA 85:8062-8066

Sherr, CJ, Rettenmier, CW, Sacca, R, Roussel, MF, Look, AT, Stanley, ER (1985) The c-*fms* proto-oncogene product is related to the receptor for the mononuclear phagocyte growth factor, CSF-1. Cell 41:665-676.

Shurtleff SA, Downing JR, Rock CO, Hawkins SA, Roussel MF, Sherr CJ (1990) Structural features of the colony-stimulating factor 1 receptor that affect its association with phosphatidylinositol 3-kinase. EMBO J 9:2415-2421

Skolnik EY, Margolis B, Mohammadi M, et al. (1991) Cloning of PI3 kinase-associated p85 utilizing a novel method for expression/cloning of target proteins for receptor tyrosine kinases. Cell 65:83-90

Stanley ER Role of colony stimulating factor-1 in monocytopoiesis and placental development. In: Mahowald AP eds (ed. 1990) Genetics of Pattern Formation and Growth Control. Wiley-Liss New York, NY pp 165-180

Tushinski RJ, Oliver IT, Guilbert LJ, Tynan PW, Warner JR, Stanley ER (1982) Survival of mononuclear phagocytes depends on a lineage-specific growth factor that the differentiated cells selectively destroy. Cell 28:71-81

Tushinski RJ, Stanley ER (1983) The regulation of macrophage protein turnover by a colony stimulating factor (CSF-1). J Cell Physiol 116:67-75

Tushinski RJ, Stanley ER (1985) The regulation of mononuclear phagocyte entry into S phase by the colony stimulating factor CSF-1. J Cell Physiol 122:221-228

Varticovski L, Druker B, Morrison D, Cantley L, Roberts T (1989) The colony stimulating factor-1 receptor associates with and activates phosphatidylinositol-3 kinase. Nature 343:699-702

Wiktor-Jedrzejczak W, Urbanowska E, Aukerman SL, et al. (1991) Correction by CSF-1 of defects in the osteopetrotic *op/op* mouse suggests local, developmental, and humoral requirements for this growth factor. Exp Hematol 19:1049-1054

Woolford J, McAuliffe A, Rohrschneider LR (1988) Activation of the feline c-fms proto-oncogene: multiple alterations are required to generate a fully transformed phenotype. Cell 55:965-977

Yeung, YG, Berg, KL, Pixley, FJ, Angeletti, RH and Stanley, ER (1992) Protein tyrosine phosphatase-1C is rapidly phosphorylated in tyrosine in macrophages in response to colony stimulating factor-1. J. Biol. Chem. 267:23447-23450

Yeung YG, Jubinsky PT, Sengupta A, Yeung DC-Y, Stanley ER (1987) Purification of the colony-stimulating factor 1 receptor and demonstration of its tyrosine kinase activity. Proc Natl Acad Sci USA 84:1268-1271

Yi T, Cleveland JL, Ihle JN (1991) Identification of novel protein tyrosine phosphatases of hematopoietic cells by polymerase chain reaction amplification. Blood 78:2222-2228

Yi T, Cleveland JL, Ihle JN (1992) Protein tyrosine phosphatase containing SH2 domains: Characterization, preferential expression in hematopoietic cells, and localization to human chromosome 12. p12-p13. Mol Cell Biol 12:836-846

THE ROLE OF P21<u>RAS</u> IN CELL SIGNALLING

Chris J. Marshall
Institute of Cancer Research
Chester Beatty Laboratories
237 Fulham Road
London SW3 6JB

The identification of p21*ras* as an oncogene suggests that the normal p21*ras* proteins play a role in cell proliferation. Inhibition of normal p21*ras* either through the microinjection of antibodies or by using a dominant interfering mutant of p21*ras* (asn17), which is thought to prevent guanine nucleotide exchange, blocks stimulation of DNA synthesis through a variety of tyrosine kinase growth factor receptors (Mulcahy et al., 1985; Feig and Cooper, 1988). Furthermore, oncogenic transformation mediated by tyrosine kinase oncogenes such as v-src is also blocked by inhibition of p21*ras* function (Smith et al., 1986). These results clearly indicate that p21*ras* is somehow required for signal transduction through tyrosine kinases. However, p21*ras* is also required for the function of tyrosine kinases mediating differentiation since inhibition of p21*ras* blocks neuronal differentiation stimulated by NGF and mesoderm induction stimulated by FGF (Thomas et al., 1992; Wood et al., 1992). In some cases stimulation of tyrosine kinases can be shown to increase the level of GTP bound to p21*ras* arguing that tyrosine phosphorylation of proteins that interact with p21*ras* can either promote GDP exchange or inhibit GTPase activity (Burgering et al., 1991). In addition, stimulation of some haemopoietic growth factor receptors which lack intrinsic tyrosine kinase activity can also lead to an increase in the amount of p21*ras* in the GTP bound form.

Activation of tyrosine kinases leads to the tyrosine phosphorylation of *ras* GAP and the promotion of interactions via its SH2 domains with p62 and p190 proteins (Ellis et al., 1990). In some cases *ras* GAP can be shown to bind to autophosphorylated tyrosine kinase receptors. Both p62 and p190 have now been identified, p62 (Wong et al., 1992) is an RNA binding protein while p190 is a multidomain protein surprisingly containing a putative GTP binding region, a GTPase activating region related to that of the GTPase activating proteins of p21*rho* and *rac* and a transcription factor domain (Settleman et al., 1992). There is a strong possibility that either or both of these proteins is involved with p21*ras* in mediating the activation of signal transduction pathways. Of great interest is the possibility that p190 may play a role in linking p21*ras* to the *rho* and *rac* proteins which regulate the cytoskeleton (Ridley et al., 1992).

NATO ASI Series, Vol. H 76
Tyrosine Phosphorylation/Dephosphorylation
and Downstream Signalling
Edited by L. M. G. Heilmeyer Jr.

What is the role of p21*ras* in mediating signal transduction from growth factor receptors? The localisation of p21*ras* to the inner surface of the plasma membrane suggests that it is involved in events at the plasma membrane. A strong possibility is that p21*ras* somehow connects plasma membrane events to cytoplasmic signal transduction pathways. The interactions of p62 and p190 with *ras* GAP may be one component of this pathway. A further clue to function of p21*ras* comes from the demonstration that p21*ras* is required for the activation of the MAP kinase pathway by tyrosine kinases (de Vries-Smits et al., 1992). MAP kinase is itself activated by phosphorylation on threonine and tyrosine by a dual specificity protein kinase: MAP kinase kinase (Nakielny et al., 1992). Furthermore MAP kinase kinase is activated by serine/ threonine phosphorylation suggesting a kinase cascade regulating MAP kinase (Gómez and Cohen, 1991). This kinase cascade usually requires extracellular signals for activation but introduction of oncogenic p21*ras* into quiescent cells rapidly activates MAP kinase in the absence of external signals (Leevers and Marshall, 1992). This suggests that the constitutively active oncogenic p21*ras* can uncouple the activation of this pathway from the requirement for extracellular signals. In addition, *ras* transformed fibroblasts (Leevers and Marshall, 1992) and human tumour cell lines containing *ras* oncogenes show constitutive activation of MAP kinases. These results strongly point to MAP kinase playing a central role in cell transformation.

Unlike tyrosine kinase oncogenes the v-raf oncogene does not require cellular p21*ras* function to induce DNA synthesis (Smith et al., 1986). This implies that the *raf* kinase may be downstream of p21*ras* in a signalling pathway. While examining transformed cell lines for MAP kinase activation we found that v-raf transformed cells contain activated MAP kinases arguing that *raf* kinase may also activate the MAP kinase pathway. Like *raf* transformation we have shown that the activation of MAP kinase by oncogenic *raf* is independent of cellular *ras* function. Furthermore the overexpression of normal p74*raf*-1 potentiates the activation of MAP kinase following stimulation by EGF or TPA (Howe et al., 1992). These results suggest that *raf* kinase may function upstream of MAP kinase in a kinase cascade. To test whether *raf* is a kinase capable of activating MAP kinase kinase, we tested whether p74*raf*-1 could reactivate inactive MAP kinase kinase. Immunoprecipitates of p74*raf*-1 from TPA stimulated cells but not from unstimulated cells activated and phosphorylated MAP kinase kinase strongly suggesting that *raf* is a MAP kinase kinase (Howe et al., 1992). These results therefore show that there is a signal transduction pathway that contains in the following order: tyrosine kinase, *ras*, *raf*, MAP kinase kinase, MAP kinase. Since we have also shown that oncogenic p21*ras* can activate protein kinase C *in vivo* (Morris et al., 1988) and that p74*raf*-1 is activated by treatment of cells with TPA, we place protein kinase C between p21*ras* and p74*raf*-1 in this pathway. This is in contrast to studies in other systems such as T-cells where protein kinase C appears to be upstream of p21*ras* (Downward et al., 1990), however, we do not find any requirement for cellular p21*ras* for the activation of MAP kinase by TPA treatment of either fibroblasts or cos-cells (de Vries-Smits et al., 1992; Howe et al., 1992) consistent with the model that protein kinase C is downstream of p21*ras*.

One potential consequence of MAP kinase activation is increased phosphorylation and activation of c-*jun* in the c-*jun*/c-*fos* AP1 transcription factor complex (Pulverer et al., 1991). Action of the AP1 appears to be central to transformation by p21*ras* since inhibition of AP1 causes *ras* transformed cells to revert to the non-transformed phenotype (Lloyd et al., 1991). Since the elements in the MAP kinase pathway are rapidly being identified in both mammal cells and yeast the analysis of this pathway is likely to throw further light on the role of p21*ras*.

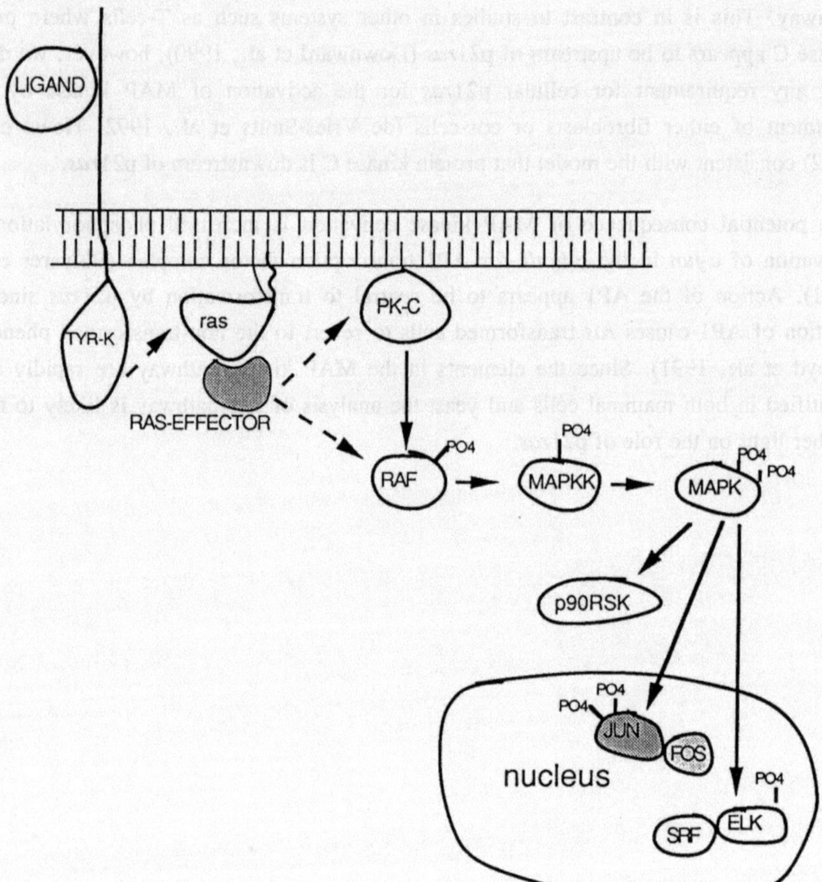

Fig. 1: Pathway of tyrosine kinase activation of MAP kinase. Dashed arrows indicate possible multiple steps. Solid arrows indicate single steps. Evidence that protein kinase C directly phosphorylates and activates *raf* is provided in Sozeri et al. (1992).

References

Burgering, B.M.Th., Medema, R.H., Maassen, J.A., van de Wetering, M.L., van der Eb, A.J., McCormick, F. and Bos, J.L. (1991), Insulin stimulation of gene expression mediated by p21*ras* activation. EMBO J. *10*, 1103-1109

de Vries-Smits, A.M.M., Burgering, B.M.Th., Leevers, S.J., Marshall, C.J. and Bos, J.L. (1992), Involvement of p21*ras* in activation of extra cellular signal-regulated kinase 2. Nature *357*, 602-604

Downward, J., Graves, J.D., Warne, P.H., Rayter, S. and Cantrell, D.A. (1990), Stimulation of p21*ras* upon T-cell activation. Nature *346*, 719-723

Ellis, C., Moran, M., McCormick, F. and Pawson, T. (1990), Phosphorylation of GAP and GAP-associated proteins by transforming and mitogenic tyrosine kinases. Nature *343*, 377-381

Feig, L.A. and Cooper, G.M. (1988), Inhibition of NIH 3T3 cell proliferation by a mutant *ras* protein with preferential affinity for GDP. Mol. Cell. Biol. *8*, 3235-3243

Gómez, N. and Cohen, P. (1991), Dissection of the protein kinase cascade by which nerve growth factor activates MAP kinases. Nature *353*, 170-173

Leevers, S.J. and Marshall, C.J. (1992), Activation of extracellular signal-regulated kinase, ERK2, by p21*ras* oncoprotein. EMBO J. *11*, 569-574

Lloyd, A., Yancheva, N. and Wasylyk, B. (1991), Transformation suppressor activity of a Jun transcription factor lacking its activation domain. Nature *352*, 635-638

Morris, J.D.H., Price, B., Lloyd, A.C., Self, A.J., Marshall, C.J. and Hall, A. (1989), Scrapeloading of Swiss 3T3 cells with *ras* protein rapidly activates protein kinase C in the absence of phosphoinositide hydrolysis. Oncogene *4*, 27-31

Mulcahy, L.S., Smith, M.R. and Stacey, D.W. (1985), Requirement for *ras* proto-oncogene function during serum-stimulated growth of NIH 3T3 cells. Nature *313*, 241-243

Nakielny, S., Cohen, P., Wu, J. and Sturgill, T. (1992), MAP kinase activator from insulin-stimulated skeletal muscle is a protein threonine/tyrosine kinase. EMBO J. *11*, 2123-2129

Pulverer, B.J., Kyriakis, J.M., Avruch, J., Nikolakaki, E. and Woodgett, J.R. (1991), Phosphorylation of c-jun mediated by MAP kinases. Nature *353*, 670-674

Ridley, A.J. and Hall, A. (1992). The small GTP-binding protein rho regulates the assembly of focal adhesions and actin stress fibres in response to growth factors. Cell *70*, 389-399

Settleman, J., Narasimhan, V., Foster, L.C. and Weinberg, R.a. (1992). Molecular cloning of cDNAs encoding the GAP-associated protein p190: Implications for a signalling pathway from ras to the nucleus. Cell *69*, 539-549

Smith, M.R., DeGudicibus, S.J. and Stacey, D.W. (1986). Requirement for c-*ras* proteins during viral oncogene transformation. Nature *320*, 540-543

Sozeri, O., Vollmer, K., Liyanage, M., Frith, D., Kour, G., Mark, G.E. and Stabel, S. (1992). Activation of the c-*raf* protein kinase by protein kinase C phosphorylation. Oncogene (in press)

Stugill, T.W., Ray, L.B., Erikson, E. and Maller, J.L. (1988). Insulin-stimulated MAP-2 kinase phosphorylates and activates ribosomal protein S6 kinase II. Nature *334*, 715-718

Thomas, S.M., DeMarco, M., D'Arcangelo, G., Halegoua, S. and Brugge, J.S. (1992). *Ras* is essential for nerve growth factor- and phorbol ester-induced tyrosine phosphorylation of MAP kinases. Cell *68*, 1031-1040

Wood, K.W., Sarnecki, C., Roberts, T.M. and Blenis, J. (1992). *Ras* mediates neve growth factor receptor modulation of three signal-transducing protein kinases: MAP kinase, *Raf*-1 and RSK. Cell *68*, 1041-1050

Wong, G., Müller, O., Clark, R., Conroy, L., Moran, M.F., Polakis, P. and McCormick, F. (1992). Molecular cloning and nucleic acid binding properties of the GAP-associated tyrosine phosphoprotein p62. Cell *69*, 551-558

THE HGF/SF RECEPTOR KINASE: STRUCTURE AND FUNCTION

Giordano S. and Comoglio P.M.
Dept of Biomedical Sciences & Oncology, University of Torino, School of Medicine
Corso M. D'Azeglio 52
10126 Torino (Italy)

ABSTRACT

Hepatocyte Growth Factor/Scatter Factor (HGF/SF) is the ligand for p190*Met*, the receptor tyrosine kinase encoded by the *MET* protooncogene. p190*MET* is a heterodimer of two disulfide-linked protein subunits, originating from glycosylation and proteolytic cleavage of a common precursor of 170 kDa. Alternative post-transcriptional processing generates two truncated *Met* proteins, lacking the cytoplasmic kinase domain. One form is soluble and released from the cells. Autophosphorylation due to HGF binding upregulates the kinase activity of the receptor, increasing the V_{max} of the phosphotransfer reaction. The major phosphorylation site has been mapped to Tyr^{1235}. Negative regulation of the receptor kinase activity occurs through phosphorylation of a critical serine residue by protein kinase C or by a calcium dependent kinase. Receptor autophosphorylation triggers the signal transduction pathway by associating *ras*GAP, phospholipase C-γ, *src*-related tyrosine kinase and phosphosphatidylinositol 3-kinase. The HGF/SF receptor is expressed in several epithelial tissues and it is often overexpressed in neoplastic cells. In some tumors of the gastro-intestinal tract it is constitutively activated, either by overexpression of the amplified *MET* oncogene or by defective post-translational processing.

NATO ASI Series, Vol. H 76
Tyrosine Phosphorylation/Dephosphorylation
and Downstream Signalling
Edited by L. M. G. Heilmeyer Jr.

Introduction

The transmembrane tyrosine kinase encoded by the *cMET* protooncogene has been the focus of renewed interest since the identification of its ligand, a molecule called Hepatocyte Growth Factor or Scatter Factor (Bottaro et al., 1991; Naldini et al., 1991 a,b). Hepatocyte Growth Factor (HGF) and Scatter Factor (SF) are identical molecules secreted by stromal fibroblasts and non-parenchimal cells of many organs (Gherardi and Stoker, 1990; Weidner et al., 1991). The factor has several activities on epithelial cells, including mitogenesis, dissociation of epithelial sheets, stimulation of cell motility, and promotion of matrix invasion (for a review see Gherardi and Stoker, 1991). It has been shown that HGF and SF are indistinguishable ligands for p190^{MET}, the receptor tyrosine kinase encoded by the *MET* proto-oncogene (Naldini et al., 1991c). This was proved by HGF/SF binding to immunopurified p190^{MET}, chemical cross-linking of radiolabelled ligand and HGF/SF-induced tyrosine phosphorylation of p190^{MET}. Formal proof of the role of p190^{MET} as the sole component of the HGF/SF receptor was obtained by reconstituting the ligand binding site in insect cells infected by a recombinant baculovirus carrying the human *MET* cDNA (Naldini et al., 1991c) and by transferring to fibroblasts the ability to respond to the factor upon trasfection of the *MET* cDNA (Giordano et al., 1993).

Structure of p190^{MET}, the tyrosine kinase encoded by the *MET* protooncogene.

The tyrosine kinase encoded by the *MET* protooncogene is the prototype of a new class of tyrosine kinase receptors because of its unique heterodimeric structure: it is a two-chain oligomer, composed of a 50 kDa (α) chain disulfide linked to a 145 kDa (β) chain in an $\alpha\beta$ complex of 190 kDa (Giordano et al., 1988, 1989b) The α chain is exposed at the cell surface and the β chain spans the plasma membrane and contains the tyrosine kinase domain (Park et al., 1987; Gonzatti Haces et al., 1988), as well as sites for tyrosine auto-phosphorylation (Ferracini et al., 1991). Both chains of the $\alpha\beta$ complex are encoded by the *MET* gene, as demonstrated by studies of the biosynthesis and the post-translational processing (Giordano et al., 1989b). The first translational product which appears after a short metabolic labelling is a single chain of 170 kDa, already glycosylated. This precursor is not exposed at the cell surface since it cannot be labelled by ^{125}I under non-permeating conditions. The 170 kDa precursor undergoes modification of intrachain disulfide bonds, as revealed by a change in the apparent molecular weight to 180 kDa in SDS-PAGE under non-reducing conditions. In cells pulsed for 15 min with radiolabelled aminoacids, the mature 50 kDa α and 145 kDa β subunits appear after 1 hr. The a and β chains originate from the proteolytic cleavage of the 170 kDa precursor and are already bound through disulfide bridges. Free β chains are never observed during pulse-chase experiments.

The p190^{MET} HGF/SF receptor is a glycoprotein that contains N-linked oligosaccharides. In the putative extracellular domains predicted from the DNA sequence there are four possible sites for N-glycosylation in the α chain and seven sites in the β chain. Digestion with endoglycosydases results in a reduction of the apparent mol. wt. of the 170 kDa precursor to 150 kd. This is the same mol. wt. of the precursor synthesized in cells in which N-linked glycosylation is inhibited by tunicamycin and matches the predicted weight calculated from aminoacids of the *MET* sequence. Glycosylation of the 150 kDa precursor is essential for further processing. Thus, treatment with tunicamycin prevents the correct cleavage and the appearance of the α and β chains (Giordano et al., 1989b).

C-terminal truncated forms of the p190^{MET} receptor

By producing a panel of monoclonal antibodies reacting with the extracellular domain of the HGF/SF receptor, two C-terminal truncated isoforms were identified: a 140 kDa trans-membrane form (p140^{MET}) and a 130 kDa soluble protein (p130^{MET}) which is released from the cell (Prat et al., 1991b). The truncated forms of the HGF/SF receptor have the same heterodimeric structure of the intact p190^{MET}, consisting of two disulfide-bonded chains (Fig.1). The α chains of the truncated forms are indistinguishable from the α chain of p190^{MET}. The β chain of p140^{MET} is about 85 kDa (p85$^{\beta}$), while the β chain of p130^{MET} is about 75 kDa (p75$^{\beta}$). Both p85$^{\beta}$ and p75$^{\beta}$ are truncated at their C-termini and lack the cytoplasmic tyrosine kinase domain. P85$^{\beta}$ and p75$^{\beta}$ share the N-terminal domain with p145$^{\beta}$, since they were all recognized by monoclonal antibodies defining four different extracellular epitopes, and yielded identical extracellular tryptic peptides (Prat et al., 1991b). The p85$^{\beta}$ chain is bound to the cell surface, while the p75$^{\beta}$ chain lacks also the transmembrane domain and it is released from the cells.

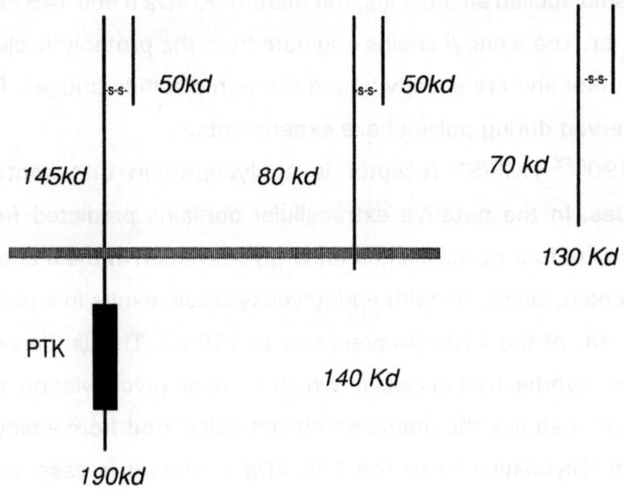

Figure 1. Schematic structure of the different HGF/SF receptor protein species.
The full-size p190^{MET}, the C-terminal truncated p140^{MET} and the soluble p130^{MET} are heterodimers composed of the same 50 kDa α chain disulfide-linked to different β chains. P190^{MET} has a transmembrane β chain of 145 kDa endowed with tyrosine kinase activity. P140^{MET} has a C-terminal truncated β chain of 85 kDa spanning the plasma membrane. P130^{MET} has a soluble C-terminal truncated β chain of 75 kDa.

The transmembrane 140^{MET} originates by post-translational proteolytic cleavage of the p170^{MET} precursor. In fact, mouse cells transfected with the full length human *MET* cDNA express p190^{MET} as well as p140^{MET} and p130^{MET}. The proteolytic event generating p140^{MET}, occurs within the endoplasmic lumen (Crepaldi et al., submitted). The soluble p130^{MET} is likely to be generated by proteolytic cleavage at the cell surface (Prat et al., 1991).

The conserved expression of the truncated forms of the HGF/SF receptor among different cell lines suggests that they play a physiological role. The soluble p130^{MET} might represent a breakdown product of the receptor. We have shown that activation of protein kinase C negatively regulates the HGF/SF receptor tyrosine kinase activity by increasing serine phosphorylation of the β chain (Gandino et al., 1990). Since TPA also increases the levels of soluble p130^{MET} (Prat et al., 1991b) it is suggested that activation of protein kinase C may induce concomitant receptor down-modulation by proteolytic activity, as described for the CSF-1 receptor (Dowing et al., 1989). Whether a similar receptor down modulation occurs also in response to the HGF/SF ligand is presently under investigation.

It is tempting to speculate that the truncated forms may also interfere with the HGF/SF receptor signal transduction pathway by competing with the intact p190^{MET} receptor for binding to the ligand. Such a negative regulatory role has already been shown for the soluble form of the IL-4 receptor (Mosley et al., 1989). Similarly, truncated EGF and PDGF receptors can inhibit wild type (i.e., full length) receptor function (Basu et al., 1989; Kashles et al., 1991, Ueno et al., 1991).

HGF/SF binding triggers the kinase activity of the p190^{MET} receptor

The earliest detectable consequence of HGF/SF stimulation of target cells is the tyrosine autophosphorylation of its receptor (Naldini et al., 1991a; 1991b; Bottaro et al., 1991; Graziani et al., 1991). Similarly to other growth factor receptors, this indicates activation of the receptor kinase. Autophosphorylation of the receptor β subunit was observed both in cells responding to HGF/SF with growth and in cells responding with increased motility. The response is time and dose dependent, and could be observed as early as 2 min after stimulation.

The tyrosine kinase activity of p190^{MET} receptor is powerfully activated by tyrosine autophosphorylation (Naldini et al., 1991a). In the presence of ATP and Mg salts, immunoprecipitated p190^{MET} incorporates phosphate into the β-subunit with first order kinetics and displays a time-dependent increase in the phosphorylation activity of a peptide substrate. Kinase activation is maximal when phosphorylation reaches the plateau. The increase in phosphorylation rate is due to a several fold increase in the V_{max} of the enzyme catalyzed phosphotransfer reaction. In the absence of ligand the K_m for ATP is not affected while a slight increase is observed in the K_m for the peptide substrate. The calculated enzymatic parameters for p190^{MET} were similar to those reported for other tyrosine kinases. In analogy with the EGF receptor a tenfold difference was observed between the K_m [ATP] for autophosphorylation and substrate phosphorylation (Weber et al., 1984; Bertics and Gill, 1985).

The kinase activity of the HGF/SF receptor is up-regulated by tyrosine phosphorylation

The major tyrosine autophosphorylation site of the p190^{MET} receptor was identified with the tyrosine residue 1235 (Ferracini et al., 1991). The Tyr^{1235} residue is embedded in a typical consensus pattern for tyrosine phosphorylation: it is located eight residues downstream from an arginine residue, and two acidic aminoacids (aspartic and glutamic acid) are present at two and five residues upstream. Tyr^{1235} is located within the tyrosine kinase domain in a segment homologous to the major autophosphorylation sites of other receptor and non-receptor kinases (Hanks et al., 1988). The sequence including Tyr^{1235} in the p190^{MET} receptor shares the highest homology with the corresponding sequence of the insulin receptor, as expected from the their phylogenetic relatedness (Hanks et al., 1988). It has been shown that autophosphorylation positively regulates the tyrosine kinase activity of insulin receptor, increasing the V_{max} of the enzyme, and modulates the biological responses to insulin. As discussed in the previous paragraph, we know that phosphorylation on tyrosine enhances the kinase activity of the HGF/SF receptor (Naldini et al., 1991a). Thus, one can assume that autophosphorylation on Tyr^{1235} positively regulates the receptor functions and triggers the cascade of events leading to the generation of intracellular signals.

Serine phosphorylation downregulates the HGF/SF receptor

Several membrane receptors have been shown to be modulated by protein kinase-C, a key enzyme involved in signal transduction (for review see Nishizuka, 1988), via phosphorylation on serine and/or threonine residues. To investigate whether the HGF/SF receptor is affected by protein kinase-C we studied its phosphorylation in response to the activation of the enzyme *in vivo*. We took advantage of the fact that in GTL-16 cells the *MET* oncogene is amplified and overexpressed and the p190^{MET} receptor is auto-phosphorylated on tyrosine, even in the absence of the ligand (Giordano et al., 1988 and 1989a). Treatment of GTL-16 cells with phorbol esters or with the membrane-permeable oleyl-acyl-glycerol, both powerful and selective activators of protein kinase-C (Rando, 1988), led to a striking reduction of the extent of tyrosine phosphorylation of the *β* subunit of the p190*MET* receptor (Gandino et al., 1990). The involvement of protein kinase-C was confirmed by the fact that the phenomenon did not occur in cells depleted of this enzyme. Phosphoaminoacid analysis revealed that the decrease of phosphorylation on tyrosine is accompanied by increased phosphorylation on serine residue, and suggests that the HGF/SF receptor itself may be a substrate for the kinase activity of the enzyme. Notably, serine was the only amino acid phosphorylated in response to protein kinase-C activation.

These data suggest that protein kinase-C negatively regulates the HGF/SF receptor through serine phosphorylation of the *β* subunit, similar to its effect on EGF and insulin receptors. The negative control by protein kinase-C seems to operate at steady-state level also in physiological conditions, i.e, in the absence of exogenous activation. In fact, depletion of protein kinase-C by prolonged treatment with TPA resulted in an increased phosphorylation of p190*MET* (Gandino et al., 1990). Similar data have been obtained in the case of the EGF-receptor in A431 cells (Sturani et al., 1988).

The normal endogenous cellular activator of protein kinase-C is 1,2 diacylglycerol which is produced transiently in the cell membrane as result of the breakdown of phosphatidyl-inositol-4,5 biphosphate by phospholipase-C. This enzyme concomitantly produces inositol-1,4,5-triphosphate (IP_3) a soluble cytoplasmic second messenger causing the release of calcium ions from intracellular stores (for a review, see Berridge, 1987). We also investigated whether both arms of the signal transduction pathway triggered by phospholipase-C modulates the kinase activity of the HGF/SF receptor. The results indicated that the increase in intracellular Ca^{2+} concentration is inhibitory.

The inhibitory effect was observed at $[Ca^{2+}]_i$ levels in the physiological range (between 200 nM and 760 nM), as determined by fluorometric analysis. Inhibition was observed both after ionophore-induced calcium influx from the extracellular compartment and after release from intracellular stores. The Ca^{2+}-induced inhibition was reversible and totally independent from protein kinase-C activation, as it took place also in protein kinase-C-depleted cells. On the other hand, protein kinase-C inhibits the p190^{MET} kinase by a mechanism which is totally independent from intracellular Ca^{2+} (Gandino et al., 1990; 1991).

Inhibition of the activity of HGF/SF receptor by Ca^{2+} is associated with serine phosphorylation of a specific phosphopeptide, implying the activation of a yet unidentified serine kinase. It is well known that calcium and calmodulin may regulate a number of protein kinases that phosphorylate serine and/or threonine residues on a wide spectrum of different substrates (for a review, see Edelman et al., 1987). Notably a unique serine residue is phosphorylated both by this Ca^{2+}-dependent kinase and by protein kinase-C (Gandino et al., in preparation). Thus two independent systems negatively modulate the HGF/SF receptor through the same molecular mechanism. In this way the signal transduced by the p190^{MET} receptor is likely to be modulated by other receptor systems coupled to protein kinase-C activation or to the induction of transient Ca^{2+} fluxes.

The HGF/SF receptor signal transduction pathway

Elucidation of the transductional pathways leading to the diverse cellular responses triggered by HGF/SF is now awaited. Cell division and motility must ultimately result from modification of distinct cellular targets. These effects may be elicited coupling the HGF/SF receptor to specific transducers or to a different combination of the same set of transducers. It has been proposed that stimulation of the tyrosine kinase activity of receptors by the appropriate ligands may result in the recruitment to the receptor (and thus to the membrane) of cytosolic enzymes responsible for the production of second messengers (Cantley et al., 1991). According to this model, the tyrosine kinase activity of receptors is primarily directed to carry out self-phosphorylation at defined tyrosine residues. These phosphotyrosines in turn represent specific "docking" sites for high affinity binding of transducers containing SH2 domains (Koch et al., 1991). We made use of recombinant HGF/SF receptor produced in insect cells to test its ability to associate a number of SH2-containing transducers. The receptor binds *in vitro*, in a phosphorylation-dependent manner, PI 3-kinase, *ras*GAP, PLC-γ, and p59*Fyn*. To date, only PDGF receptor is known to associate with such a wide spectrum of transducers *in vivo* (Cantley et al., 1991). This coincidence is perhaps not surprising, since PDGF receptor is itself known to mediate a pleiotropic response including cell movement and mitogenesis.

The HGF/SF receptor associates PI 3-kinase also *in vivo,* in intact cells stimulated by nanomolar concentrations of HGF/SF (Graziani et al., 1991). The 85 kDa subunit of PI-3 kinase, coprecipitates with the $p190^{MET}$ in immunocomplexes made with anti-receptor antibodies from HGF/SF stimulated cells. PI 3-kinase activity was present in these immunocomplexes, as shown by generation of the D-3 phosphorylated inositol lipids in the presence of the appropriate substrates and $[^{32}P]$-γ-ATP. It is conceivable that the D-3 polyphosphoinositides generated by PI 3-Kinase play a role in the signalling pathway triggered by HGF/SF. This factor induces different biological responses in epithelial cells, including mitogenesis and stimulation of cell motility. Interestingly, D-3 polyphosphoinositides are generated in response to other extracellular signals inducing either cell growth (Auger et al., 1989; Varticovski et al., 1989; Ruderman et al., 1990; Endemann et al., 1990) or chemotaxis (Traynor-Kaplan et al., 1989).

Expression of the HGF/SF receptor in normal human tissues

Study of the expression of the gene in a variety of human tissues at both RNA and protein levels allowed us to obtain information about the physiological role of the HGF/SF receptor. Northern blot analysis showed that in normal tissues, the only *MET* transcript detectable is a 9Kb mRNA, having the same size as the major transcript described in cultured cell lines (Park et al., 1986; Giordano et al., 1989a). High levels of specific mRNA were found in the liver and kidney. The transcript was also found in tissues of the gastrointestinal tract, including stomach, ileum, colon and rectum, and in prostate, seminal vesicles and breast. In lung, uterus, ovary, skin, thyroid and skeletal muscles specific mRNA was barely detectable while other organs, such as adrenal glands, bone marrow and spleen, were negative (Di Renzo et al., 1991).

By Western blot analysis, high levels of $p190^{MET}$ were found in liver, kidney, ovary, in endometrium and in tissues of the gastrointestinal tract including stomach, ileum, caecum, colon, sigma and rectum (Di Renzo et al., 1991). In these samples also the 170kDa *MET* precursor was revealed by the antiserum. Lower levels of $p190^{MET}$ were found in the lung, prostate, seminal vesicles, skin and breast. The *MET* protein is also expressed in endothelial cells and it has been shown that the interaction with HGF/SF can stimulate these cells to proliferate and migrate (Bussolino et al., 1992).

The cellular localization of the *MET* protein was studied with indirect immunofluorescence on frozen sections of a wide panel of normal adult human organs, using a monoclonal antibody directed against the extracellular domain of the $p190^{MET}$ (Prat et al., 1991a). The antibody reacted with the epithelial cells of the parenchymal liver, the major biliary ducts, the stomach, and of the small and the large intestine. The epithelium of the endometrium and the epithelial component of the ovary were also positive. Basal keratinocytes of the skin and of the oesophagus were stained as well. In hepatocytes and endometrial cells antibody reactivity was preferentially localized at the level of the plasma membrane, with a homogeneous pattern. Recently it has been shown that the HGF/SF receptor is expressed also in the microglial cells of the human central nervous sistem (Di Renzo et al., 1992).

The tissue distribution of the HGF/SF receptor reflects the pattern of expression of its ligand. HGF/SF transcripts and immunoreactive HGF/SF were found in several organs of the gastrointestinal tract, including the liver, thyroid, salivary glands and some areas of thẹ brain. HGF/SF was localized in either stromal components and non-parenchymal cells by some authors (Rubin et al., 1991), or in epithelial cells by others (Tạshiro et al., 1990; Zarnegar et al., 1990). Although the precise cellular localization of HGF/SF is not clear at present, the expression of the HGF/SF receptor in the same organs suggests a paracrine relationship between adjacent cells. HGF/SF (Scatter Factor) is produced by stromal fibroblasts of the organs listed above and it has a scattering effect exclusively on epithelial cells (Stoker et al., 1987; Gherardi et al., 1989; Weidner et al., 1990).

Activation of the *MET* proto-oncogene

The *MET* proto-oncogene has been found amplified and overexpressed and the protein activated in a gastric carcinoma cell line (Giordano et al., 1989a). In this cell line, the primary structure of the *MET* protein is indistinguishable from that found in other normal cells, suggesting that overexpression alone is most probably sufficient for the activation of the *MET* kinase. A point mutation or a minor rearrangement could however escape detection in the protein analysis. This hypothesis was tested by directly sequencing a series of overlapping cDNAs obtained from GTL-16 cells mRNA. This analysis revealed that the *MET* cDNA obtained from GTL-16 cells reflects the sequence of the most abundant transcript in human tissues and that activation of the *MET* kinase in GTL-16 cells is associated with amplification and overexpression of the proto-oncogene (Ponzetto et al., 1991). Such a mechanism of activation has already been shown also in the case of several tyrosine kinase receptors such as the Epidermal Growth Factor receptor (Di Fiore et al., 1987) and the Colony Stimulating Factor receptor (Roussel et al., 1987).

We have also found an other tumor cell line in which the *MET* receptor is activated, that is constitutively phosphorylated on tyrosine. A human colon carcinoma cell line (LoVo) expresses at the cell surface an uncleaved *MET* receptor ($p190^{NC}$) endowed with de-regulated tyrosine kinase activity (Mondino et al., 1991). As described, under physiological conditions the *MET* mRNA is translated into a precursor with an apparent M_r of 170 kDa which is then cleaved by a protease. This process originates the 50 kDa α and the 145 kDa β subunits, already joined by disulphide bonds into the correctly processed $p190^{\alpha\beta}$ heterodimer (Giordano et al., 1989b). In LoVo cells the 170 kDa precursor is normally produced and glycosylated, but not cleaved. The proteolytic cleavage site is conserved, as it was detected in the cDNAs derived from this line. These data showed that a defective post-translational mechanism is responsible for the lack of cleavage. The intracellular compartment where the *MET* precursor undergoes maturation has not yet been identified, but it is likely to be the Golgi apparatus.

$P190^{NC}$ is a functional tyrosine kinase and is constitutively phosphorylated in the colon carcinoma line studied. Since in LoVo cells the *MET* gene is neither amplified nor overexpressed and does not present any point mutation, we concluded that the "structural" defect leading to kinase activation is the lack of cleavage between α and β chains. The detailed molecular mechanisms by which HGF/SF triggers the β chain kinase activity are still to be clarified. However, by analogy with the other growth factor receptors, it is likely that ligand-induced conformational change(s) of the extracellular domain should be critical to stabilize the intracellular kinase in an active state. We propose that the uncleaved extracellular domain of $p190^{NC}$ mimics the activating conformation reached by $p190^{\alpha\beta}$ upon binding to HGF/SF. If this is true, the *MET* oncogene expressed by the colon carcinoma cell line studied provides an example of activation due to a post-translational mechanism.

Expression of HGF/SF receptor in human tumors

The expression of the *MET* oncogene is conserved or increased in epithelial tumors (Di Renzo et al., 1991). Out of the three major transcripts described (9, 7, and 5.2kb; Park et al., 1986; Giordano et al., 1989b), the 9kb is consistently present, as in the normal tissue counterparts. In some fresh tumors the 7 and 5.2kb transcripts were also detected. An increased amount of the *MET* mRNA and protein was observed in the majority of carcinomas of the gastrointestinal tract. A hundred-fold increase of HGF/SF receptor was found in thyroid tumors of the papillary histotype, derived from the follicular epithelium (Di Renzo et al., 1992). *MET* gene expression was not increased in non-neoplastic thyroid diseases, in adenomas and in anaplastic carcinomas. The abnormal expression of the HGF/SF receptor in papillary thyroid carcinomas is intriguing. It is noteworthy that among the tumors of this histotype one half expresses either an activated *TRK* or an activated *RET* oncogenes (Fusco et al., 1987). The finding of an abnormal amount of HGF/SF receptor suggests that oncogenes encoding tyrosine kinase receptors may be relevant in the pathogenesis of tumors originated from the thyroid follicular epithelium.

From this survey it is concluded that the expression of the *MET* oncogene in human carcinomas confers a selective advantage to neoplastic cells.

Acknowledgements

The experimental work reviewed in this paper is the result of the collaborative effort of our colleagues T.Crepaldi, F.Di Renzo, G.Gaudino, L. Gandino, A. Graziani, A. Mondino, L. Naldini, C.Ponzetto, M.Prat, who are gratefully acknowledged. The manuscript has been written with the excellent assistance of A.Cignetto. These researches are supported by grants from the Associazione Italiana Ricerche Cancro (AIRC) and from the Italian National Research Council (CNR), progetto finalizzato ACRO.

REFERENCES

Auger, K.R., Serunian, L.A., Soltoff, S.P., Libby, P., and Cantley, L.C. (1989) PDGF- dependent tyrosine phosphorylation stimulates production of novel phosphoinositides in intact cells Cell 57: 167-175.

Bardelli, A., Maina, F., Gout, I., Fry. M.J., Waterfield, M.D., Comoglio, P.M., and Ponzetto, C. (1992) Autophosphorylation promotes complex formation of recombinant Hepatocyte Growth Factor Receptor with cytoplasmic effectors containing SH2 domains. Oncogene 7: 1973-1978

Basu, A., Raghunath, M., Bishayee, S. and M. Das (1989) Inhibition of tyrosine kinase activity of the epidermal growth factor (EGF) receptor by a truncated receptor form that binds to EGF: role for interreceptor interaction in kinase regulation. Mol. Cell. Biol. 9: 671-677.

Berridge, M.J. (1987) Inositol trisphosphate and diacylglycerol: two interacting second messengers Ann. Rev. Biochem. 56: 159-193.

Bertics, P.J. and G. N. Gill (1985) Self-phosphorylation enhances the protein-tyrosine kinase activity of the epidermal growth factor receptor. J. Biol. Chem. 260: 14642-14647.

Bottaro, D.P., Rubin, J.S., Faletto, D.L., Chan, A.M.L., Kmiecick, T.E., Vande Woude, G.F., and Aaronson, S.A. (1991) Identification of the hepatocyte growth factor receptor as the c-*met* proto-oncogene. Science 251: 802-804

Bussolino, F., Di Renzo, M.F., Ziche, M., Bocchietto, E., Olivero, M., Naldini, L., Gaudino, G., Tamagnone, L., Coffer, A., and Comoglio, P.M. (1992) Hepatocyte Growth Factor is a potent angiogenic factor which stimulates endothelial cell motility and growth. J. Cell Biol. 119: 629-641

Cantley, L.C., Auger, K., Carpenter, C., Duckworth, B., Graziani, A., Kapeller, R., and Soltoff, S. (1991) Oncogenes and signal transduction. Cell 64: 281-302

Crepaldi T., Prat M., Giordano S. and Comoglio P.M. 1993 Post traslational processing of the HGF/SF receptor precursor. Mol. Cell. Biol. submitted.

Di Renzo, M.F., Narsimhan, R.P., Olivero, M., Bretti, S., Giordano, S., Medico, E., Gaglia, P., Zara, P., and Comoglio, P.M. (1991) Expression of the *Met*/HGF receptor in normal and neoplastic human tissues. Oncogene 6: 1997-2003

Di Renzo, M.F., Bertolotto, A., Olivero, M. Crepaldi, T., Schiffer, D., Pagni C.A. and Comoglio, P.M. (1992a) Selective expression of the *MET*/HGF receptor in the human central nervous system. Oncogene, 7: 000-000

Di Renzo, M.F., Olivero, M., Ferro, S., Prat, M., Bongarzone, I., Pilotti, S., Pierotti M. and Comoglio P.M. (1992b) Overexpression of the c-*MET*/HGF receptor gene in human thyroid carcinomas. Oncogene 7: 2549-2553

Di Fiore, P.P., Pierce, J.H., Fleming, T.P., Hazan, R., Ullrich, A., King, C.R., Schlessinger, J. and Aaronson, S.A. (1987) Overexpression of the human EGF receptor confers an EGF-dependent transformed phenotype to NIH 3T3 cells. Cell 51: 1063-1070.

Downing, J.R., Roussel, M.F., and C.J. Sherr (1989) Ligand and protein kinase C downmodulate the Colony-Stimulating Factor 1 receptor by independent mechanisms. Mol. Cell. Biol. 9: 2890-2896.

Edelman, A.M., Blumenthal, D.K. and Krebs, E.G. (1987) Protein serine/threonine kinases. Ann. Rev. Biochem. 56: 567-613.

Endemann, G., Yonezawa, K., and Roth, R.A. (1990) Phosphatidylinositol kinase or an associated protein is a substrate for the insulin receptor tyrosine kinase J. Biol. Chem. 265: 396-400

Ferracini, R., Longati, P., Naldini, L., Vigna, E., and Comoglio, P.M. (1991) Identification of the major autophosphorylation site of the *Met*/Hepatocyte Growth Factor receptor tyrosine kinase. J. Biol. Chem. 266: 19558-19564

Fusco, A., Grieco, M., Santoro, M., Berlingieri, M.T., Pilotti, S., Pierotti, M.A., Della Porta, G. and Vecchio, G. (1987) A new oncogene in human thyroid papillary carcinomas and their lymph-nodal metastases. Nature 328: 170-172.

Gandino, L., Munaron, L., Naldini, L., Magni, M., and Comoglio, P.M. (1991) Intracellular calcium regulates the tyrosine kinase receptor encoded by the *MET* oncogene. J. Biol. Chem. 266: 16098-16104

Gandino, L., M.F. Di Renzo, S. Giordano, F. Bussolino and P.M. Comoglio (1990) Protein kinase-c activation inhibits tyrosine phosphorylation of the c-*met* protein. Oncogene 5: 721-725

Gherardi E. and Stoker M. (1991) Cancer Cells 3: 227-232

Gherardi, G., Gray, J., Stoker, M., Perryman, M., and Furlong, R. (1989). Purification of scatter factor, a fibroblast-derived basic protein that modulates epithelial interaction and movement. Proc. Natl. Acad. Sci. USA 86: 5844-5848

Gherardi, E., and Stoker, M. (1990) Hepatocytes and scatter factor Nature 346: 228

Giordano, S., Di Renzo, M.F., Ferracini, R., Chiado'Piat, L. and Comoglio, P.M. (1988) P145 a protein with associated tyrosine kinase activity in a human gastric carcinoma cell line. Mol. Cell. Biol. 8: 3510-3517.

Giordano, S., Ponzetto, C., Di Renzo, M.F., Cooper, C.S. and Comoglio, P.M. (1989a) Tyrosine kinase receptor indistinguishable from the c-*MET* protein. Nature 339: 155-156

Giordano, S., Di Renzo, M.F., Narsimhan, R., Cooper, C.S., Rosa, C., and Comoglio, P.M. (1989b) Biosynthesis of the protein encoded by the *c-met* proto-oncogene. Oncogene 4: 1383-1388

Giordano S., Zhen Z., Medico E., Gaudino G., Galimi F. and Comoglio P.M. (1993) Transfer of the motogenic and invasive response to scatter factor/hepatocyte growth factor by transfection of the human cMET protooncogene. Proc Natl. Acad. Sci. USA 90: 000-000

Graziani, A., Gramaglia, D., Cantley, L.C., and Comoglio, P.M. (1991) The tyrosine phosphorylated Hepatocyte Growth Factor/Scatter Factor receptor associates with phosphatidylinositol 3-kinase. J. Biol. Chem. 266: 22087-22090

Hanks, S.K., Quinn, A.M. and T. Hunter (1988) The protein kinase family: conserved features and deduced phylogeny of the catalytic domains. Science 241: 42-52.

Kashles, O., Yarden, Y., Fisher, R., Ullrich, A. and Schlessinger, J. (1991) A dominant negative mutation suppresses the function of normal epidermal growth factor receptors by heterodimerization. Mol. Cell. Biol. 11: 1454-1463.

Koch, C.A., Anderson, D., Moran, M.F., Ellis, C., and Pawson, T., (1991) SH2 and SH3 domains: elements that control interactions of cytoplasmic signaling proteins. Science 252: 668-674

Mondino, A., Giordano, S., and Comoglio P.M. (1991) Defective posttranslational processing activates the tyrosine kinase encoded by the *mET* proto-oncogene (Hepatocyte Growth Factor Receptor) Mol. Cell. Biol. 11: 6084-6092

Mosley, B., Beckmann, M.P., March, C.J., Idzerda, R.L., Gimpel, S.D., VandenBos, T., Friend, D., Alpert, A., Anderson, D., Jackson, D., Wignall, J.M., Smith, C., Gallis, B., Sims, J.E., Urdal, D., Widmer, M.B., Cosman, D. and Park, L.S. (1989) The murine interleukin-4 receptor: molecular cloning and characterization of secreted and membrane bound forms. Cell 59: 335-348.

Naldini, L., Vigna, E., Ferracini, R., Longati, P., Gandino, L., Prat, M., and Comoglio, P.M. (1991a) The tyrosine kinase encoded by the *MET* proto-oncogene is activated by autophosphorylation. Mol. Cell. Biol. 11: 1793-1803

Naldini, L., Vigna, E., Narshiman, R.P., Gaudino, G., Zarnegar, R., Michalopoulos, G., and Comoglio, P.M. (1991b) Hepatocyte Growth Factor (HGF) stimulates the tyrosine kinase activity of the receptor encoded by the proto-oncogene c-*MET*. Oncogene 6: 501-504

Naldini, L., Weidner, M., Vigna, E., Gaudino G., Bardelli A., Ponzetto, C., Narsimhan, R., Hartmann, G., Zarnegar, R., Michalopoulos, G., Birchmeier, W., and Comoglio, P.M. (1991c) Scatter Factor and Hepatocyte Growth Factor are indistinguishable ligands for the *Met* receptor. EMBO J. 10: 2867-2878

Nishizuka, Y. (1988) The molecular heterogeneity of protein kinase C and its implications for cellular regulation Nature 334: 661-665.

Park, M., Dean, M., Cooper, C.S., Schmidt, M., O'Brian, S.J., Blair, D.G. and Vande Woude, G.F. (1986) Mechanism of met oncogene activation. Cell 45: 895-904.

Park. M., Dean, M., Kaul, K., Braun, M.J., Gonda, M.A., and Vande Woude, G. (1987) Sequence of *MET* protooncogene cDNA has features characteristic of the tyrosine kinase family of growth-factor receptors. Proc. Natl. Acad. Sci. USA 84: 6379-6383

Ponzetto, C., Giordano, S., Peverali, F., Della Valle, G., Abate, M., Vaula, G. and Comoglio, P.M. (1991) c-*met* is amplified but not mutated in a cell line with an activated *met* tyrosine kinase. Oncogene 6: 553-559

Prat, M.P., Narsimhan, R.P., Crepaldi, T., Nicotra, M.R., Natali, P.G., and Comoglio, P.M. (1991) The receptor encoded by the human c-*MET* oncogene is expressed in hepatocytes, epithelial cells and solid tumors. Int. J. Cancer 49: 323-328

Prat, M., Crepaldi, T., Gandino, L., Giordano, S., Longati, P., and Comoglio, P.M. (1991) C-terminal truncated forms of Met, the Hepatocyte Growth Factor receptor Mol. Cell. Biol. 11: 5954-5962.

Rando, R.R. (1988) Regulation of protein kinase C activity by lipids FASEB 2: 2348-2355.

Roussel, M.F., Dull, T.J., Rettenmier, C.W., Ralph, P., Ullrich, A. and Sherr, C.J. (1987) Transforming potential of the c-*fms* proto-oncogene (CSF-1 receptor) Nature 325: 549-552.

Rubin, J.S., Chan, A.M.L., Bottaro, D.P., Burgess, W.H., Taylor, W.G., Cech, A.C., Hirschfield, D.W., Wong, J., Miki, T., Finch, P.W., and Aaronson, S.A. (1991) A broad-spectrum human lung fibroblast-derived mitogen is a variant of hepatocyte growth factor. Proc. Natl. Acad. Sci. U.S.A. 88: 415-419

Ruderman, N.B., Kapeller, R., White, M.F., and Cantley, L.C. (1990) Activation of phosphatidylinositol 3-kinase by insulin Proc. Natl. Acad. Sci. U.S.A. 87: 1411-1415.

Stoker, M., Gherardi, E., Perryman, M., and Gray, J. (1987) Scatter Factor is a fibroblast-derived modulator of epithelial cell mobility. Nature 327: 239-242

Sturani, E., Zippel, R., Toschi, L., Morello, L., Comoglio, P.M., and Alberghina, L. (1988) Kinetics regulation of the tyrosine phosphorylation of epidermal growth factor receptor in intact A_{431} cells Mol. Cell. Biol. 8: 1345-1351.

Tashiro, K., Hagiya, M., Nishizawa, T., Seki, T., Shimonishi, M., Shimizu, S., and Nakamura, T. (1990). Deduced primary structure of rat hepatocyte growth factor and expression of the mRNA in rat tissues. Proc. Natl. Acad. Sci. USA 87: 3200-3204.

Traynor-Kaplan, A.E., Thompson, B.L., Harris, A.L., Taylor, P., Omann, G.M., and Sklar, L.A. (1989) Transient increase in phosphatidylinositol 3,4-bisphosphate and phosphatidylinositol trisphosphate during activation of human neutrophils. J. Biol. Chem. 264: 15668-15673

Ueno, H., Colbert, H., Escobedo, J.A. and Williams, L.T. (1991) Inhibition of PDGF β receptor signal transduction by coexpression of a truncated receptor. Science, 252: 844-848.

Varticovski, L., Drucker, B., Morrison, D., Cantley, L., and Roberts, T. (1989) The colony stimulating factor-1 receptor associates with and activates phosphatidylinositol-3 kinase Nature 342: 699-702

Weber, W., P.J. Bertics and G. N. Gill (1984) Immunoaffinity purification of the epidermal growth factor receptor. Stoichiometry of binding and kinetics of self-phosphorylation. J. Biol. Chem. 259: 14631-14636.

Weidner, K.M., Arakaki, N., Vandekerchove, J., Weingart, S., Hartmann, G., Rieder, H., Fonatsch, C., Tsubouchi, H., Hishida, T., Daikuhara, Y., and Birchmeier, W. (1991) Evidence for the identity of human Scatter Factor and Hepatocyte Growth Factor. Proc. Natl. Acad. Sci. U.S.A. 88: 7001-7005

Weidner, K.M., Behrens, J., Vandekerckove, J. and Birchmeier, W. (1990) Scatter Factor: molecular characteristics and effect on the invasiveness of epithelial cells. J. Cell Biol. 111: 2097-2108

Zarnegar, R., Muga, S., Rahdija, R. and Michalopoulos, G. (1990) Tissue distribution of hepatopoietin-A: A heparin-binding polypeptide growth factor for hepatocytes. Proc. Natl. Acad. Sci USA 87: 1252-1256.

RET/ptc AND TRK ONCOGENES IN PAPILLARY THYROID CARCINOMA

M.A. Pierotti, E. Arighi, I. Bongarzone, M.G. Borrello, G. Butti, A. Greco, C. Mariani, M. Miozzo, C. Miranda, G. Sozzi

Division of Experimental Oncology A, Istituto Nazionale Tumori - Via G. Venezian 1 - 20133 Milan - Italy

INTRODUCTION

From the evidence that the oncogenes of tumor viruses could be routinely detected by their ability to induce neoplastic transformation of cells in culture, there arose the possibility that tumor cells might also contain transforming genes detectable in the same way. Gene transfer techniques (transfection) were therefore used to test the DNA of tumor cells for the presence of genes that, when introduces into cells in culture, can elicit a transformed phenotype and the ability to form tumors in animals (tumorigenicity). The first successful application of this concept was obtained with mouse fibroblasts transformed by a chemical carcinogen (Shih et al., 1979), but was then applied to a number of tumor cell lines and tumor specimens (Weinberg, 1982; Varmus, 1984; Barbacid, 1987). A transforming activity can be detected in approximately 20% of all specimens tested with some types of tumors rarely, or ever, containing transfection detectable transforming genes and others displaying consistent frequencies of activation of one or more oncogenes. The great majority of the transforming genes detected by transfection with the DNA of human cancers belongs to the family of the RAS proto-oncogenes (Barbacid, 1987).

A remarkable exception is represented by the papillary type of the thyroid carcinomas where in about 50% of the case assayed by transfection the activation of the tyrosine kinase proto-oncogenes TRK and RET has been detected (Fusco et al., 1987; Bongarzone et al., 1989; Grieco et al., 1990; Greco et al., 1992; Sozzi et al., 1992). This peculiarity may be related to the fact that the thyroid epithelium is simple and homogeneous in its organization, behaviour and differentiation. In addition, its growth is regulated by a single trophic hormone, TSH. These features are in contrast with the complexity of differentiation and growth control of epithelia of more common tumors such as those from breast, colon and lung, and make it possible that a significant percentage of carcinomas originating from the

NATO ASI Series, Vol. H 76
Tyrosine Phosphorylation/Dephosphorylation and Downstream Signalling
Edited by L. M. G. Heilmeyer Jr.

thyroid epithelium and belonging to the same histotype, can be characterized by the presence of a common genetic lesion(s).

Oncogene activation in papillary thyroid carcinoma: RET/ptc

The papillary carcinoma provided the first example of oncogene activation in thyroid tumors. Using the transfection assay on NIH3T3 fibroblasts the occurrence of a common transforming sequence in 5 out 20 papillary thyroid carcinomas was reported by the laboratory of G. Vecchio (University of Naples) and ourselves (Fusco et al., 1987). Since no homology was found either by hybridization with the available probes related to known oncogenes or by comparison of the Alu pattern of the transfectants with that of several known oncogenes, it was suggested that the transforming sequence could represent a new human oncogene that was designated as ptc (from Papillary Thyroid Carcinoma). Subsequently, when ptc was cloned and sequenced, it was found that it originated by the fusion of an unknown 5' sequence (designated as H4 or D10S170) to the tyrosine kinase domain of the RET proto-oncogene (Grieco et al., 1990). The RET oncogene was isolated by Takahashi et al., (1985) following transfection of NIH3T3 cells with human T cell lymphoma DNA and was shown to originate from the fusion of two separate human DNA genes: the RET proto-oncogene, encoding a receptor for an unknown ligand and with an associated tyrosine kinase activity (Takahashi and Cooper, 1987; Takahashi et al., 1991; Lanzi et al., 1992), and a gene coding for a putative zinc finger protein, rfp (Takahashi et al., 1988). However, this rearrangement had occurred in vitro during the procedures of the transfection assay (RET = REarranged during Transfection) since no evidence of such rearrangement was found in the DNA of the original T cell lymphoma.

On the contrary, the rearrangements detected in our cases of papillary thyroid carcinoma occurred somatically in all tumors whose DNA was found active in the transfection assay because of the ptc oncogenic sequence (Fusco et al., 1987; Bongarzone et al., 1989; Grieco et al., 1990). We have mapped to chromosome 10q the two genes forming ptc:RET (10q11.2) and H4/D10S170 (10q21), and have established their order on that chromosome as cen-RET-H4-qter (Donghi et al., 1989; Sozzi et al., 1991).

The mechanism responsible for this tumor specific genetic event, has been revealed following a cytogenetic analysis of several cases of papillary thyroid carcinomas performed in our laboratory and in that of R.B. Jenkins at the Mayo Clinic (Rochester, Minnesota, USA). In four cases the results indicated the presence of a paracentric 10q inversion, inv(10)(q11.2q21), with breakpoints characterization of these cases, performed in our and in G. Vecchio laboratories, demonstrated the activation of ptc, we concluded that a chromosome 10q inversion provides the structural basis for the H4-RET fusion that forms the chimeric transforming sequence ptc (Sozzi et al., 1992; Pierotti et al., 1992).

More recently, in our laboratory we have identified a second oncogenic rearrangement of RET (Bongarzone et al., 1993). We cloned a biologically active RET-related cDNA from NIH3T3 cells transformed by a papillary thyroid tumor DNA that was found to contain a somatically rearranged RET alleles. The analysis of its sequence has revealed that it contained 717 nucleotides of RI gene, followed by 1150 nucleotides of the RET proto-oncogene, including its entire tk domain. RI gene encodes one isoform of the regulatory subunit of cAMP-dependent protein kinase A (cAMP-PKA). The latter is a holoenzyme complex consisting of a regulatory subunit dimer and two catalitic subunits (Sandberg et al., 1990). In higher eukaryotes, most of the physiological responses to cAMP are mediated by the activation of cAMP-PKA (Dumont et al., 1989). Recently, RI gene has been mapped to human chromosome 17q23q24 (Boshart et al., 1991; Jones et al., 1991).

These findings led us to designate RET/ptc1 the first identified rearrangement between H4/D10S170 and RET and RET/ptc2 this novel rearrangement involving the RI of PKA.

The analysis of the RET-related sequences of RET/ptc2 rearrangement (Bongarzone et al., 1993) revealed two major points of interest in comparison with the homologous sequences displayed by the previously reported RET/ptc1 rearrangement (Grieco et al., 1990). First, although two different genes provide the 5' end of the RET transforming sequences, the breakpoint of proto-RET was identical, at the cDNA level, in the two different cases, confirming the previously reported finding that tumor specific RET rearrangements all occur within the same intronic region (Grieco et al., 1990). Second, the 3' end sequences of the two oncogenic rearrangements of RET were found, on the contrary, to differ as a consequence of an alternative splicing, as originally proposed by Ishizaka et al. (1989).

By comparing the carboxytermini of the cloned RET/ptc1 and RET/ptc2 transforming sequences, we found that they were derived from the longer and shorter isoforms, respectively (Fig. 1). As a consequence, we found that both RET/ptc1 and RET/ptc2 oncogenes encode two isoforms of the transforming proteins able to form both homo- and heterodimers (Bongarzone et al., 1993).

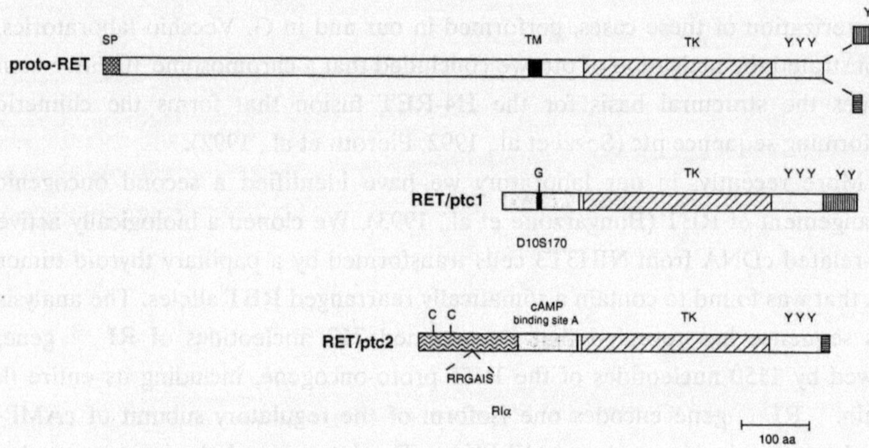

Fig. 1 - Schematic structure of the product of the RET proto-oncogene and of its oncogenic versions RET/ptc1 and RET/ptc2. Shown are the signal peptide (SP), transmembrane (TM) and tyrosine kinase (TK) domains, a glycine stretch in D10S170 (G), cystein (C) and tyrosine (Y) residues, a catalitic domain binding sequence (RRGAIS), and cAMP binding sites in RI . p51 and p9 indicate the different COOH termini of the two RET isoforms.

A biochemical analysis of RET/ptc1 and RET/ptc2 products has then revealed that they are constitutively phosphorylated on tyrosine, display an "in vitro" autophosphorylation activity, are located in the cytoplasm and have intrinsic tyrosine phosphorylation that is not modulated by TPA treatment of the transformed cells (Lanzi et al., 1992). These features contrast with those displayed by the normal product of the RET proto-oncogene as expressed in neuroblastoma cell lines, which is a membrane receptor (Takahashi et al., 1991) and is not phosphorylated on tyrosine under the culture conditions used (Lanzi et al., 1992).

Finally, we have identified other genes rearranged with RET whose characterization is underway (Table I).

Oncogenes activation in papillary thyroid carcinoma: TRK

The human TRK oncogene was first identified in a colon carcinoma, following transfection and focus formation in NIH3T3 cells (Martin-Zanca et al., 1986). The analysis of the mechanism of its malignant activation indicated that TRK oncogenicity was generated in tumor DNA by a somatic rearrangement that fused seven exons of the tropomyosin gene with the tyrosine kinase (tk) domain of the TRK proto-oncogene (Martin-Zanca et al., 1986). The isolation of the TRK proto-oncogene cDNA revealed that it encodes a cell surface receptor protein for a ligand recently identified as the nerve growth factor (Klein et al., 1991; Hempstead et al., 1991). Proto-TRK transcripts have been detected exclusively in peripheral nervous ganglia indicating that the gene may have a role in the nervous system (Barbacid et al., 1991).

We have mapped the two genes forming the oncogenic TRK sequence, proto-TRK or neurotrophin tyrosine kinase receptor 1 (NTRK1) to chromosome 1q32q41 (Miozzo et al., 1990) and the isoform of non muscle tropomyosin gene (TPM3) to the band q31 of the same chromosome (Radice et al., 1991).

A second case of TRK activation, designated as TRK-2h, was reported by Kozma et al.: it was generated by the fusion of the TRK tk domain with the L7a ribosomal protein gene. However, the formation of the chimeric transforming gene occurred during the transfection of DNA from a human breast carcinoma cell line (Barbacid et al., 1991). In vitro activation of the TRK oncogene following transfection procedure occurs with a high frequency, through different modalities: gene fusion, point mutations, in frame deletions and duplication of the tk domain (Barbacid et al., 1991).

By DNA transfection/focus formation assay we have detected activation of the TRK oncogene in 8/51 cases of human papillary thyroid carcinomas (Table I). TRK activation was demonstrated in the DNA both from the primary and the metastatic tumors but not from normal cells of the same patients, thus representing a tumor specific event (Bongarzone et al., 1989).

Southern blot analysis showed the involvement of the tropomyosin gene (TPM3) in the generation of three TRK-thyroid oncogene (Sozzi et al., 1992; Bongarzone et al., unpublished results), as for the original TRK oncogene (Martin-Zanca et al., 1986). The cytogenetic analysis of two TPM3/TRK positive papillary thyroid carcinomas has revealed rearrangements of chromosome 1 that, although not sufficient to indicate the precise mechanism of fusion of the two sequences, are demonstrating that also TRK is generated by an intrachromosomal rearrangement between an activating gene at 5' end (TPM3) and the tyrosine kinase domain of a growth factor receptor (NTRK1) that provides the 3' end of the transforming sequence (Sozzi et al., 1992). In the other three cases analysed no involvement of tropomyosin was detected. Therefore, we propose to designate as TRK-T the thyroid TRK oncogenes activated by foreign sequences different from tropomyosin. A first of these cases TRK-T1 was cloned and characterized (Greco oncogene produced by an intrachromosomal rearrangement that linked the tyrosine kinase domain of the TRK proto-oncogene to the 5' region of the TPR gene (Park et al., 1986) both mapping to chromosome 1 (Miozzo et al., 1990; Park et al., 1986). TRK-T1 oncogene encodes a 55 KD protein reacting with antibodies against the 14 carboxy-terminal aminoacids of the proto TRK protein. We demonstrated the involvement of TPR sequences in the generation of another different TRK-T oncogene, designated as TRK-T2 (Greco et al., 1992), whereas one case (TRK-T3) did not show an involvement of either TPM3 or TPR. TPR was first identified as part of the MET oncogene, a human transforming gene isolated from HOS cells treated with MNNG and created by the fusion of TPR with the tk domain of a novel receptor (Park et al., 1986) subsequently identified as the hepatocytes growth factor receptor (HGFr). Since in papillary thyroid carcinomas TRK activation occurred in the original tumors, TRK-T1 and TRK-T2 oncogenes represent the first evidence of "in vivo" oncogenic rearrangement involving TPR sequences, and indicate that also the tumor-specific activation of TRK oncogene may involve the rearrangement with different 5' end donor genes (Fig. 2).

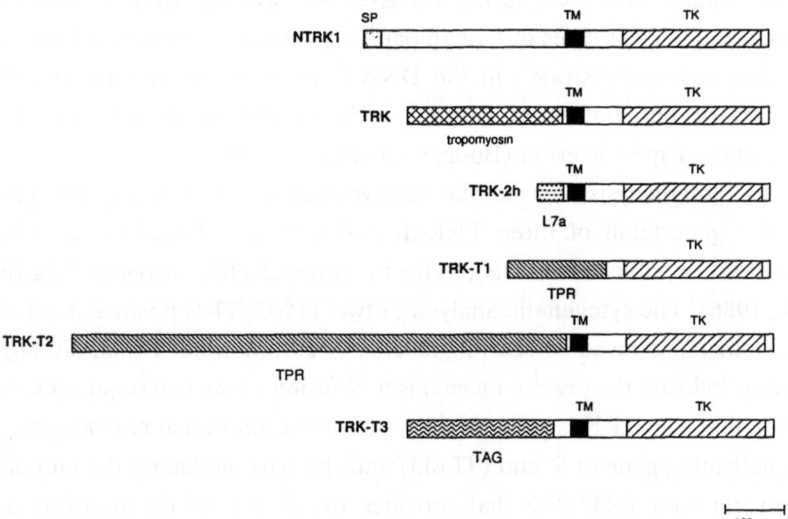

Fig. 2 - Schematic structure of the product of the TRK proto-oncogene (NTRK1) and of its oncogenic versions. TRK-T indicates the oncogenes derived from human papillary thyroid carcinomas. Signal peptide (SP), transmembrane

Table I - RET AND NTRK1 ACTIVATION IN HUMAN PAPILLARY THYROID CARCINOMAS

Activated proto-oncogene	Positive/ Total cases	Unrelated sequences fused to TK domain of proto-oncogene		Designation of chimeric transforming sequence
RET	18/51 (35%)	9/18	H4/D10S170 (uncharacterized gene)	ptc-1
		2/18	RI (regulatory subunit of cAMP-dependent PKA)	ptc-2
		7/18	neither H4 nor RI	ptc...
NTRK1	8/51 (15%)	3/8	TPM3 (non muscle tropomyosin gene)	TRK
		3/8	TPR (Translocated Promoter Region)	TRK-T1 (T2 and T4)
		1/8	TAG (TRK Activating Gene, uncharacterized gene)	TRK-T3
		1/8	not yet determined	--

CONCLUSION

The molecular analysis of human papillary thyroid carcinomas has revealed some peculiar features. These tumors show an high frequency of tyrosine kinase oncogene activation since RET/ptc and TRK together are found activated in about 50% of the cases analysed (Table I). Beside the RAS gene family this is the highest frequency of oncogene activation detected in solid tumors. In addition, the mechanism of RET and TRK activation involves the truncation of their 5' extracellular domain which is replaced by sequences provided by different genes as depicted in Figure 3.

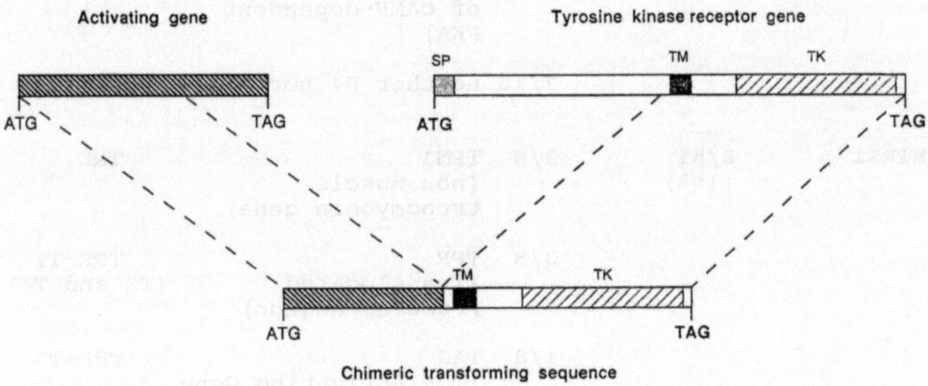

Fig. 3 - Schematic representation of the molecular rearrangements between different genes leading to the generation of chimeric transforming sequences. The activating gene provides the 5' end sequences regulating the expression of the oncogenic chimera while its transforming activity is due to part of a gene encoding a receptor molecule with a tyrosine kinase enzymatic activity.

The gene fusions are mediated by chromosome rearrangements that in the majority of the cases involve genes located on the same chromosomes where RET and TRK map, chromosome 10q and 1q, respectively (Fig. 4).

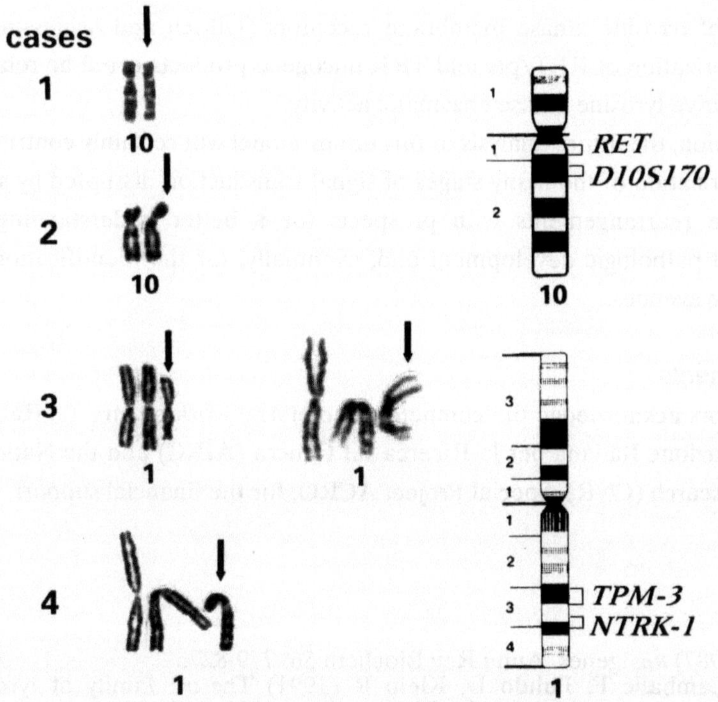

Fig. 4 - Intrachromosomal rearrangements activating RET (cases 1 and 2) and NTRK1 (cases 3 and 21) by formation of the chimeric transforming sequences RET/ptc1 and TRK.

These findings make the papillary thyroid carcinoma one of the first type of human non-hematophoietic tumor where the molecular basis underlying the neoplastic transformation are emerging (Editorial, Nature Genetics, 1992). In fact, this high frequency of gene rearrangements would suggest the existence, in thyroid epithelial cells, of an unscheduled activity of certain recombinases. In this respect, the analysis of the sequences involved in the different RET/ptc and TRK breakpoints, would elucidate whether they arise from common or different mechanisms, that is whether they could involve or not the same recombination machinery.

Finally, also the products of the activated oncogenes share several unique features. The chimeric transforming proteins show a constitutive phosphorylation on tyrosine and display a constitutive enzymatic activity. In addition, the different sequences providing the 5' end of the different RET/ptc and TRK oncogenes are ubiquitously expressed and are demonstrated (TPR and RI) or predicted the activation of tyrosine kinase membrane receptors (Ullrich and Schlessinger, 1991), the dimerization of RET/ptc and TRK oncogene products could be related to their constitutive tyrosine kinase enzymatic activity.

In conclusion, the future analysis of this tumor model will certainly contribute to the characterization of the many stages of signal transduction disrupted by such oncogenic gene rearrangements with prospects for a better understanding of physiologic and pathologic development and, eventually, for the identification of new therapeutic avenues.

Acknowledgments

The authors acknowledge the competent secretarial work of Mrs. G. Raineri and the Associazione Italiana per la Ricerca sul Cancro (AIRC) and the National Counsil for Research (CNR), Special Project ACRO, for the financial support.

References

Barbacid M (1987) *Ras* genes. Annu Rev Biochem 56: 779-827.

Barbacid M, Lamballe F, Pulido D, Klein R (1991) The *trk* family of tyrosine protein kinase receptors. Biochim Biophys Acta, 1072: 115-127.

Bongarzone I, Pierotti MA, Monzini N, Mondellini P, Manenti G, Donghi R, Pilotti S, Grieco M, Santoro M, Fusco A, Vecchio G, Della Porta G (1989) High frequency of activation of tyrosine kinase oncogenes in human papillary thyroid carcinoma. Oncogene 4:1457-1462.

Bongarzone I, Monzini N, Borrello MG, Carcano C, Ferraresi G, Arighi E, Mondellini P, Della Porta G, Pierotti MA (1993) Molecular characterization of a thyroid tumor specific transforming sequence formed by the fusion of ret tyrosine kinase and the regulatory subunit RI of c-AMP protein kinase. A Mol Cell Biol, in press.

Boshart M, Weih F,, Nichols M, Schutz G (1991) The tissue-specific extinguisher locus TSE1 encodes a regulatory subunit of cAMP-dependent protein kinase. Cell, 66: 849-859.

Donghi R, Sozzi G, Pierotti MA, Biunno I, Miozzo M, Fusco A, Grieco M, Santoro M, Vecchio G, Spurr NK, Della Porta G (1989) The oncogene associated with human papillary thyroid carcinoma (PTC) is assigned to chromosome 10q11-q12 in the same region as multiple endocrine neoplasia type 2A (MEN2A).Oncogene 4: 521-523.

Dumont JE, Jauniaux J-C, Roger PP (1989) The cyclic AMP-mediated stimulation of cell proliferation. Trends Biochem Sci 14: 67-71.

Editorial (1992) Keeping track of the translocations. Nature Genet, 2: 85-86.

Fusco A, Grieco M, Santoro M, Berlingieri MT, Pilotti S, Pierotti MA, Della Porta G, Vecchio G (1987) A new oncogene in human thyroid papillary carcinomas and their lymph nodal metastases. Nature 328: 170-172.

Greco A, Pierotti MA, Bongarzone I, Pagliardini S, Lanzi C, Della Porta G (1992) *TRK*-T1 is a novel oncogene formed by the fusion of *TPR* and *TRK* genes in human papillary thyroid carcinomas. Oncogene 7:237-242.

Grieco M, Santoro M, Berlingieri MT, Melillo RM, Donghi R, Bongarzone I, Pierotti MA, Della Porta G, Fusco A, Vecchio G (1990) PTC is a novel

Hempstead BL, Martin-Zanca D, Kaplan DR, Parada LF, Chao MV (1991) High affinity NGF binding requires coexpression of the *trk* proto-oncogene and the low affinity NGF receptor. Nature 350: 678-683.

Ishizaka Y, Ochiai M, Tahira T, Sugimura T, Nagao M (1989) Activation of the *ret-II* oncogene without a sequence encoding a transmembrane domain and transforming activity of two *ret-II* oncogene products differing in carboxy-termini due to alternative splicing. Oncogene, 4: 789-794.

Jones KW, Shapero MH, Chevrette M, Fournier REK (1991) Subtractive hybridization cloning of a tissue-specific extinguisher: TSE1 encodes a regulatory subunit of protein kinase A. Cell 66: 861-872.

Klein R, Jing S, Nanduri V, O'Rourke E, Barbacid M (1991) The *trk* proto-oncogene encodes a receptor for nerve growth factor. Cell, 65: 189-197.

Lanzi C, Borrello MG, Bongarzone I, Migliazza A. Fusco A, Grieco M, Santoro, R.A. Gambetta, F. Zunino, G. Della Porta, and M.A. Pierotti. 1992. Identification of the product of two oncogenic rearranged forms of the *RET* proto-oncogene in papillary thyroid carcinomas. Oncogene, 7: 2189-2194.

Martin-Zanca D, Hughes SH, Barbacid M (1986) A human oncogene formed by the fusion of truncated tropomyosin and protein tyrosine kinase sequences. Nature (London), 319: 743-748.

Miozzo M, Pierotti MA, Sozzi G, Radice P, Bongarzone I, Spurr NK, Della Porta G (1990) Human TRK proto-oncogene maps to chromosome 1q32-q41. Oncogene, 5: 1411-1414.

Park M, Dean M, Cooper CS, Schmidt M, O'Brien SJ, Blair DG, Vande Woude GF (1986) Mechanism of *met* oncogene activation. Cell, 45: 895-904.

Pierotti MA, Santoro M, Jenkins RB, Sozzi G., Bongarzone I, Grieco M, Monzini N, Miozzo M, Herrmann MA, Fusco A, Hay I.D., Della Porta G, Vecchio G (1992) Characterization of an inversion on the long arm of chromosome 10 juxtaposing *D10S170* and *RET* and creating the oncogenic sequence *RET/PTC*. Proc Natl Acad Sci USA, 89: 1616-1620.

Radice P, Sozzi G, Miozzo M, De Benedetti V, Cariani T, Bongarzone I, Spurr NK, Pierotti MA, Della Porta G (1991) The human tropomyosin gene involved in the generation of the TRK oncogene maps to chromosome 1q31. Oncogene 6: 2145-2148.

Sandberg M, Skalhegg B, Jahnsen T (1990) The two mRNA forms for the type I regulatory subunit of cAMP-dependent protein kinase from human testis are due to the use of different polyadenylation site signals. Biochem Biophys Res Comm, 167: 323-330.

Shih C, Shilo BZ, Goldfarb MP, Dannenberg A, Weinberg RA (1979) Passage of phenotypes ofchemically transformed cells via transfection of DNA and chromatin. Proc Nat Acad Sci 76: 5714-5718.

Sozzi G, Pierotti MA, Miozzo M, Donghi R, Radice P, De Benedetti V, Grieco M, Santoro M, Fusco A, Vecchio G, Mathew CGP, Ponder BAJ, Spurr NK, Della Porta G (1991) Refined localization to contiguous regions on chromosome 10q of the two genes (H4 and RET) that form the oncogenic sequence PTC. Oncogene, 6: 339-342.

Sozzi G, Bongarzone I, Miozzo M, Cariani CT, Mondellini P, Calderone C, Pilotti S, Pierotti MA, Della Porta G (1992) Cytogenetic and molecular genetic characterization of papillary thyroid carcinomas. Genes Chrom Cancer, 5:1-7.

Takahashi M, Ritz J, Cooper GM (1985) Activation of a novel human transforming gene, ret, by DNA rearrangement. Cell 42: 581-588.

Takahashi M, Cooper GM (1987) ret transforming gene encodes a fusion protein homologous to tyrosine kinases. Mol Cell Biol 7, 1378-1385.

Takahashi M, Inaguma Y, Hiai H, Hirose F (1988) Developmentally regulated expression of a human "Finger"-containing gene encoded by the 5' half of the ret transforming gene. Mol Cell Biol 8: 1853-1856.

Ullrich A, Schlessinger J (1991) Signal transduction by receptors with tyrosine kinase activity. Cell, 51: 203-212.
Varmus HE (1984) The molecular genetics of cellular oncogenes. Annu Rev Genet 18: 553-612.
Weinberg RA (1982) Oncogenes of spontaneous and chemically induced tumors. Adv Cancer Res 36: 149-163.

Structure-Function Analysis of Hepatocyte Growth Factor and its Tyrosine-Kinase Receptor c-Met

Nathalie A. Lokker, Melanie R. Mark and Paul J. Godowski
Department of Cell Genetics, Genentech, Inc.
460 Point San Bruno Blvd.
So. San Francisco, CA 94080, U.S.A.

Abstract

Hepatocyte growth factor (HGF) exhibits mitogenic and/or motogenic activities for a variety of cells (see for review Matsumoto and Nakamura, 1992). Structurally, it has similarities to kringle-containing serine-proteases, although it does not possess proteolytic activity. The HGF receptor (HGFr) is the product of the c-met proto-oncogene, a membrane-spanning tyrosine kinase receptor. The 190-kDa precursor is proteolytically processed within the extra-cellular domain (ECD) to a heterodimer consisting of a 50-kDa α subunit disufide-linked to a 145-kDa ß subunit. A structure-activity relationship analysis of both HGF and its receptor was performed by functional analysis of ligand and receptor variants.

I. Analysis of HGF variants was accomplished by defining their ability to induce DNA synthesis on hepatocytes in primary culture and to compete with wild-type HGF for binding to a soluble form of the HGFr (sHGFr: ECD fused to the hinge and constant region of human IgG heavy chain). Our results show that 1) Pro-HGF can bind to, but not activate the HGF receptor (HGFr); 2) the HGF ß-chain is not required for receptor binding but contributes to receptor activation; 3) a primary receptor binding determinant is located within the N-terminal 272 residues.

II. Like the mature form of the HGFr, sHGFr is a heterodimer which arises by proteolytic processing. The putative cleavage site for the HGFr is 303-RKKKRS-308. Mutational analysis of this cleavage site suggests that HGFr maturation is mediated by furin or a furin-like protease. Finally, our data indicate that processing of the sHGFr into α/ß form is not required for high affinity binding to either pro- or mature HGF.

Results and Discussion

Part I. Structure-function analysis of HGF

A schematic representation of the α and ß subunits of huHGF is shown in Fig. 1. The α-chain contains a hairpin and four kringle domains, whereas the ß-chain has homology to serine-proteases (38 % to the plasminogen serine-protease domain; Nakamura et al., 1989). However, HGF does not appear to be an active protease because some of the residues proposed to be essential for proteolytic activity are changed.

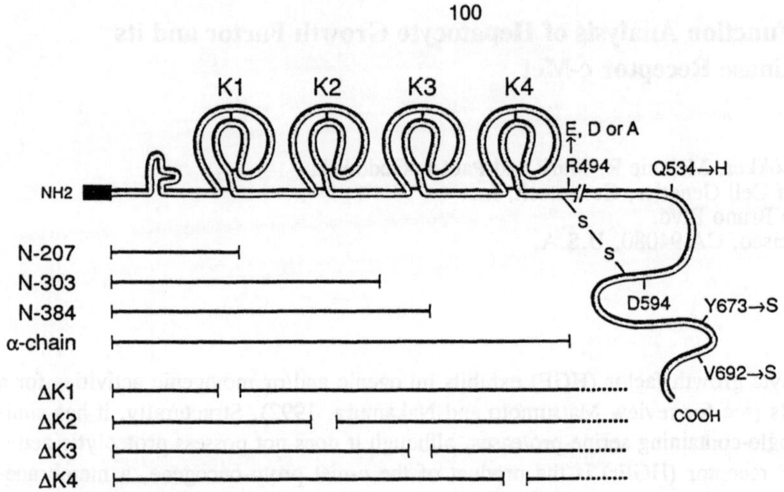

Fig. 1: Schematic representation of the secondary structure prediction of huHGF. Indicated are the point mutations which have been introduced. Depicted below are the mature forms of the C-terminal truncation variants (from Lokker et al., 1992, with permission).

Mutations were introduced into the HGF cDNA in a CMV based expression plasmid and conditioned media from stable populations of human 293 cells expressing each variant were assayed by Western blotting to monitor the size and expression level of the HGF variants (Lokker et al., 1992). The mitogenic activity on liver cells in primary culture and ability to bind to a soluble form of the HGFr was determined. For this, the ECD of the HGFr was fused to the hinge and constant region (Fc) of an human IgG heavy chain and binding was performed in solution. The affinity of HGF for purified sHGFr is identical to the affinity of HGF for endogenous HGFr on A549 cells.

Analysis of cleavage site, protease, C-terminal and kringle deletion variants

We chose to try to block cleavage of HGF by replacing the R494 with either D, E, or A. Each of these mutations appeared to block processing of HGF (Lokker et al., 1992). These variants were defective in inducing the proliferation of hepatocytes in primary culture (Table I). In binding assays, a 2- to 10-fold higher K_d when compared to wild-type (wt) rhuHGF was observed (Table I). These results argue strongly that cleavage of HGF into the two-chain form is required for mitogenic activity, i.e. that single-chain HGF is a promitogen.

All protease variants (as single, double and triple mutations, Fig. 1) exhibited a reduced specific activity (SA) except for variant Q534H (Table I). Interestingly, these variants showed a K_d similar to wt rhuHGF. Thus, it appears that these mutants are defective in an activity subsequent to receptor binding.

A number of C-terminal truncations of HGF were made by deleting either the ß-chain or the ß-chain in addition to a progressive number of kringles as depicted in Fig. 1. The data indicate that the N-terminal 272 residues of HGF (the mature form of variant N-303) are sufficient for high affinity binding to the HGFr (Table I).

Results from deleting each kringle domain are shown in Table I. Deletion of the first kringle (variant ΔK1) of HGF affected biological activity and binding most, showing at least a 100-fold reduction both in SA and K_d. Deletion of all other kringles (variants ΔK3,

ΔK3 or ΔK4) also induces severely reduced mitogenic activity (Table I). However, the K_d of these deletion variants remained close to that observed with wt rhuHGF.

Variants (var)	SAvar/SAwt +/- S.D.	Kdwt/Kdvar +/- S.D.
Single chain		
R494A	<0.03	0.32 +/- 0.18
R494D	<0.03	0.51 +/- 0.21
R494E	<0.02	0.31 +/- 0.13
Protease		
Q534H	1.19 +/- 0.44	1.48 +/- 0.85
Y673S	0.27 +/- 0.07*	1.35 +/- 0.72
V692S	0.08 +/- 0.04*	1.02 +/- 0.13
Q534H, Y673S	<0.03	2.24 +/- 1.11
Y673S, V692S	<0.02	1.76 +/- 0.63
Q534H, Y673S, V692S	<0.02	1.91 +/- 1.28
C-terminal truncation		
N-303	<0.05	0.23 +/- 0.03
N-384	<0.05	0.25 +/- 0.02
α-chain	<0.04	0.25 +/- 0.03
Kringle deletion		
ΔK1	<0.002	<0.03
ΔK2	<0.05	0.41 +/- 0.18
ΔK3	<0.03	0.56 +/- 0.36
ΔK4	<0.07	0.86 +/- 0,46

Table I: Residual biological activity and receptor binding capacity of single-chain, protease, C-terminal and kringle deletion rhuHGF variants compared to wild-type rhuHGF.
Ratios are mean +/- S.D. from at least three assays. Protein contents from supernatants were determined in two different ELISAs and specific activity (SA) and K_d were calculated from mitogenic activity and inhibition curves as described in Materials and Methods. When a < sign is used, this corresponds to the highest K_d or the lowest SA which could be defined. (*) indicate that the plateau of the mitogenic activity curve of the variant was low compared to wild-type (wt) rhuHGF (from Lokker et al., 1992, with permission).

Induction of tyrosine-phosphorylation of the HGF receptor

We determined if variants R494E or Y673S, V692S, which bind the HGFr in vitro but are defective for mitogenic activity, could stimulate tyrosine-phosphorylation of the HGFr in A549 cells. Both variants exhibit a reduced ability to induce phosphorylation of the HGFr (Fig. 2) indicating that these mutations affect a function of HGF subsequent to binding, e.g. dimerization of the HGFr.

The putative cleavage site for the HGFr is 303-KRKKR-307 which fits the consensus sequence for cleavage by furin (RXK/RR; Watanabe et al., 1992). We could show by SDS-PAGE and Western blotting that replacement of the P1 or P4 arginine, but not the P3 lysine, with alanine inhibited conversion to the α/ß heterodimer. The precessing of variants P2A and P5A, containing Lys to Ala substitutions at positions 306 and 303, respectively, was partially inhibited. This result strongly suggests that this is the *bona fide* cleavage site of the HGFr and that processing of the HGFr is mediated by furin or a furin-like protease (Mark et al., in press).

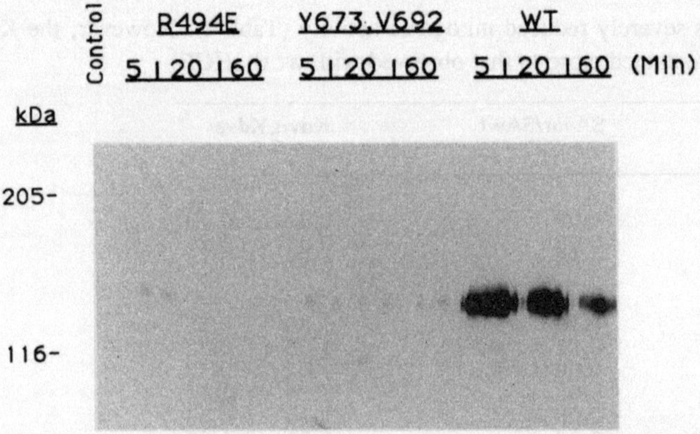

Fig. 2: Time course of ligand-induced tyrosine phosphorylation on the 145 kDa ß-subunit of the HGFr by wild-type rhuHGF (WT), single-chain (R494E) or protease (Y673S, V692S) variants. Lysates from A549 cells incubated for 5, 20 or 60 min with 200 ng/ml of HGF or variants were prepared and immunoprecipitated with an anti-HGF receptor antibody and blotted with anti-phosphotyrosine antibodies.

We determined if processing of the HGFr ECD to the α/ß heterodimer is required for binding to pro (single-chain) or mature (two-chain form) HGF (Fig. 3). These data show that cleavage of the HGFr is not necessary for binding, because the P1A and P4A sHGFr variants can bind with high affinity to both wild-type and the single-chain HGF variant R494E.

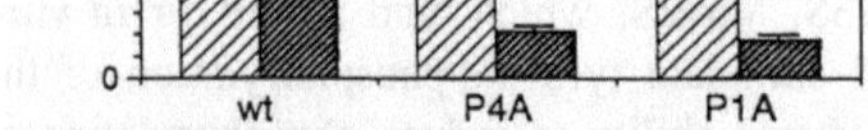

Fig. 3: Effect of cleavage of the sHGFr on binding to wild-type or single-chain HGF.
Data is expressed as the ratio of the K_d of the wt sHGFr for mature HGF divided by the K_d of wt sHGFr or the P1A or P4A variant of either mature two-chain HGF or R494E pro-HGF variant times 100 %. Values are mean +/- S.D. from three independent assays (from Mark et al., in press, with permission).

The unique structure of the HGF molecule suggests that there may be multiple events that regulate the biological activity of this growth factor. A first step of regulation may be the conversion of pro HGF to mature HGF which subsequently may control receptor activation.

References

Lokker N.A., Mark, M.R., Luis, E.A., Bennett, G.L., Robbins, K.A., Baker, J.B., and Godowski, P.J. (1992) EMBO J. 11, 2503-2510

Matsumoto, K. and Nakamura, T. (1992) Crit. Rev. Oncol. 3, 27-54

Mark, M.R., Lokker, N.A., Zioncheck, T.F., Luis, E.A. and Godowski, P.J. (1992) J. Biol. Chem. (in press)

Nakamura, T., Nishizawa, T., Hagiya, M., Seki, T., Shimonishi, M., Sugiura, A., Tashiro, K. and Shimizu, S. (1989) Nature 342, 440-443

Watanabe, T., Nakagawa, T., Ikemizu, J., Nagahama, M., Murakami, K. and Nakayama, K. (1992) J. Biol. Chem. 267, 8270-8274

ONCOGENIC ACTIVATION OF THE MET/HGF RECEPTOR TYROSINE KINASE IS PROMOTED BY LEUCINE ZIPPER MEDIATED DIMERIZATION

Gerard A. Rodrigues and Morag Park
Molecular Oncology Laboratory
Departments of Oncology and Medicine
McGill University
687 Pine Ave. West
Montreal, Canada
H3A 1A1

ABSTRACT

The *met* receptor tyrosine kinase is oncogenically activated by a genomic rearrangement that results in the production of a 65 kd protein containing *tpr* sequences fused directly to the *met* receptor kinase domain. *Tpr* was found to be essential for cell transformation mediated by *tpr-met*. Two leucine zipper motifs have been identified in *tpr*, one of which is essential for dimerization of *tpr-met* and cell transformation. From analysis of mutant *tpr-met* molecules we propose that the leucine zipper region of *tpr* oncogenically activates the *met* receptor by promoting constitutive dimerization of the kinase domain. In support of this hypothesis, a kinase negative mutant of *tpr-met* was found to suppress transformation in a dominant negative fashion.

INTRODUCTION

The *met* protooncogene, a receptor tyrosine kinase (RTK) (Park et al., 1987; Gonzatti-Haces et al., 1988; Tempest et al., 1986), is the receptor for hepatocyte growth factor/scatter factor (HGF/SF)(Bottaro et al., 1991; Naldini et al., 1991). The *met* oncogene was first identified in a human osteogenic sarcoma cell line (HOS) treated *in vitro* with N-methyl-N-nitronitrosoguanidine (MNNG)(Cooper et al., 1984). The oncogenic activation of *met* results from a genomic rearrangement which places sequences from chromosome 1, designated *tpr* (translocated promoter region) into the *met* locus on chromosome 7 (Park et al., 1986). Consequently, the extracellular, transmembrane and a portion of the cytoplasmic domain of the *met* receptor are deleted in the oncogene product (Park et al., 1987). In addition,142 novel amino acids are fused directly to the *met* kinase domain (Chan et al., 1987) resulting in the production of a 65 kd hybrid protein in MNNG-treated HOS cells (Gonzatti-Haces et al., 1988). Unlike the *met* receptor, the *tpr-met* protein is constitutively phosphorylated on tyrosine

NATO ASI Series, Vol. H 76
Tyrosine Phosphorylation/Dephosphorylation and Downstream Signalling
Edited by L. M. G. Heilmeyer Jr.

residues consistent with its proposed role in cell transformation (Gonzatti-Haces et al., 1988). Significantly, similar to the *met* oncogene, *tpr* sequences are fused to the kinase domain of the *trk*/nerve growth factor RTK in oncogenes isolated from human thyroid papillary tumors (Greco et al., 1992) and to the *raf* serine/threonine kinase in an oncogene generated by transfection of hepatocellular carcinoma DNA into NIH 3T3 cells (King et al., 1988). This frequent rearrangement of *tpr* with loci encoding kinases suggests that *tpr* may contribute to oncogenic activation of these genes by a common molecular mechanism.

RESULTS

Characterization of the *tpr-met* protein product

Whereas the *met* RTK is a transmembrane product, in a transformed Fischer Rat 3T3 cell line expressing a *tpr-met* cDNA a cytoplasmic product of 65kd is detected by SDS-PAGE under reducing conditions (data not shown). This is consistent with previous results obtained with MNNG-HOS cells (Gonzatti-Haces et al., 1988). However, under non-reducing conditions a protein with a molecular weight of 130 kd is observed (data not shown).The size of the larger molecular weight species is consistent with the predicted size of a *tpr-met* dimer stabilized by disulfide bonds *in vitro*.

To demonstrate if a direct physical association exists between *tpr-met* molecules *in vivo*, sucrose gradient fractionation of lysates prepared from *tpr-met* expressing cells was conducted. Fractions were collected and assayed for the presence of *tpr-met*. *Tpr-met* was detected in fractions corresponding to a molecular weight of 130 kd, consistent with that of a dimer (data not shown). This result provides evidence for the presence of *tpr-met* dimers *in vivo*.

Tpr contains leucine zipper motifs required for dimerization and for cellular transformation by the *tpr-met* oncogene

The absence of any mutations within the *met* portion of the oncogene suggests that deletion of the extracellular domain of *met* alone is not sufficient to oncogenically activate *tpr-met*. To formally test this, a deletion mutant, dl*tpr*, was constructed in which all *tpr* sequences are removed (Figure 1).The transforming capacity of this mutant was assessed in the focus forming assay using Fr3T3 cells. Although the dl*tpr* product has autokinase activity similar to *tpr-met* (data not shown), the dl*tpr* mutant had no transforming activity (Figure 1). Thus, *tpr* is essential for cell transformation.

The nucleotide sequence for the *tpr-met* cDNA clone was examined for the presence of potential dimerization motifs.Two leucine zipper motifs designated leucine zipper A and B were identified in *tpr* (Figure 1). To test whether these leucine zipper motifs mediate dimerization of *tpr-met* and whether this is important for transformation, mutations were introduced into these regions by substituting the first three of the four leucines in leucine zippers A and B with valines individually or together (Figure 1). Mutations in *tpr* do not affect the *met* catalytic domain as all mutants were found to be enzymatically active *in vitro* (data not shown). When tested in the focus forming assay the leucine zipper A region was found to be essential for transformation. Mutation of this region alone or together with mutations in the leucine zipper B region abolished transformation. In contrast, mutation of the leucine zipper B region alone was only slightly deleterious (Figure 1). To test for dimerization of these mutants, a carboxy terminal deletion mutant of *tpr-met* was generated. The *tpr-met*ΔC mutant can be distinguished from the wild type protein both on the basis of size (60 kd) and because it has lost the epitope recognized by the antibody Ab144 raised against a peptide at the C-terminus of the *met* receptor (Rodrigues et al., 1991). This mutant was co-transfected into COS-1 cells with *tpr-met* mutants and immunoprecipitations of proteins from cells metabolically labelled with [^{35}S] methionine were carried out with Ab144. The *tpr-met*ΔC protein can be coimmunoprecipitated when cotransfected with *tpr-met* or with the leucine zipper B mutant but not with dl*tpr* or the leucine zipper A mutant (summarized in Figure 1). These results demonstrate that the leucine zipper A region of *tpr* mediates dimerization and that this event is related to the capacity of the *tpr-met* products to cause cell transformation.

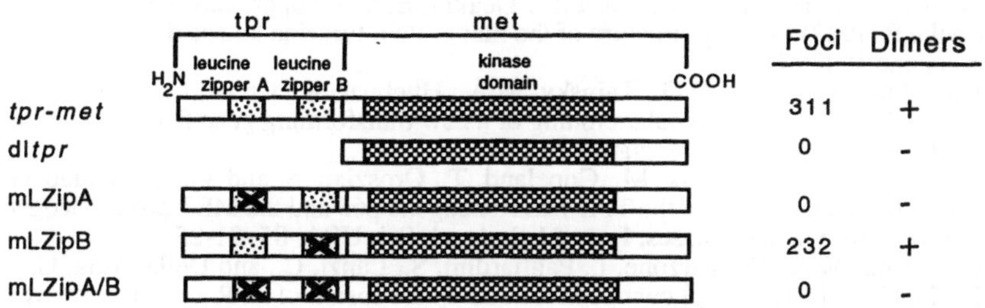

Figure 1. Schematic representation of *tpr-met* mutants. Mutations in the leucine zipper region are indicated by an X. The transforming and dimerization abilities are indicated on right. The number of foci were scored two weeks following transfection of 5µg of plasmid DNA.

A kinase defective mutant of *tpr-met* acts as a dominant negative suppressor of cell transformation

One prediction to arise from our model is that a kinase defective mutant would suppress cell transformation through the formation of inactive heterodimers with *tpr-met*. The *tpr-met*ΔC mutant was rendered kinase defective by substituting the highly conserved lysine residue in the ATP binding pocket of the *met* kinase domain with an alanine thereby generating the mutant K241AΔC. The K241AΔC mutant was inserted into a retroviral vector to achieve higher levels of expression than that of *tpr-met*. When equal amounts of this vector and *tpr-met* in pXM were co-transfected into Fr3T3 cells a dramatic reduction (up to 85%) in the number of foci was observed.

DISCUSSION

Dimerization has been shown to be required for the ligand-stimulated mitogenic response mediated by RTKs (Ullrich and Schlessinger, 1990). The hypothesis that activation of the *met* oncogene is a consequence of *tpr* mediated dimerization is supported by the finding that mutations within the leucine zipper A region eliminates the ability of *tpr-met* to form dimers *in vivo* and abolishes the transforming ability of *tpr-met*, and that a kinase defective mutant can suppress transformation in a dominant negative fashion. We have noted the presence dimerization motifs in a number of other activated RTKs that have undergone DNA rearrangement, suggesting the model we propose for oncogenic activation of *tpr-met* may be applicable to many members of this class of RTKs.

REFERENCES

Bottaro, D.P., Rubin, J.S., Faletto, D.L., Chan, A.M.-L., Kmiecik, T.E., Vande Woude, G.F., and Aaronson, S.A. (1991). Identification of the HGF receptor as the c-*met* proto-oncogene product. Science 251: 802-804.

Chan, A.M.-L., King, H.W.S., Tempest, P.R., Deakin, E.A., Cooper, C.S. and Brooke,P.(1987) Primary structure of the met protein tyrosine kinase domain. Oncogene 1: 229-233.

Cooper, C.S., Park, M., Blair, D.G., Tainsky, M.A., Huebner, K., Croce, C.M., and Vande Woude, G.F. (1984). Molecular cloning of a new transforming gene from a chemically transformed human cell line. Nature 311: 29-33.

Gonzatti-Haces, M., Seth, A., Park, M., Copeland, T., Oroszlan, S. and Vande Woude, G.F. (1988). Characterization of the TPR-MET oncogene p65 and the MET proto-oncogene p140 protein tyrosine kinases. Proc. Natl. Acad. Sci. USA. 85: 21-25.

Greco, A., Pierotti, M.A., Bongarzone, I., Pagliardini, S., Lanzi, C., and Della Porta, G. (1992). TRK-T1 is a novel oncogene formed by the fusion of TPR and TRK genes in a human papillary thyroid carcinomas. Oncogene 7:237-242.

King, H.W.S., Tempest, P.R., Merrifield, K.R., and Rance, A.J. (1988). Tpr homologues activate *met* and *raf*. Oncogene 2:617-619.

Naldini, L., Vigna, E., Narsimhan, R.P., Gaudino, G., Zarnegar, R., Michalopoulos, G.A., and Comoglio, P.M. (1991). Hepatocyte growth factor (HGF) stimulates the tyrosine kinase activity of the receptor encoded by the proto-oncogene c-*MET*. Oncogene 6:501-504.

Park, M., Dean, M., Cooper, C.S., Schmidt, M., O'Brien, S.J., Blair, D.G., and Vande Woude, G.F. (1986). Mechanism of *met* oncogene activation. Cell 45: 895-904.

Park, M., Dean, M., Kaul, K., Braun, M.J., Gonda, M.A., and Vande Woude G.F. (1987). Sequence of *met* proto-oncogene cDNA has features characteristic of the tyrosine kinase family of growth factor receptors. Proc. Natl. Acad. Sci. USA. 84:6379-6383.

Rodrigues, G.A., Naujokas, M.A., and Park, M. (1991). Alternative splicing generates isoforms of the *met* receptor tyrosine kinase which undergo differential processing. Mol. Cell. Biol. 11: 2962-2970.

Tempest, P.R., Cooper, C.S., and Major, G.N. (1986). The activated human *met* gene encodes a protein tyrosine kinase. FEBS lett 209: 357-361.

Ullrich, A. and Schlessinger, J. (1990). Signal transduction by receptors with tyrosine kinase activity. Cell 61: 203-212.

SPECIFICITY OF THE TRUNCATED FORM OF HUMAN T-CELL PROTEIN TYROSINE PHOSPHATASE TOWARD PHOSPHORYLATED SYNTHETIC PEPTIDES.

Maria RUZZENE, James A. LORENZEN[1], Arianna DONELLA-DEANA, Oriano MARIN, John W. PERICH[2] , Edmond H. FISCHER[1] and Lorenzo A. PINNA
Dipartimento di Chimica Biologica e Centro per lo Studio della Fisiologia Mitocondriale del Consiglio Nazionale delle Ricerche, Università di Padova, Via Trieste 75, 35121 Padova, Italy

INTRODUCTION

The family of protein tyrosine phosphatases (PTPs) includes a striking moltitude and diversity of members, all apparently committed with a single post-translational modification quantitatively almost negligible. PTPs should be therefore endowed with a marked substrate specificity.
Despite such a theoretical expectation, very little is known about the factors determining the substrate and site recognition by PTPs.
In this study we characterize the site specificity of a human recombinant T-cell PTP, expressed in its active truncated form of 37 kDa, devoided of the carboxyl terminal regulatory domain of 11 kDa (Cool, D.E. et al., 1989; Cool, D.E. et al., 1990; Zander, N.F. et al., 1991).
The analysis has been undertaken with a number of phosphorylated synthetic peptides that reproduce, with suitable modifications, the phosphoacceptor sites of a variety of phosphotyrosine proteins.

RESULTS AND DISCUSSION

The structure of the phosphopeptides deeply influences the dephosphorylation rate determined at submicromolar substrate concentration (TAB.1) : the best substrate is the 20 residues long peptide *src*[514-533] including the C terminal physiological phosphoacceptor site of pp60$^{c\text{-}scr}$ (Tyr527).

[1] Department of Biochemistry, SJ-70, University of Washington, Seattle, WA 98195, USA
[2] Biochemical and Molecular Biology Unit, University of Melbourne, Australia

TABLE 1. Dephosphorylation of synthetic [^{32}P]phosphopeptides by truncated TC-PTP.
Peptides were synthesized as previously described (Marin, O. et al., 1991) and phosphorylated in the presence of [γ-^{32}P]ATP by spleen tyrosin kinase TPK-III (Brunati, A.M. and Pinna, L.A., 1988). Phosphatase assay was carried out as elsewhere described (Cool, D.E. et al., 1990).
Activity is expressed as percentage of dephosphorylation after 10 min of incubation in the presence of 0.1 µM phosphopeptides and 1.7 ng/ml enzyme. Dephosphorylation was kept within the limit of linearity with respect to the time and enzyme concentration.

Peptide substrate	PTP activity
AFLEDYFTSTEPQYpQPGENL (*src*[514-533])	20.5
AFLED<u>F</u>FTSTEPQYpQPGENL (*src*[514-533]Phe^{6})	19.7
NIDGDGEVNYpEE (CaM (loop IV))	16.0
EDNEYpTA (*src*[412-418])	9.4
KDDEYpNPA (*fgr*[408-415])	7.7
EPQYpQPA (*src*[524-530]Ala^{7})	6.0
DEN(YY)pKA (*syk*[515-521])	3.7
DKDGNGYpISAAE (CaM (loop III))	1.1

The calmodulin dodecapeptide reproducing the calcium binding loop IV is a substrate almost as good as the *src*[514-533] peptides, while the peptide reproducing the calcium binding loop III of calmodulin is nearly unaffected. A shortened derivative of *src*[514-533], namely the heptapeptide *src*[524-530] is still efficiently dephosphorylated albeit less rapidly than the parent peptide. The dephosphorylation rate of *src*[524-530] is somewhat lower than that of the heptapeptide EDNEYpTA (c-*src*[412-418]), reproducing the main autophosphorylation site of p60^{src}(Tyr416), which in turn is comparable to that of the octapeptide KDDEYpNPA, reproducing the c-*fgr* sequence homologous to c-*src*[412-418]. The N-terminal acidic residues, with special reference to that located at position -3 and -4 relative to P-Tyr, play a relevant role as positive determinant of dephosphorylation by TC-PTP, as shown by substitutions of the *src*[412-418] peptide (TAB. 2).
The kinetic constants of some representative phosphopeptides are shown in TAB. 3. It should be noted that the values for the best peptide substrates are comparable to that obtained for protein substrates.

TABLE 2. Dephosphorylation of the *src*[412-418] peptide EDNEYpTA and its substituted derivatives by truncated TC-PTP.
Experimental details are described in Tab. 1.

Peptide substrate	PTP activity
EENEYpTA	13.9
EDNEYpTA	11.7
EDNAYpTA	10.0
EDAAYpAA	8.3
EANEYpTA	6.5
ENNEYpTA	5.6
QNAAYpTA	1.3

TABLE 3. Kinetic constants of truncated TC-PTP for phosphopeptide and phosphoprotein substrates.
Determinations were carried out with 1.7 ng/ml enzyme. Values for RCML (reduced, carboxiamidomethylated and maleylated lysozyme) and MBP (myelin basic protein) are drawn from Zander, N.F. et al.,1991.

Substrate	Km(μM)	Vmax(μmol/min/mg)
EDNEYpTA	0.60	16.80
AFLEDFFTSTEPQYpQPGENL	0.40	24.15
NIDGDGEVNYpEE	0.21	15.10
DRVYpIHPF	3.50	11.90
RCML	0.05	26.00
MBP	1.25	4.70

Phosphoseryl and phosphothreonyl peptides cannot be dephosphorylated by TC-PTP (not shown). The absolute specificity of the enzyme for Tyr-P residues was also demonstrated by synthesizing (Perich, J.W., 1991a/b) the phosphoseryl and phosphotyrosyl derivatives of the *src* peptide

NEYTA, i.e. NESpTA and NEYpTA, respectively, and by comparing their capability to compete against the enzymatic dephosphorylation of the radiolabelled *src* peptide EDNEY32pTA. While the IC_{50} value for the phosphotyrosyl peptide is around 3 μM, the phosphoseryl homologue is totally ineffective, providing the evidence that not only it cannot be dephosphorylated by TC-PTP, but it cannot even bind to the enzyme (not shown).

From the data presented here, it can be concluded that:

a) *Conditio sine qua non* for substrate recognition by TC-PTP appears to be the presence of a phosphotyrosyl residue;

b) the enzyme displays markedly variable activities toward peptides that reproduce different phosphotyrosyl sites;

c) its dephosphorylation efficiency can be deeply altered by deletions and/or aminoacid substitutions in the parent substrate.

The peptide length is relevant, but is not *per se* a sufficient condition.

The presence of acidic residues at positions –3 and/or –4 relative to Tyr-P appears to be a positive determinant of dephosphorylation. The best peptide substrates for TC-PTP share indeed the common feature of bearing an acidic residue at position –3.

REFERENCES

Brunati, A.M. & Pinna, L.A. (1988) Characterization of four tyrosine protein kinases from the particulate fraction of rat spleen. Eur. J. Biochem. 172, 451-457

Cool, D.E., Tonks, N.K., Charbonneau, H., Walsh, K.A., Fischer, E.H. & Krebs, E.G. (1989) cDNA isolated from a human T-cell library encodes a member of the protein-tyrosine-phosphatase family. Proc. Natl. Acad. Sci. USA *86*, 5257-5261

Cool, D.E., Tonks, N.K., Charbonneau, H., Fischer, E.H. & Krebs, E.G. (1990) Expression of a human T-cell protein-tyrosine-phosphatase in baby hamster kidney cells. Proc. Natl. Acad. Sci. USA 87, 7280-7284

Marin, O., Donella-Deana, A., Brunati, A.M., Fischer, S. and Pinna, L.A. (1991) Peptides reproducing the phosphoacceptor sites of $pp60^{c-src}$ as substrates for TPK-IIB, a splenic tyrosine kinase devoid of autophosphorylation activity. J. Biol. Chem 266, 17798-17803

Perich, J.W. (1991a) Synthesis of O-phosphoserine- and O-phosphothreonine-containing peptides. Meth. Enzymol. 201, 225-233

Perich, J.W. (1991b) Synthesis of O-phosphotyrosine-containing peptides. Meth. Enzymol. 201, 234-245

Zander, N.F., Lorenzen, J.A., Cool, D.E., Tonks, N.K., Daum, G., Krebs, E.G. & Fischer, E.H. (1991) Purification and characterization of a human recombinant T-cell protein-tyrosine-phosphatase from a baculovirus expression system. Biochemistry 30, 6964-6970

REGULATION OF TYROSINE PROTEIN PHOSPHATASES BY BINDING TO PHOSPHOLIPIDS AND COVALENT MODIFICATION

Hans Werner Hofer and Christine Stader
University of Konstanz, Faculty of Biology, D-7750 Konstanz, Germany

The regulation of the phosphorylation state of tyrosine residues in proteins is of pivotal importance for the transduction of growth factor signals, cellular growth and differentiation as well as for the suppression of tumour formation. Protein phosphatases are involved in these processes in a dual mode by regulating the activity of some serine/threonine- and tyrosine-specific protein kinases and by counteracting phosphorylation of proteins by tyrosine kinases. The majority of the tyrosine phosphatases so far identified exhibit sequence homology within their catalytic domains and can be attributed to two different subfamilies (reviewed in 1,2): The members of the receptor phosphatase subfamily contain a transmembrane domain and extracellular domains with similarities to cell adhesion molecules or immunoglobulins and (with one known exception) two catalytic domains (see Ref. 3). Most, if not all, members of this subfamily are integral membrane proteins. A second subfamily is formed by tyrosine phosphatases consisting of a single catalytic domain. Though they are not comprising a transmembrane sequence, they have nevertheless also been reported to associate with membranes or other particulate cellular compartments (4-6).

We have previously reported (7) that a 56 kDa tyrosine phosphatase (PTP) isolated from a particulate fraction of spleen is strongly inhibited by RNA (I_{50} 15 μg/ml) and DNA (I_{50} 150 μg/ml). The inhibition by these intracellular macromolecules resembles the inhibition by other polyanionic substances like heparin and poly(TyrGlu). Association with nucleic acids in ribosomes or in the nucleus, therefore, would be a potential mechanism for the attenuation of tyrosine phosphatase activity.

Here, we report that the same tyrosine phosphatase is strongly inhibited by low concentrations of anionic phospholipids. Fig. 1 depicts the diagrammatic representation of the influence of micellar lipids on the isolated tyrosine phosphatase. The presence of diacylglycerol or of lipids like PE*) and PC had no effect on the

*)Abbreviations: PC, phosphatidyl choline; PE, phosphatidyl ethanolamine; PI, phosphatidyl inositol; PS phosphatidyl serine; RCML, reduced carboxymethylated and maleylated lysozyme.

phosphatase activity. Anionic lipids like PI and PS, however, were inhibitory when present in the assays in micromolar concentrations. The IC_{50} values (6.0 ± 0.7 μM for PI and 3.7 ± 0.8 μM for PS) were similar for both of these lipids. Evidently, the phosphatase was not selective for the inositol or serine moieties of the lipids indicating that the phosphate (and/or its negative charge) may have played a major role in the interaction with the inhibitory site of the phosphatase. The inhibition by PI or PS was not influenced by the simultaneous presence of 100 μM PE (data not shown).

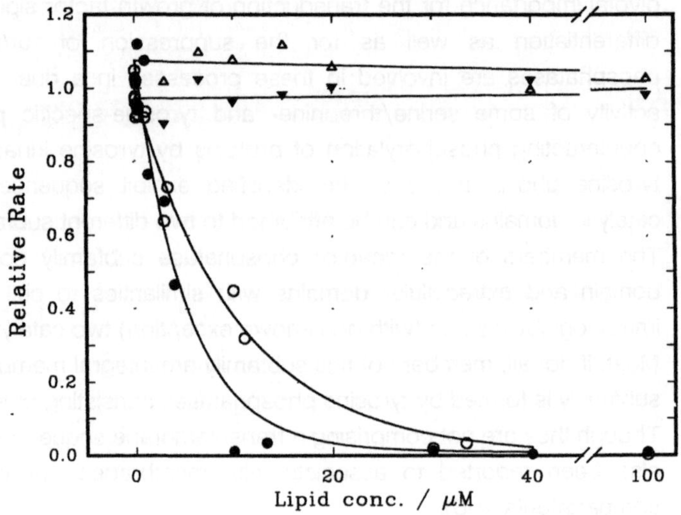

Fig. 1
Influence of various phospholipids on the activity of a lienal tyrosine phosphatase. The assays were performed using RCM-lysozyme as substrate and the activities expressed relative to that measured in the absence of effectors. Hollow circles: PI; full circles: PS; hollow triangles: PE; full triangles PC.

The inhibition by anionic phospholipids deserves special interest because of the fact that the phosphatase was isolated from a particulate fraction of spleen and, therefore, is likely located on membranes though apparently is not an integral membrane protein. Its kinetic and immunologic (see below) properties suggest relationship to the T-cell phosphatase (8). This is in accordance with the preference for an anionic substrate such as RCM-lysozyme, whereas receptor phosphatases generally exhibit higher activity with myelin basic protein. Localization at the membranes of the endoplasmic reticulum has also been found for full-length tyrosine phosphatase 1B (6) and the T-cell phosphatase has been reported to be strongly attached to particle fractions when over-expressed as a full-length clone (4).

PS (and to lesser extent) PI are major phospholipid fractions in cellular membranes and expose their polar head groups to the cytoplasm leading to high concentrations of negative charges on the membrane surface. Considering the *in vitro* effects, inhibition of the membrane-bound tyrosine phosphatase by anionic phospholipids in its vicinity does not appear unreasonable. The problem whether phosphorylation of PI might alter the inhibitory properties of the phospholipid and thus possibly subject the phosphatase activity under the control of PI kinases and phosphatases was not addressed in this study.

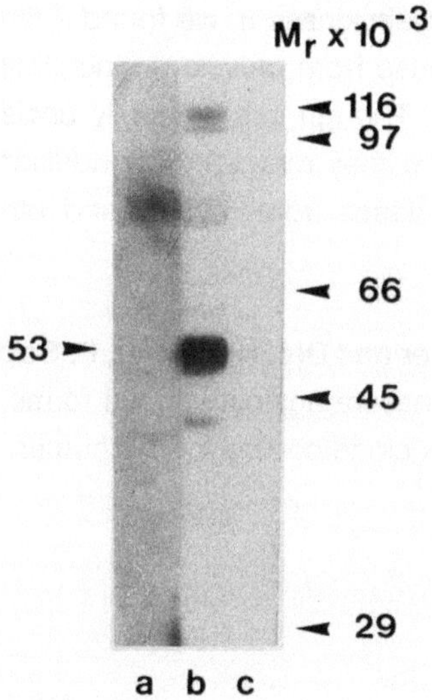

Fig. 2
Immunoblot of the lienal tyrosine phosphatase probed with antisera against tyrosine phosphatases. The antisera were raised against the following tyrosine phosphatase forms a: human receptor phosphatase (9), b: human T-cell phosphatase (8), c: human phosphotyrosine phosphatase 1B (10).

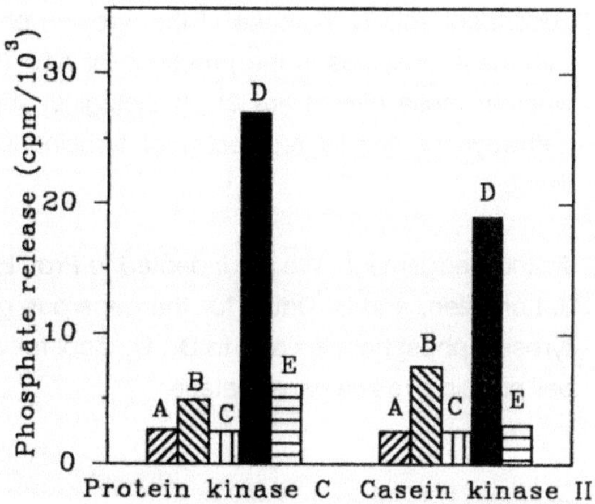

Fig. 3
Influence of protein kinase C and casein kinase II on the activity on a tyrosine phosphatase from rat skeletal muscle. The assays contained: (A) tyrosine phosphatase, (B) A + ATP, (C) B + effectors which were Ca^{2+} and brain phospholipids for protein kinase C and 125 mM KCl for casein kinase II, (D) C + protein kinase, (E) D minus phosphatase.

The data indicate that translocation of an active soluble form of tyrosine phosphatase to cellular structures containing either nucleic acids (nucleus or ribosomes) or phospholipids (plasma membrane or intracellular membranes) potentially leads to transient inhibition of the phosphatase and may, therefore, serve as a control mechanism to stimulate tyrosine phosphorylation.

In order to elucidate the nature of the lienal tyrosine phosphatase employed for these studies, electrophoretic gels transferred to nitrocellulose were probed with several antisera against tyrosine protein phosphatases generously provided by Prof. E. Fischer. Fig. 2 shows that positive immunoreaction was only obtained with an an antiserum against human T-cell phosphatase (8), whereas the reaction with antisera against human RPTR (9) and PTP-1B (10) was negative.

We also isolated a 44 kDa tyrosine phosphatase from 100,000•g supernatants from rat skeletal muscle. While the activity of the lienal phosphatase was not distinctly influenced by treatment with protein kinase C and casein kinase II, we found a time-dependent activity increase of the tyrosine phosphatase from muscle on incubation with these enzymes in the presence of ATP (Fig. 3). Though it is presently unclear whether these effects are of physiological importance they may suggest additional pathways of activity regulation of tyrosine phosphatases from muscle and other tissues.

Acknowledgement. We are indebted to Prof. E. Fischer and Drs. N. Zander, F. Solca, J. Lorenzen, and G. Daum for the generous gift of antisera against various forms of tyrosine phosphatases and to Dr. D. Cool for a cDNA clone coding for the human T-cell protein tyrosine phosphatase.

References:

1. Fischer, E.H., Charbonneau, H., Tonks, N.K. (1991) Science 253, 401-406.
2. Pot, D.A., Dixon, J.A. (1992) Biochim.Biophys.Acta 1136, 35-43.
3. Krueger, N.X., Saito, H. (1992) Proc. Natl. Acad. Sci.U.S.A. 89, 7417-7421.
4. Cool, D.E., Tonks, N.K., Charbonneau, H., Fischer, E.H., Krebs, E.G. (1990) Proc.Natl.Acad.Sci.U.S.A. 87, 7280-7284.
5. Brautigan, D.L., Pinault, FM. (1991) Proc.Natl.Acad.Sci.U.S.A. 88, 6696-6700.
6. Frangioni, J.V., Beahm, P.H., Shifrin, V., Jost, C.A., Neel, B.G. (1992) Cell, 68, 545-560.
7. Stader, C., Dierig, S., Tidow, N., Kirsch, S., Hofer, H.W. (1991) in: Cellular Regulation by Protein Phosphorylation (ed. Heilmeyer, L.M.G.) NATO ASI Series Vol. H 56, Springer, Berlin Heidelberg.
8. Cool, D.E., Tonks, N.K., Charbonneau, H., Walsh, K.A., Fischer, E.H., Krebs, E.G. (1989) Proc. Natl.Acad. Sci. U.S.A. 86, 5257-5261.
9. Daum, G., Zander, N.F., Morse, B., Hurwitz, D., Schlessinger, J., Fischer, E.H. (1991) J.Biol.Chem. 266, 12211-12215.
10. Tonks, N.K., Diltz, C.D., Fischer, E.H. (1988) J.Biol.Chem. 263, 6722-6730; 6731-6737.

LOCALIZATION - DEPENDENT REGULATION OF LYMPHOCYTE TYROSINE PHOSPHATASES.

Christine Stader and Hans Werner Hofer
University of Konstanz, Faculty of Biology, D-7750 Konstanz, Germany.

A soluble tyrosine phosphatase isolated from pig spleen was strongly inhibited by nucleic acid and anionic phospholipids (1,2). Potentially, therefore, the enzyme could be regulated by reversible binding to membranes or nucleic acid - containing structures. In order to examine this hypothesis we studied the localization of tyrosine phosphatases in lymphoid cells.

Cell were isolated from spleen and the splenocyte suspensions gently lysed with digitonin and fractionated into nuclear, mitochondrial and plasma membrane, microsomal, and cytosolic fractions by differential centrifugation. The highest PTP activity was found in the 500*g* pellet which mainly contained the nuclear fraction (corresponding to the 500 *g* pellet in Fig. 1) followed by the 100,000 *g*

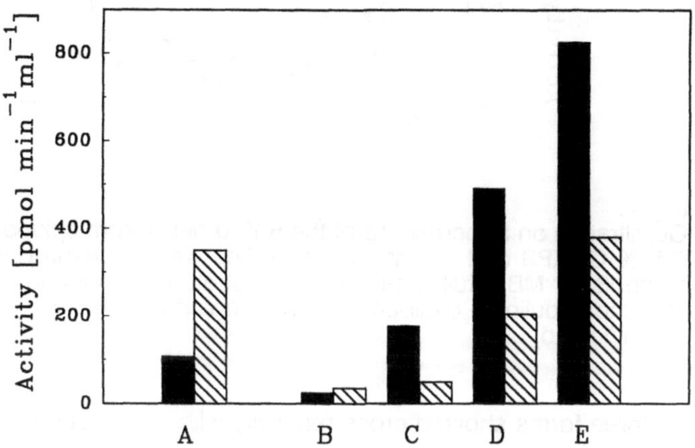

Fig. 1.
Activities of PTP in the 100,000 *g* supernatant (A) and in the 500 *g* pellet (B-E) from splenocytes after extraction with buffer (B), 0.5 M KCl (C), 0.5% CHAPS (D), and 0.5% CHAPS plus 0.5 M KCl (E). Enzyme activities were determined with RCML[1] (full bars) and MBP (hatched bars) as substrates.

supernatant. The high proportion of dephosphorylation of MBP in this supernatant indicates that a substantial amount of receptor phosphatases was solubilized by the

[1] Abbreviations: CHAPS, 3-[(3-cholamidopropyl)dimethylammonio]-2-hydroxy-1-propanolsulfonat; HMC, high molecular weight complex; MBP, myelin basic protein; PTP phosphotyrosine protein phosphatase; RCML, reduced carboxymethylated and maleylated lysozyme.

cell lysis procedure which employed digitonin. The amounts of phosphatase activity extracted from the 500*g* pellet by low-ionic strength buffer, 0.5 M KCl, and by detergent were additive suggesting that the tyrosine phosphatase activities were relieved from different modes of association with cellular structures.

Fig. 2 shows the distribution of activity on Superose 12 gel permeation chromatography of the PTP extracted from the 500*g* pellet by the combined application of salt and detergent. Three different M_r of RCML-dephosphorylating PTP were found: a HMC[1])-form migrating in the exclusion volume and two forms of M_r ca. 53,000 and 37,000, respectively.

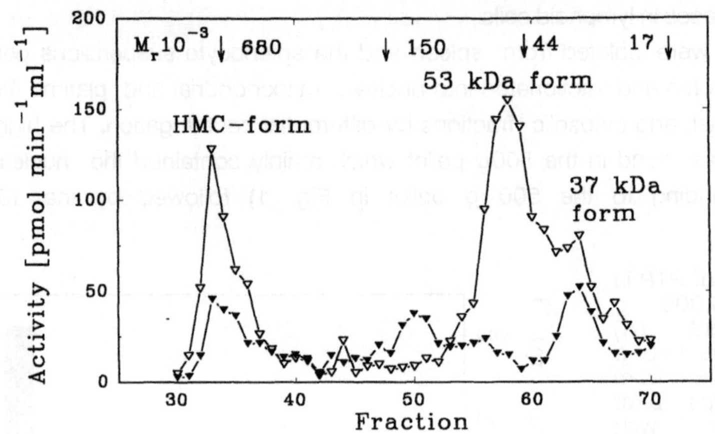

Fig. 2.
Gel filtration on Superose 12 of the 500 *g* pellet from splenocytes after extraction with 0.5 % CHAPS and 0.5 M KCl. Activities were measured with 2 μM RCML (hollow symbols) or MBP (full symbols) as substrates. The elution positions of thyreoglobulin, immunoglobulin G, ovalbumin, and myoglobin used for the calibration of the column are indicated.

All three forms showed cross-reactivity with an antiserum against human T-cell PTP (3, the serum was generously provided by Prof. E. Fischer) on immunoblots. It should be noted, however, that the 53 kDa form did not cross-react with an antiserum against a N-terminal peptide of PTP-1B (4) and a human receptor PTP (5) (data not shown). The kinetic properties of the HMC and the 53 kDa forms were also in agreement with those of the T-cell PTP (cf Table 1). The HMC form mainly differed from the 53 kDa form by lower inhibition by Zn^{2+} and unusually strong activation by (unphosphorylated) MBP (cf Ref. 6 for comparison).

Table 1.
Influence of various effectors on the HMC and 53 kDa forms of PTP from splenocytes.

Effector		HMC PTP	53 kDa PTP
nil		100 %	100 %
Vanadate	100 µM	18 %	12 %
Molybdate	10 µM	21 %	28 %
$ZnCl_2$	100 µM	76 %	10 %
EDTA	5 mM	110 %	87 %
Heparin	10 µM	26 %	16 %
poly(Tyr,4Glu)	10 µM	61 %	59 %
Spermine	2 mM	410 %	73 %
MBP	50 µM	1475 %	138 %

Activities were determined with RCML (2 µM) as substrate and expressed as percent of the activity measured in the absence of effector.

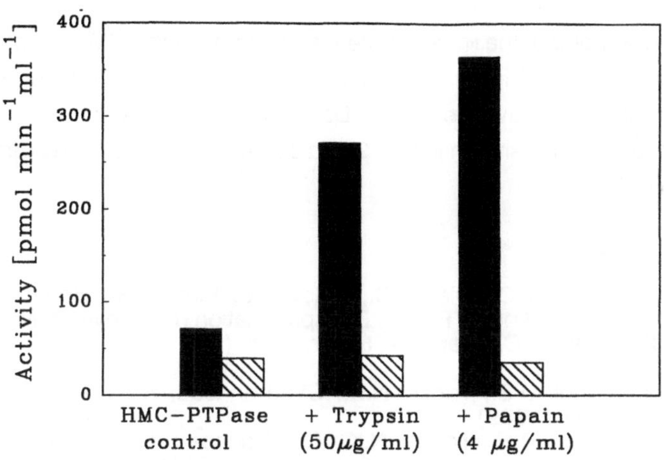

Fig. 3. Limited proteolysis of the HMC form in the presence of trypsin or papain. The PTP activities were determined with RCML (full bars) or MBP (hatched bars) as substrates.

A similar extent of activation of the HMC form was observed in the presence of low concentrations of trypsin and papain (Fig. 3). The stimulation, however, was restricted to an increase in RCML dephosphorylation while the rate of MBP dephosphorylation was unaffected. This different response on protease treatment either suggests the presence of different PTP (including a protease unsensitive PTP specific for MBP) or claims for a change in substrate specificity (see also Ref. 7).

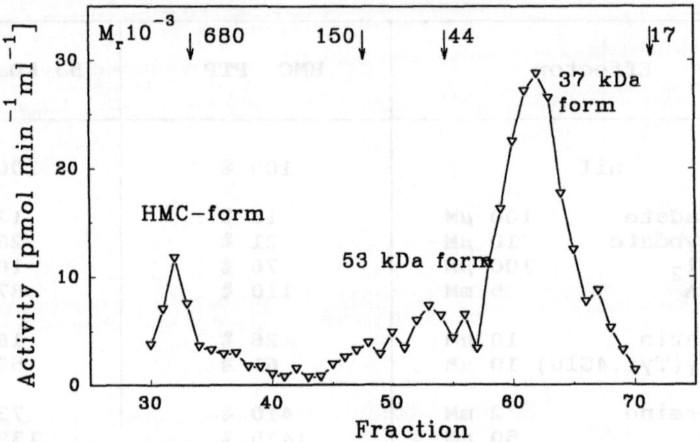

Fig. 5. Gel filtration on Superose 12 of the HMC form of the PTP after limited proteolysis with trypsin (50 μg/ml, 5 min at 30 °C). The proteolysis hjad been terminated by the additon of a 3-fold excess of soybean trypsin inhibitor. The activity was determined with RCML (2 μM) as substrate.

As expected, the limited protease treatment of the HMC form of PTP leads to a change in its M_r (c.f. Fig. 5). A small amount of a 53 kDa PTP is formed. The most prominent product, however, is a a 37 kDa form of the PTP which may correspond to the PTP of the same M_r shown in Fig. 2 and similar proteolytic forms described by other authors (7, 8).

References:

1. Stader, C., Dierig, S., Tidow, N., Kirsch, S., Hofer, H.W. (1991) in: Cellular Regulation by Protein Phosphorylation (Heilmeyer, L.M.G. ed.) pp. 355-360.
2. Stader, C., Dierig, S., Hofer, H. W. (1991) Adv. Protein Phosphatases 6 (Suppl.), 89-90.
3. Cool, D.E., Tonks, N.K., Charbonneau, H., Walsh, K.A., Fischer, E.H., Krebs, E.G. (1989)
4. Cool, D.E., Tonks, N.K., Charbonneau, H., Fischer, E.H., Krebs, E.G. (1990) Proc.Natl.Acad.Sci.U.S.A. 87, 7280-7284.
5. Daum, G., Zander, N.F., Morse, B., Hurwitz, D., Schlessinger, J., Fischer, E.H. (1991) J.Biol.Chem. 266, 12211-12215
6. Boylan, J.M., Brautigan, D.L., Madden, J., Raven, T., Ellis, L., Gruppuso, P.A. (1992) J. Clin. Invest. 90, 174-179.
7. Zander, N.F., Lorenzen, J.A., Cool D.E., Tonks, N.K., Daum, G., Krebs, E.G., Fischer, E.H. (1991) Biochemistry 30, 6964-6970.
8. Brautigan, D.L., Pinault, FM. (1991) Proc. Natl. Acad.Sci. U.S.A. 88, 6696-6700.

PROTEIN TYROSINE PHOSPHATASES IN THE NEMATODE *ASCARIS SUUM*

Brigitte Schmid and Hans Werner Hofer

University of Konstanz, Faculty of Biology, D-7750 Konstanz, Germany

Phosphorylation of tyrosine residues as secondary modification of proteins has been demonstrated on all levels of eukaryotic organization. Nevertheless, special forms of protein tyrosine phosphatases have also been demonstrated in bacteria and as translation products of the vaccinia virus genome (reviewed in Ref. 1). By polymerase chain reaction using degenerated oligonucleotides of conserved PTP[1] domains 27 distinct sequences have been identified indicating that phosphatases related to the enzymes so far mainly characterized from mammalians and insects are also present in lower animals (2).

We used muscle and developing embryonal cells from the parasitic nematode *Ascaris suum* for a comparative study of the localization and properties of tyrosine phosphatases in lower animals. This organism is an attractive tool for biochemical and developmental studies because of its well characterized simplicity, size and availability.

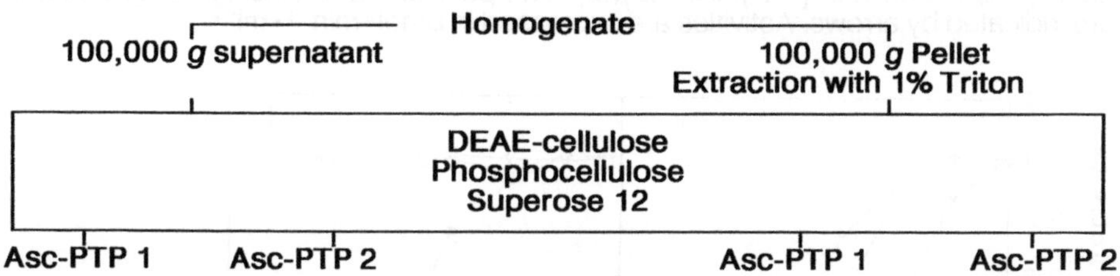

Fig. 1. Schematic representation of the isolation of two forms of PTP from Ascaris muscle.

Muscle tissue from Ascaris was fractionated into soluble and particulate compartments according to the scheme shown in Fig. 1. The 100,000 *g* supernatant was subjected to subsequent chromatographies on DEAE-cellulose and phosphocellulose and two peaks of tyrosine phosphatase activity were obtained which were operationally designated Asc-PTP 1 and Asc-PTP 2. Re-chromatographies of the separately pooled peaks on a Superose 12 column are shown in Fig. 2a and b.

Asc-PTP 1 exhibited a single peak of activity corresponding to a native M_r of 45-50,000, whereas most of the activity found in the Asc-PTP 2 sample was found in a

1) Abbreviation: PTP, phosphotyrosine phosphatase.

peak of M_r ca. 180,000. Interestingly, each of the two tyrosine phosphatases was specific for one of the two experimental substrates. Asc-PTP 1 only dephosphorylated RCML and Asc-PTP 2 was specific for MBP.

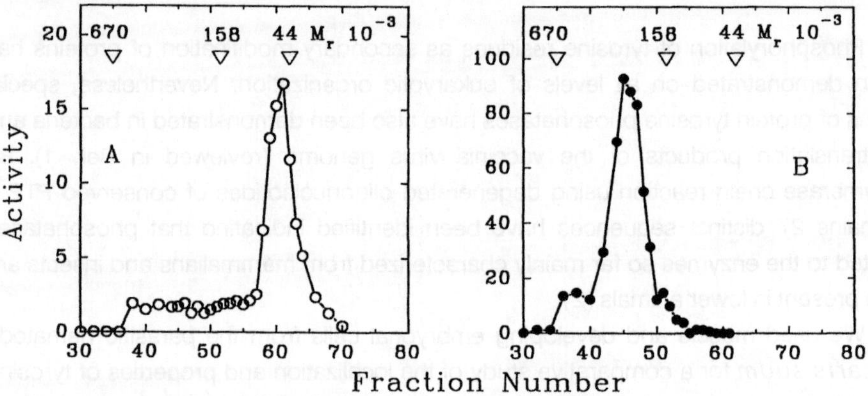

Fig. 2. Gel permeation chromatography of two forms of PTP from Ascaris muscle on Superose 12
(A) Asc-PTP 1, activity measured with [^{32}P]RCM-lysozyme (2 μM). (B) Asc-PTP 2, activity measured with [^{32}P]MBP (2 μM). The positions of molecular weight markers are indicated by arrows. Activities are expressed in pmol•min^{-1}•ml^{-1}.

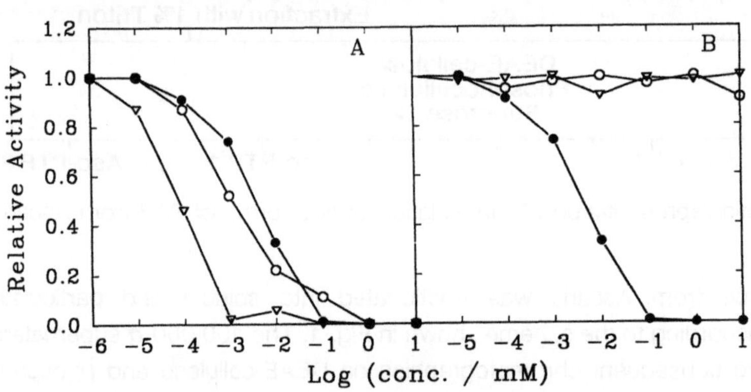

Fig. 3. Effects of inorganic ions on Asc-PTP 1 (A) and Asc-PTP 2 (B). Zn^{2+} (Full circles), molybdate (hollow circles), vanadate (triangles).

A comparison of some kinetic properties of the tyrosine phosphatases showed that Asc-PTP 2 was not inhibited by vanadate and molybdate and was only slightly inhibited by high concentrations of heparin and poly(Tyr,4Glu). Asc-PTP 1 was inhibited by the polyanions and also by the classical tyrosine phosphatase inhibitors vanadate and molybdate as well as by Zn^{2+}. The complete inhibition of this enzyme by EDTA suggests the requirement for a divalent cation.

Table 1. Influence of various effectors on the activities of Asc-PTP 1 and Asc-PTP 2.

Effector		Activity (% of control)	
		Asc-PTP 1	Asc-PTP 2
Heparin	100 µM	0	78
	10 µM	32	100
	1 µM	51	100
poly(Tyr,4Glu)	100 µM	0	56
	10 µM	38	100
	1 µM	57	100
Iodoacetamide	1 mM	0	0
EDTA	5 mM	0	98
NaF	10 mM	100	100
Tetramisol	10 mM	100	100
Microcystin	6 nM	100	100

The activity of Asc-PTP 1 was determined with [^{32}P]RCM-lysozyme, the activity of Asc-PTP 2 with [^{32}P]MBP and expressed as the percentage of the control activity measured in the absence of effectors.

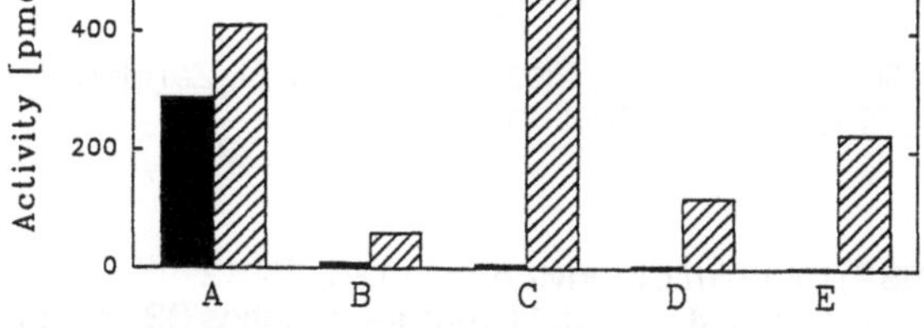

Fig. 4. Activities of Asc-PTP 1 and Asc-PTP 2 in subcellular fractions from Ascaris muscle. A: 100,000 *g* supernatant; B - C: Pellet fraction re-extracted with low-ionic strength buffer (B), 1.5% Triton X-100 (C), 0.7 M KCl (D), and 1.5% Triton plus 0.7 M KCl (E).

Asc-PTP 1 and Asc-PTP 2 were different not only with respect to molecular weight, substrate specificity and inhibitory properties but also with respect to localization. Asc-PTP 1 was only found in the 100,000 *g* supernatant whereas most of

Asc-PTP 2 was extracted by detergent from the pellet fraction but was also present in the supernatant which comprised about 30% of the total Asc-PTP 2 activity.

A number of the properties observed for Asc-PTP 1 and Asc-PTP 2 were reminiscent to the properties of the non-receptor and receptor forms of mammalian protein tyrosine phosphatases. The most notable difference was the absence of inhibition by vanadate and molybdate of the Asc-PTP 2 form of the enzyme. The close relationship between mammalian and non-mammalian phosphatases is also expressed by the cross-reaction of the enzymes from Ascaris with a polyclonal antiserum against the human T-cell phosphatase (3) on immunoblots after denaturing electrophoresis. Asc-PTP 1 exhibited a single band migrating according to a M_r 56,000. The migration of this band was only slightly slower than the migration of the phosphatase from splenocytes (Stader & Hofer, this volume). The pattern obtained by immunostaining of Asc-PTP 2 was more complex: a band representing M_r 80,000 was consistently visualized with the same intensity as a band of M_r ca. 40,000. The relationship between these bands is unknown. Yet, the smaller protein could be derived from the bigger protein by proteolytic processes. Both molecular weights, however, were distinctly lower than those of the known receptor forms of PTP (4,5).

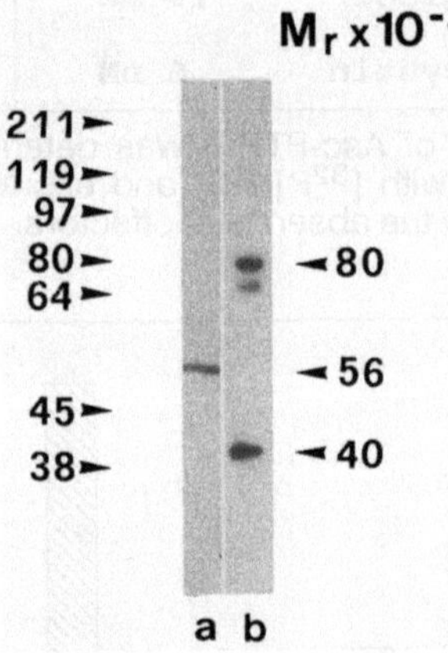

Fig. 5. Immunoblots of Asc-PTP 1 (Lane a) and Asc-PTP 2 (Lane b) with an antiserum raised against recombinant T-cell protein phosphatase (3). The proteins were blotted from a 10% polyacrylamide gel to nitrocellulose after electrophoresis in the presence of sodium dodecylsulfate.

Acknowledgement. We are indebted to Prof. E. Fischer and Drs. N. Zander, F. Solca, J. Lorenzen, and G. Daum for the generous gift of antisera.

References

1. Pot, D.A., Dixon, J.A. (1992) Biochim.Biophys.Acta 1136, 35-43.
2. Mathews, R.J., Flores, E., Thomas M.L. (1991) Immunogenetics 33, 33-41.
3. Cool, D.E., Tonks, N.K., Charbonneau, H., Walsh, K.A. Fischer, E.H., Krebs, E.G. (1989) Proc. Natl. Acad. Sci.U.S.A 86, 5257-5261.
4. Tonks, N.K., Charbonneau, H., Diltz, C.D., Fischer, E.H., Walsh, K.A. (1988) Biochemistry 27, 8695-8701.
5. Krueger, N.X., Saito, H. (1992) Proc. Natl. Acad. Sci.U.S.A. 89, 7417-7421.

Non-radioactive determination of phosphotyrosine phosphatase (PTPase) activity

Joachim Kremerskothen and Angelika Barnekow
Dept. of Exp. Tumorbiology
University of Muenster
Badestraße 9
W-4400 Muenster
Germany

A.Introduction

Using phosphotyrosine as an artificial substrate, a colorimetric assay to determine PTPase activity has been developed. This very sensitive and simple assay is based on the complex formation between malachite green and phosphomolybdate in the presence of 1 M H_2SO_4. Determination of intracellular PTPase activity in crude cell extract from chicken embryo fibroblasts (CEF) transformed by a temperature-sensitive mutant of the Rous sarcoma virus (RSV), NYts68 (Weber and Friis, 1979), was carried out by a modification of the method of Baykov et al. (1988). Our results are comparable to those obtained with a commonly used radioisotopic phosphatase assay described by Nelson and Branton (1984). PTPase activity is increased shortly after RSV induced CEF transformation induced by the very active tyrosine kinase pp60$^{v\text{-}src}$ encoded by the RSV oncogene, v-*src*, or by the resulting increased phosphotyrosine levels in the cells (Sefton et al., 1980).

B. Methods and Results

Primary CEF were prepared and cultured as described earlier (Branton and Landry-Magnan, 1979). After infection of CEF with NYts68, we were able to switch from an inactive form of pp60$^{v\text{-}src}$ and a normal CEF phenotype at the non-permissive temperature (40°C) to an active transforming pp60$^{v\text{-}src}$ at the permissive temperature (35°C). Cell extracts were prepared by homogenization in 50 mM Tris/HCl, pH 7.5, using an electric potter.

NATO ASI Series, Vol. H 76
Tyrosine Phosphorylation/Dephosphorylation and Downstream Signalling
Edited by L. M. G. Heilmeyer Jr.

Cell lysates were clarified by centrifugation for 1 h at 6000 x g (4°C) and desalted on PD10 columns. For the PTPase assay 30 μg (30 μl) of cytosolic protein were incubated in 20 μl of a reaction buffer containing 50 mM Tris/HCl, pH 7.5 and 10 mM phosphotyrosine as the substrate. After a 30 min. incubation period at 37°C, the reaction was terminated by the addition of 20 μl 10% BSA (w/v) and 60 μl trichloric acid (25%). The probes were centrifuged at 6000 x g for 10 min. and 10 μl of the supernatant were mixed with 190 μl of 50 mM Tris/HCl, pH 7.5 and 50 μl of the malachite green colour reagent (Baykov et al., 1988). After a 10 min. incubation period at room temperature, the extinction was measured at 655 nm in microtiter plates (Fig.1).

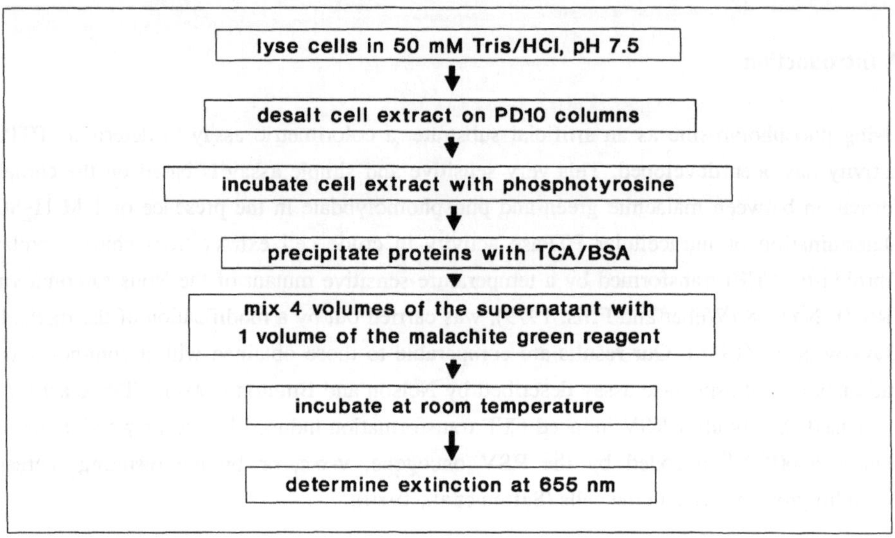

Fig.1: Protocol for the determination of PTPase activity in crude cell extracts using the malachite green method

Using a standard graph ranging from 1.0 to 5.0 nmol inorganic phosphate (P_i) we were able to calculate specific PTPase activity in normal and transformed CEF (Fig.2). The PTPase activity in NYts68 infected cells cultured at the non-permissive temperature was comparable to that in uninfected CEF. Shifting the NYts68 infected cells to the permissive temperature led to an increase in PTPase activity and to cellular transformation. Reshifting the transformed cells led to a normalization of PTPase activity and to a normal phenotype.

Using the radioactive method (Foulkes et al., 1981) to determine PTPase activity, we obtained similar results (data not shown).

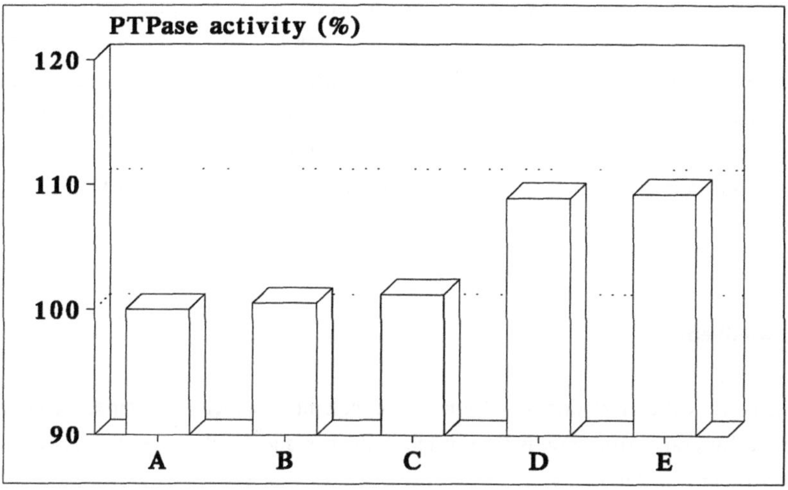

Fig.2: Comparison of PTPase activities in extracts of uninfected (A) and NYts68 infected CEF cultured at the non-permissive temperature (B) and for 30 min. (C), 1 h (D) and 3 h (E) at the permissive temperature

To control the selectivity of the malachite green assay we determined the effects of well-known PTPase inhibitors (Tab.1). The phosphate analogues ortho-vanadate and molybdate as well as zinc chloride, when used in μM concentrations, displayed an expected strong inhibition of PTPase activity, whereas mM concentrations of the phosphothreonine/phosphoserine specific inhibitor sodium fluoride had only slight inhibitory effects.

Tab.1: Effects of inhibitors on PTPase activity in NYts68 transformed CEF

Inhibitor	Concentration	PTPase activity (%)
None	-	100
molybdate	2 μM	13
o-vanadate	2 μM	15
zinc chloride	2 μM	17
sodium fluoride	2 mM	93

C. Conclusion

Using the malachite green assay for the detection of inorganic phosphate within the nM range, we were able to develop a very fast, selective, simple and convenient procedure for the determination of PTPase activity. Our results are comparable to those obtained from radioactive assays and showed a transformation specific increase in PTPase activity in RSV infected CEF. To characterize this increase of PTPase activity and its role in signal transduction and transformation processes, it will be necessary to purify distinct PTPases from CEF.

Acknowledgements

We thank E. Ossendorf and U. Wend for excellent technical assistance. This work was supported by grants from Fonds Chemische Industrie and Deutsche Forschungsgemeinschaft (Ba 876/1-2) to A. Barnekow.

References

Baykov, A.A., Evtushenko, O.A. and Avaeva, S.M. (1988). A malachite green procedure for orthophosphate determination and its use in alkaline phosphatase-based enzyme immunoassay. *Anal. Biochem.* **171**, 266-270

Branton, P.E. and Landry-Maglan, J. (1979) Plasma membrane phosphoproteins in normal and Rous sarcoma virus transformed chick embryo fibroblasts: characterization by in vitro phosphorylation. *J. Cell. Physiol.* **100**, 159-168

Foulkes, J.G., Howard, R.F. and Ziemiecki, A. (1981) Detection of a novel mammalian protein phosphatase with activity for phosphotyrosine. *FEBS Letters* **130**, 197-204

Nelson, R.L. and Branton, P.E. (1984). Identification, purification and characterization of phosphotyrosine-specific protein phosphatases from cultured chicken embryo fibroblasts. *Mol. Cell. Biol.* **4**, 1003-1012

Sefton, B.M., Hunter, T., Beemon, K. and Eckhart, N. (1980). Evidence that the phosphorylation of tyrosine is essential for cellular transformation by Rous sarcoma virus. *Cell* **20**, 207

Swarup, C., Cohen, S. and Garbers, D.L. (1982). Inhibition of membrane phosphotyrosyl-protein phosphatase activity by vanadate. *Biochem. Biophys. Res Commun.* **107**, 1104-1109

Weber, M.J. and Friis, R.R. (1979) Dissociation of transformation parameters using temperature-conditional mutants of Rous sarcoma virus. *Cell* **16**, 25-32

NEW PROTEIN TYROSINE PHOSPHATASES FROM VASCULAR SMOOTH MUSCLE CELLS: TRANSCRIPT LEVELS IN TISSUES AND IN CULTURED CELLS

L. Ghira Borges
D. Bowen-Pope
Department of Pathology
University of Washington
Seattle, Washington 98195, USA

KEYWORDS/ABSTRACT: protein tyrosine phosphatase/ smooth muscle cell/ vascular / transcript / culture cells / tissue / cell density

Partial cDNA sequences of 5 new protein tyrosine phosphatases (PTPase) were obtained after PCR amplification of cDNA from vascular smooth muscle cells. On Northern blots, these sequences hybridize to messages ranging from 3.5 to 11.2 Kb. At the transcript level, the 5 PTPase genes are differentially expressed both in cultured cells and in various rat tissues. In cultured cells, one of the PTPase genes, VPTP alpha has a very narrow pattern of expression. In vivo, every PTPase gene is expressed in more than one tissue, and they all have high transcript levels in the brain. Four of the five PTPase genes show a very strong transcript up-regulation in vascular newborn smooth muscle cells plated at increased cell densities.

INTRODUCTION

Vascular smooth muscle cells (SMC) have been implicated in various pathologies affecting blood vessels. A large fraction of the cells that form the advanced lesions of arteriosclerosis are SMCs, which have undergone massive cell proliferation during plaque development (Ross,1990). Plaque growth leads to enlargement of the vessel wall area and narrowing, or even complete occlusion, of the lumen of the artery.

NATO ASI Series, Vol. H 76
Tyrosine Phosphorylation/Dephosphorylation
and Downstream Signalling
Edited by L. M. G. Heilmeyer Jr.

In the vessel wall, smooth muscle cells, endothelial cells, and leukocytes are all capable of producing a wide variety of growth factors. Most of these growth factors interact through receptors, which initiate signal transduction events by phosphorylation of signaling proteins at tyrosine residues. Modulation of these signal transduction pathways may also be achieved by PTPases. Since no PTPases have been identified in SMCs, we decided to clone new PTPases from these cells, and to look at the regulation of these genes in various pathological conditions that affect blood vessels. We report here the isolation of 5 PTPase sequences from rat smooth muscle, and we examine their transcript levels in different cell types and tissues, and in SMCs cultured at different cell densities.

MATERIAL AND METHODS

cDNA cloning. PTPase sequences were PCR amplified from cDNA made from mRNA extracted from newborn rat SMCs. They were cloned in a bacterial vector and sequenced.

Cultured cells. All of the cells used in this study were derived from rat tissues. They included: newborn and adult rat SMCs isolated from the aortas of 12-day newborn and 3-month old rats; medial SMCs and injured medial SMCs derived from uninjured and injured rat carotids; skin fibroblasts derived from newborn and adult rats, tendon fibroblasts, muscle myoblasts, heart myoblasts, hepatocytes, ileum epithelial cells, kidney fibroblasts, and kidney epithelial cells.

RNA isolation and Northern blot analysis. Total RNA was isolated by the guanidinium isothiocyanate method (Chomczynski and Sacchi, 1987). For the Northern blots, 10 to 15 micrograms of total RNA was loaded in each lane.

Phosphorimager analysis of the Northern blots. Northern blots that had been hybridized to [^{32}P]-dCTP labeled PTPase probes were exposed to phosphorimaging plates and analyzed in a Molecular Dynamics PhosphorImager.

Cell density experiment. For this experiment only newborn and adult rat SMCs were used. On day zero, cells were plated at different cell densities in 150mm plates. Two days after plating, cells were lysed and the total RNA extracted.

RESULTS AND DISCUSSION

We have cloned partial PTPase cDNA sequences, after PCR amplification of cDNA from smooth muscle cells cultured from the aorta of newborn rats. The transcript sizes for each PTPase are as follows: 8.7 Kb for VPTP alpha, 6.2 Kb for VPTP beta, 3.5 and 6.4 Kb for VPTP gamma, 6.5 and 10.2 Kb for VPTP delta, and 6.2 and 11.2 Kb for VPTP epsilon. The cDNA sequences were used as probes for Northern blot analysis of expression of the various PTPase genes both in vivo and in cultured cells. In addition, we looked at the modulation of PTPase transcript levels in cells plated at different densities.

In cultured cells, VPTP alpha is strongly expressed in aortic SMC from newborn rats. In every other cell type (see Materials and Methods) the mRNA levels are extremely low. These include other types of muscle cells (muscle and heart myoblasts) and even SMCs derived from the aorta and carotid of adult rats. VPTP beta and VPTP epsilon show the highest mRNA expression in newborn SMCs but, unlike VPTP alpha, their transcript levels are also relatively high in other cultured cells. VPTP gamma and VPTP delta are strongly expressed in various cells including adult SMCs, muscle and heart myoblasts, ileum and kidney epithelial cells.

In vivo, all 5 PTPase sequences are expressed in various tissues at levels that vary from sequence to sequence. VPTP alpha is expressed at low levels in muscle tissues (heart, skeletal muscle, diaphragm, stomach and intestine) and is strongly expressed in the cerebellum, brain cortex, lung, and kidney cortex. Interestingly, every PTPase gene shows a strong expression in at least one of the brain regions that we looked at. These brain regions include the cerebellum, brain cortex and brain stem. Other studies have implicated the brain as a rich source of PTPases. Using o-phospho-L-tyrosine as a substrate for in situ localization of PTPase activity, Yoshioka et al. (1991)

detected strong phosphatase activity in the hippocampus. In Drosophila embryos, Tian et al. (1991) reported the isolation of 7 different PTPase cDNAs, of which three were primarily expressed in the central nervous system.

Besides the brain, the lung consistently showed high levels of PTPase expression for each of the sequences studied. Esophagus (VPTP beta, VPTP gamma, VPTP delta, and VPTP epsilon), heart (VPTP beta, VPTP gamma, and VPTP delta), and kidney cortex (VPTP gamma, VPTP delta, VPTP epsilon, and VPTP alpha) are other tissues that are relatively rich in PTPase transcripts.

In cultured SMCs, four of the 5 PTPases (VPTP alpha, VPTP beta, VPTP gamma, and VPTP epsilon) show a very strong mRNA up-regulation when cells are shifted from low to high density. Interestingly, this modulation of transcript levels is seen in SMCs derived from aortas of newborn rats, but not in the same type of cells derived from the aorta of adult rats. SMCs derived from newborn and adult rats have distinct phenotypes in culture. The newborn cells are flattened epithelioid cells that grow in culture as a monolayer. The adult cells are spindle-shaped fibroblast-like cells that grow in culture in overlapping layers. As cell density is increased, transcript levels are upregulated by about 5-fold (VPTP beta and VPTP epsilon) and 21-fold (VPTP alpha and VPTP gamma). The possibility that PTPases might be involved in cell-adhesion has been raised by several authors and is based on the observation that some PTPases have on their extracellular domain, fibronectin type-III and immunoglobulin-like motifs (Fisher et al.,1991). The same type of motifs has been shown to be present in cell-adhesion molecules, such as N-CAM (Cunningham et al.,1987).

REFERENCES

Chomczynski, P. and Sacchi, N.(1987) Anal.Biochem. 162, 156-159.
Cunningham, B.A., Hemperly, J.J., Murray, B.A., Prediger, E.A., Brackenbury, R., Edelman, G.M. (1987) Science 236, 799-806.
Fisher, E.H., Charbonneau, H. and Tonks, N.K. (1991) Science 253, 401-406.
Ross, R. (1990) Adv. Nephrol. 19, 79-86.
Tian, S.-S., Tsoulfas, P. and Zinn, K. (1991) Cell 67, 675-685.
Yoshioka, T., Tanaka, O., Otani, H. and Shinohara, H. (1991) Brain Res. 555, 177-179.

RECOMBINANT CSK EXPRESSED IN E.COLI IS PHOSPHORYLATED ON TYROSINE RESIDUE(S) AND UNDERGOES IN VITRO PHOSPHORYLATION

C. Bougeret, S. Fischer and R. Benarous
Institut Cochin de Génétique Moléculaire,
INSERM U332, Université Paris V
22 rue Méchain, 75014 Paris
FRANCE

ABSTRACT

The C-terminal Src Kinase (CSK) is responsible for the phosphorylation of the negative regulatory carboxy terminal tyrosine of several members of the Src family. CSK is unique among the members of the protein tyrosine kinase family because it lacks the conserved tyrosine autophosphorylation site. Using the glutathione S-transferase (GST) bacterial expression system (Smith & Johnson, 1988), we have produced large amounts of a chimeric rat CSK protein. We have determined that the GST-CSK fusion protein isolated from bacteria is phosphorylated on tyrosine residue(s) and is capable of undergoing phosphorylation on tyrosine residue(s) *in vitro*. Such autophosphorylation of CSK might have a role in the regulation of its kinase activity.

INTRODUCTION

Protein tyrosine kinases (PTKs) are involved in cell proliferation and differentiation. The activity of Src-PTKs is regulated by phosphorylation of critical tyrosine residues which are located in the catalytic domain and conserved in all members of the family. For example, tyrosine 416 of c-Src is involved in positive regulation of the kinase activity upon autophosphorylation. At the same time, the phosphorylation of the C-terminal tyrosine 527 of c-Src is implicated in negative modulation of the activity.

Recently the PTK responsible for the phosphorylation of the carboxy terminal tyrosine (527) of $p60^{c\text{-}src}$, designated CSK (C-terminal Src Kinase), has been purified and cloned from rat brain (Okada & Nakagawa, 1989; Nada et al., 1991), from chicken brain (Sabe et al., 1992a) and from the human erythroleukemia cell line K562 (Partanen et al., 1991). CSK is expressed in a large number of tissues and has been shown, *in vitro*, to specifically

NATO ASI Series, Vol. H 76
Tyrosine Phosphorylation/Dephosphorylation and Downstream Signalling
Edited by L. M. G. Heilmeyer Jr.

phosphorylate the negative regulatory sites of p60$^{c\text{-}src}$ (Okada & Nakagawa, 1989), p56^{lyn}, p59^{fyn} (Okada et al., 1991) and p56^{lck} (Bergman et al., 1992). *In vivo*, overexpression of CSK causes reversion to normal phenotype of transformed cells overexpressing v-Crk and c-Src, but not of cells transformed by v-Src or c-Src in which tyrosine 527 was changed to phenylalanine. These results strongly suggest that CSK acts on tyrosine 527 of c-Src and suppresses c-Src kinase activity *in vivo* (Sabe et al., 1992b).

CSK contains SH3 and SH2 domains. CSK has been previously described as being devoid of autophosphorylation activity (Okada et al., 1991; Bergman et al., 1992). Furthermore, the deduced sequence of CSK from cDNA showed that its catalytic domain contains all the features of PTK except for the conserved tyrosine autophosphorylation site, and the regulatory carboxy terminal tyrosine phosphorylation site (Nada et al., 1991; Partanen et al., 1991; Sabe et al., 1992a).

RESULTS

1/ CSK was expressed and purified as a GST-CSK fusion protein.

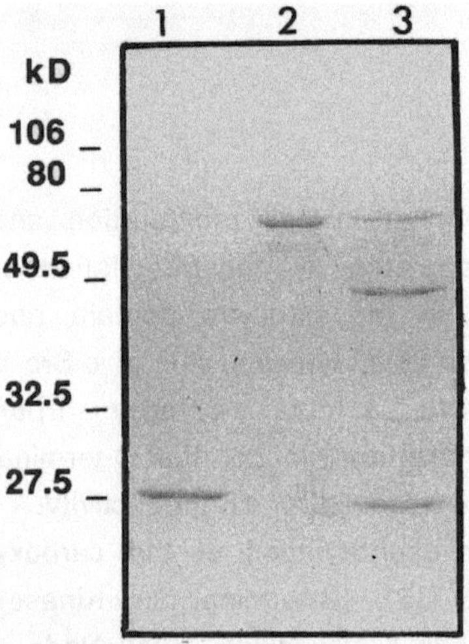

Figure1 : Purified GST (1), purified GST-CSK untreated (2) or treated (3) with thrombin, were analyzed on 12% SDS-PAGE followed by Coomassie blue staining.

2/ GST-CSK synthesized in *E.coli* is reactive with a monoclonal antiphosphotyrosine antibody.

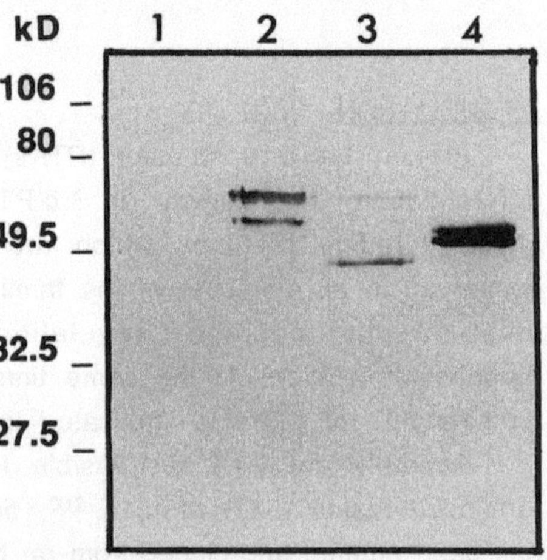

Figure 2 : purified GST (1), purified GST-CSK untreated (2) or treated (3) with thrombin, and lysate of Sf9 cells overexpressing p56^{lck} (4), were resolved on 12% SDS PAGE, blotted and probed with a monoclonal antiphosphotyrosine antibody.

3/ GST-CSK is autophosphorylated *in vitro* on tyrosine residue(s).

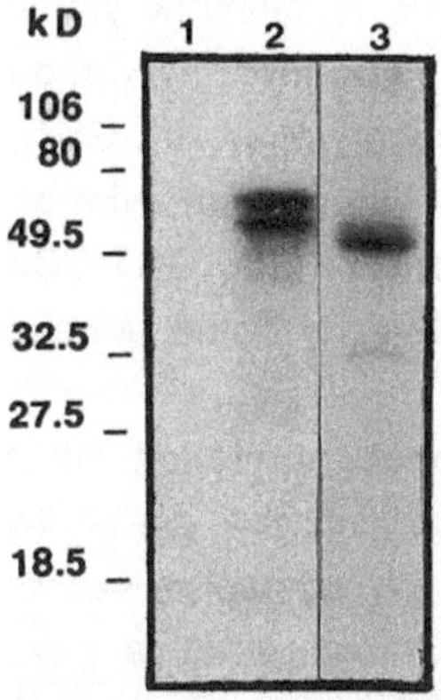

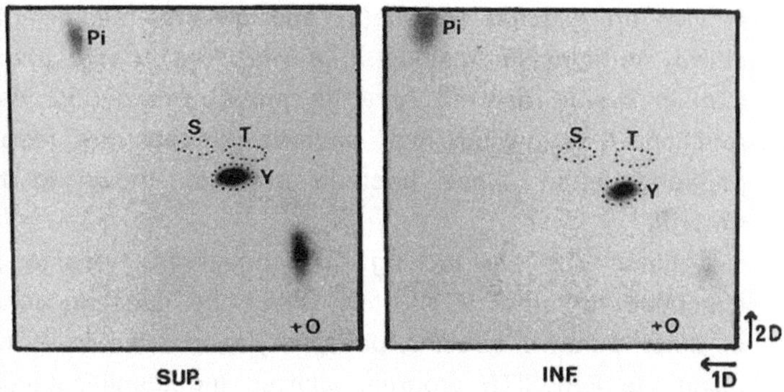

Figure 3A : purified GST (1), purified GST-CSK (2), lysate of Sf9 cells overexpressing $p56^{lck}$ (3), were labeled *in vitro* with (γ-^{32}P) ATP and analyzed on 12% SDS-PAGE followed by autoradiography.

Figure 3B : The superior and inferior bands of labeled GST-CSK were excised and hydrolysed. Resulting phosphoaminoacids were resolved by 2D electrophoresis and detected by autoradiography. Position of phosphoaminoacids (S : P-serine, T : P-threonine, Y : P-tyrosine), free PO_4 (Pi) and origin (O) are marked.

4/ Tyrosine kinase activity of GST-CSK is determined on exogenous substrate.

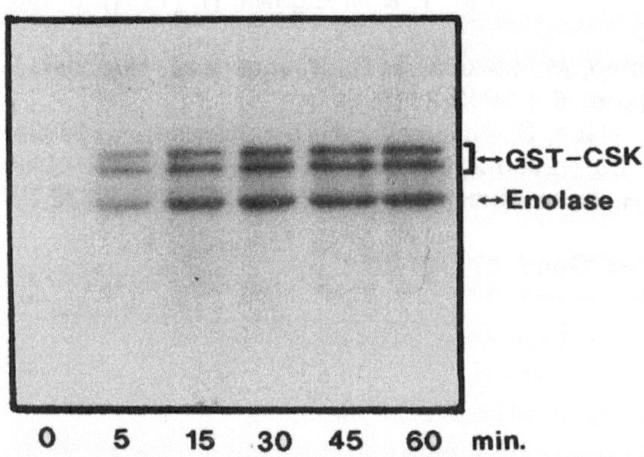

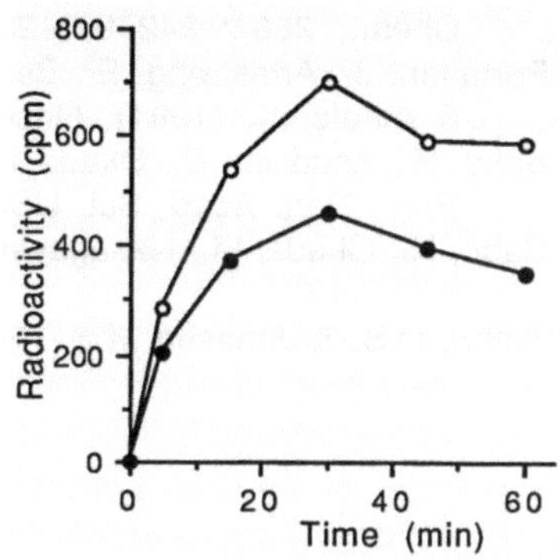

Figure 4A : Purified GST-CSK (300 ng) was incubated with a reaction mixture containing 5 µM of (γ-^{32}P) ATP and 5 µg of acid-denatured enolase. After various times at 30°C, the reaction was resolved on 12% SDS-PAGE followed by autoradiography.

Figure 4B : Labeled proteins, GST-CSK (open circles) and enolase (closed circles), were excised from the gel and radioactivity was counted.

CONCLUSION

We have demonstrated that bacterially produced CSK is phosphorylated *in vivo* on tyrosine residue(s) and displays significant autophosphorylation activity on tyrosine residue(s) *in vitro.* The *in vivo* phosphorylation of GST-CSK in *E.coli* reported here, is probably occuring via autophosphorylation reaction. It is unlikely that bacterial kinases are responsible for GST-CSK phosphorylation, since bacteria are not known to have tyrosine kinase activity.

Since CSK is lacking the conserved tyrosine phosphorylation sites described for other PTKs, this raises the question concerning the possibility of other putative tyrosine phosphorylation sites in CSK. A computer search using the PROSITE program allowed the identification of two putative sites for tyrosine phosphorylation, namely tyrosine 97 and 145. Interestingly, both tyrosines are located in the SH2 domain of CSK. Such phosphorylation might have an effect on the binding properties of this SH2 domain, and consequently on the ability of CSK to phosphorylate its substrates.

REFERENCES

Bergman, M., Mustelin, T., Oetken, C., Partanen, J., Flint, N.A., Amrein, K.E., Autero, M., Burn, P. & Alitalo, K. (1992). EMBO J., 11 : 2919-2924.

Nada, S., Okada, M., Mac Auley, A., Cooper, J.A. & Nakagawa, H. (1991). Nature, 351 : 69-72.

Okada, M. & Nakagawa, H. (1989). J. Biol. Chem., 264 : 20886-20893.

Okada, M., Nada, S., Yamanashi, Y., Yamamoto, T. & Nakagawa, H. (1991). J. Biol. Chem., 266 : 24249-24252.

Partanen, J., Armstrong, E., Bergman, M., Mäkelä, T. P., Hirvonen, H., Huebner, K. & Alitalo, K. (1991). Oncogene, 6 : 2013-2018.

Sabe, H., Knudsen, B., Okada, M., Nada, S., Nakagawa, H. & Hanafusa, H. (1992a). Proc. Natl. Acad. Sci. USA, 89 : 2190-2194.

Sabe, H., Okada, M., Nakagawa, H. & Hanafusa, H. (1992b). Mol. Cell. Biol., 12 : 4706-4713.

Smith, D.B. & Johnson, K.S. (1988). Gene, 67 : 31-40.

POLYLYSINE ACTIVATES A SPLENIC TYROSINE PROTEIN KINASE ENCODED BY THE *fgr* PROTOONCOGENE

A. Donella-Deana, A. M. Brunati, M. Ruzzene and L.A. Pinna
Dipartimento di Chimica Biologica and Centro per lo Studio della Fisiologia Mitocondriale del CNR, Universita' di Padova, Via Trieste 75, Padova, Italy

Introduction.

A tyrosine protein kinase previously isolated from rat spleen, conventionally termed TPK-III, displaying autophosphorylation activity [Brunati and Pinna, 1988], has been purified to homogeneity. The enzyme is immunologically related to the tyrosine kinase encoded by the *fgr* gene, a member of the *src* family [Brunati et al., manuscript in preparation].
Different authors have previously demonstrated that basic polycations and basic proteins can activate several tyrosine protein kinases either cytosolic or associated with membrane receptors. We report here the positive effect of polylysine on *fgr*-TPK activity.

Results and Discussion.

Figure 1A shows the dose dependency of polylysine effect on either enzyme autophosphorylation or phosphorylation of the peptide EDNEYTA. This peptide reproduces the main autophosphorylation site (Tyr-416 of $pp60^{c\text{-}src}$) of most tyrosine protein kinases belonging to the *src* family and seems to be responsible for the increase of tyrosine kinase activity (Cooper, 1990). The maximal enzyme activation obtained with very low concentration of effector suggests that polylysine would essentially interact with the enzyme (70 nM) rather than with the substrate (200 μM). This hypothesis is consistent with the finding that the nanomolar polylysine concentration giving the maximal stimulation

NATO ASI Series, Vol. H 76
Tyrosine Phosphorylation/Dephosphorylation and Downstream Signalling
Edited by L. M. G. Heilmeyer Jr.

of the enzyme activity is not significantly changed by using different kinds and concentrations of substrates (data not shown).

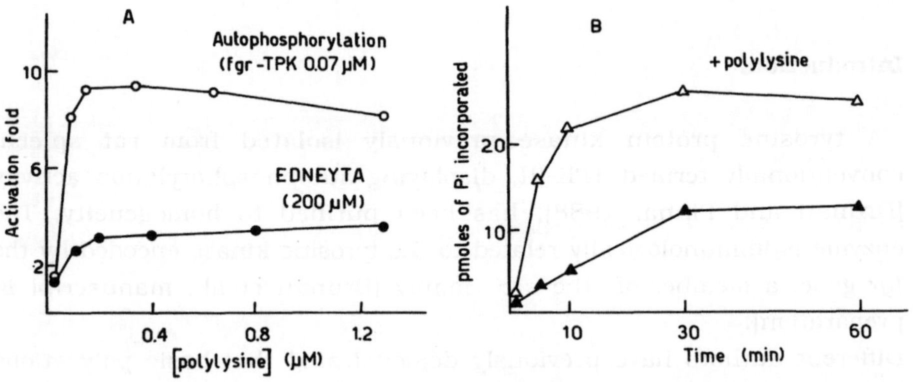

Fig.1 A) Activation by polylysine of *fgr*-TPK autophosphorylation and EDNEYTA phosphorylation by the enzyme. B) Time courses of *fgr*-TPK autophosphorylation in the absence and in the presence of polylysine. Enzyme autophosphorylation was performed as elsewhere described (Donella-Deana et al., 1992).

The time courses of *fgr*-TPK autophosphorylation, obtained in the presence or absence of polylysine, show that the effector increases both the rate and the degree of phosphorylation of the enzyme, which, under these conditions, incorporates about three moles of Pi/mole of *fgr*-TPK (Fig. 1B).

Polylysine also modifies the dose-dependent effect of divalent cations (Mg^{2+} or Mn^{2+}) on the *fgr*-TPK activity. While the maximal increase of the basal protein kinase activity is found at about 10 mM cation concentration, with Mg^{2+} showing higher activatory effect than Mn^{2+}, in the presence of polylysine the maximal substrate phosphorylation is obtained at 1.2-4 mM Mg^{2+} and 0.4-1 mM Mn^{2+}. The requirements of a cation concentration closer to the physiological level, observed in the presence of polylysine, suggests that this compound can substitute for the high concentrations of Mg^{2+}/Mn^{2+} that seems to be necessary for the activation of *fgr*-TPK at a second regulatory and/or catalytic site(s).

Table I. Kinetic constants for *fgr*-TPK with synthetic peptides related to the putative autophosphorylation site of *abl*-TPK. Underlining denotes substitutions relative to the parent peptide. Phosphorylation of peptides was performed as previously described (Donella-Deana et al., 1992).

Peptides	Kcat		Km (mM)	
	-polyLys	+polyLys	-polylys	+polyLys
GDTYTAHA	47	130	1.1	1.2
GDAYAAHA	14	66	1.2	2.3
GETYTAEA	47	90	3.8	1.6
GKTYTAKA	37	71	14.2	13.9
GATYTAAA	41	120	6.2	6.1
GDTYTAEA	43	130	1.1	1.4
GATYTAHA	36	98	3.3	3.3
GDTYTAAA	29	104	1.2	1.4
GKTYTAHA	16	71	3.5	3.2

The stimulation of *fgr*-TPK by polylysine occurs primarily due to 5-fold increase in Vmax with no significant effect on the Km for ATP (data not shown). Table I shows the kinetic constants for *fgr*-TPK with different synthetic peptides related to the main putative autophosphorylation site of the tyrosine protein kinase encoded by the oncogene of Abelson murine leukemia virus. In spite of different kinetic values obtained with different peptide substrates the favourable effect of polylysine is always accounted for by an increase of Vmax; Km values being almost unaffected. These data suggest that the effect of polylysine does not correlate with the different electrolytic nature of the peptides used as substrates, but it is plausible that polylysine directly interacts with an acidic region of *fgr*-TPK inducing the appropriate conformational changes required for the enzyme activation at low divalent cation concentration. In this regard it is of interest to note that the *fgr*-TPK molecule contains the highest negative charge density in the C-terminal part of the protein, which presents 10 acidic groups over a 32 amino acids stretch (Cooper, 1990). This acidic region, highly conserved in the tyrosine kinases belonging to the *src* family, contains also the residue (homologous to Tyr-527 of pp60$^{c\text{-}src}$) which can be phosphorylated by heterologous protein tyrosine kinase(s) (Nada et al., 1991), and can consequently induce the folding of this domain into the SH-2 domain of the same molecule in a head-to-tail self-association, that inhibits tyrosine kinase activity (Matsuda et al., 1990).

The conclusion that polylysine could exert its positive effect by interacting with this acidic region and preventing the self-association-linked inhibitory effect is consistent with the fluorescence spectra obtained with an eicosapeptide reproducing the C-terminal part of *src*-TPK, highly homologous to *fgr*-TPK. A marked increment of fluorescence emission of the tyrosyl residue is found in the presence of polylysine suggesting that the peptide interacts with the polyelectrolyte and changes the environment in which the phenolic group of tyrosyl residue is situated.

Acknowledgments. This work was supported by AIRC and Italian M.U.R.S.T. and C.N.R. (target project on Biotechnology and Bioinstrumentation and A.C.R.O.)

References

Brunati, A.M. and Pinna, L.A. (1988). Characterization of four tyrosine protein kinases from the particulate fraction of rat spleen. *Eur. J. Biochem.* 172, 451-457

Cooper J.A. (1990). The *src* family of protein-tyrosine kinases. in *Peptides and protein phosphorylation* (B.E.Kemp ed.) pp. 85-113, CRC Press, Boca Raton, FL

Donella-Deana A., Marin O., Brunati A.M. and Pinna L. (1992). Selective effect of poly(lysine) on the enhancement of the *lyn* tyrosine protein kinase activity: increased specificity toward *src* peptides. *Eur. J. Biochem*. 204, 1159-1163

Matsuda M., Mayer B.J., Fukui Y. and Hanafusa H. (1990). Binding of transforming protein p47$^{gag\text{-}crk}$ to a broad range of phosphotyrosine-containing proteins. *Science* 248, 1537-1539.

Nada, S., Okada, M., MacAuley, A., Cooper, J.A. & Nakagawa, H. (1991). Cloning of a complementary DNA for a protein-tyrosine kinase that specifically phosphorylates a negative regulatory site of p60$^{c\text{-}src}$. *Nature* 351, 69-72

PROTEIN PHOSPHORYLATION IN THE SIGNAL TRANSDUCTION OF THE NEUROPEPTIDE BOMBESIN IN SWISS 3T3 CELLS

J. Van Lint, P. Agostinis, W. Merlevede and J.R. Vandenheede
Afdeling Biochemie, Faculteit Geneeskunde
Katholieke Universiteit te Leuven
Campus Gasthuisberg
B-3000 Leuven
Belgium

INTRODUCTION

Neuropeptides (e.g. bombesin) are small regulatory peptides localized in the neurons of the central and peripheral nervous system as well as in neuroendocrine cells. They can act as neurotransmitters, paracrine or endocrine hormones but also as growth factors, which shows that the demarcation line between hormones, cytokines, neuroendocrine effectors and growth factors is becoming increasingly blurred. Bombesin is a "sole mitogen" for Swiss 3T3 cells, which means that bombesin can initiate all the necessary events for cell division on its own (Rozengurt E, Sinnett-Smith J 1990). This is quite surprising since bombesin signals through a G-protein linked receptor, whereas all other sole mitogens for Swiss 3T3 cells (PDGF, FGF) signal through a tyrosine kinase containing receptor (Rozengurt E, Sinnett-Smith J 1990). We therefore set out to investigate whether bombesin could stimulate any of the protein kinases in Swiss 3T3 cells, that are known to be activated in the signaling pathways initiated by tyrosine kinase containing receptors.

RESULTS AND DISCUSSION

MBP/MAP2 kinases

Bombesin rapidly stimulates two MBP kinases which by their elution position on Mono Q anion exchange chromatography, their sensitivity to PP2A phosphatase action, and by their reaction with antiphosphotyrosine-, antiphosphothreonine- and antiERK antibodies can be classified as ERK2 and ERK1 respectively. The activation of MBP/MAP2 kinases appears to be a very

NATO ASI Series, Vol. H 76
Tyrosine Phosphorylation/Dephosphorylation and Downstream Signalling
Edited by L. M. G. Heilmeyer Jr.

general extracellular signal related phenomenon, hence their classification as ERKs (Boulton et al. 1991). Activation of both the 42 kDa and the 44 kDa kinases requires phosphorylation at tyrosine as well as at threonine residues; removal of either the threonine-phosphate (by the type 2A phosphatases) or the tyrosine-phosphate (by the CD45 tyrosine phosphatase) results in their inactivation (Boulton et al. 1991). ERK activation may constitute an early intracellular message which primes the cell for the more specific information that comes along the signal transduction pathway to dictate what kind of response is needed.

S_6 peptide kinases

The S_6 peptide used as a kinase substrate in our studies corresponds to the phosphorylation site of the 40S ribosomal protein S_6 located at the C-terminus, which is phosphorylated after a variety of mitogenic stimuli (Sturgill et al. 1991, Ahn et al. 1990, Erikson RL 1991, Agostinis et al. 1992). We reasoned that if bombesin is a sole mitogen for Swiss 3T3 fibroblasts and if S_6 phosphorylation is crucial for cell division, than bombesin should stimulate S_6 kinase(s).

Indeed, using the S_6 peptide as a substrate to screen for S_6 kinases, we found a group of five bombesin stimulated S_6 peptide kinases. Their time course of activation, substrate specificity, chromatographic behaviour and their sensitivity to phosphatase action relates them to the EGF activated S_6 peptide kinases described by Ahn *et al*. 1990. Several studies have indicated that the early and late phases of the growth factor induced S_6 kinase activations are brought about by distinct but converging signaling pathways (Sturgill et al. 1991, Erikson RL 1991). The *in vivo* phosphorylation of the S_6 protein is catalyzed by two main classes of enzymes: the 90 kDa S_6 kinases, collectively called RSK (<u>R</u>ibosomal <u>S_6</u> <u>K</u>inases), and the 70 kDa S_6 kinases. Members of the RSK family seem to participate mainly in the early phase of the growth factor response and these enzymes have the ability to phosphorylate the Kemptide substrate as well as the G subunit of the type 1 protein phosphatase at the PK-A site 1. The bombesin stimulated S_6 peptide kinases eluting in the breakthrough and early in the Mono Q gradient, display a similar time dependent activation and substrate specificity so that they might be related to the 90 kDa RSK enzymes (Sturgill et al 1991, Ahn et al. 1990, Agostinis et al. 1992). It is tempting to speculate that the bombesin stimulated S_6 peptide kinase eluting at 0.2 M NaCl in the Mono Q gradient may be related to the S_6 kinase activated in the late responses of Swiss 3T3 cells to EGF and insulin (Sturgill et al. 1991, Agostinis et al. 1992).

Casein kinase 2 (CK-2)

When bombesin stimulated cytosols were tested for CK-2 activity, no clear stimulation was observed with the specific CK-2 peptide substrate $R_3E_3SE_3$ (Agostinis et al. 1992). Surprisingly, when cytosolic kinases were resolved by Mono Q anion exchange chromatography, a several fold stimulation was seen at the 0.35 M NaCl elution position, with all CK-2 substrates. These peptide kinase activities were all inhibited by heparin and polyclonal antibodies to CK-2 revealed the presence of the α, α' and ß subunits of the enzyme at the 0.35 M NaCl elution place. We have no evidence for a regulation of the CK-2 enzyme by a direct phosphorylation: both the type 2A and alkaline phosphatase where without any effect on the basal and stimulated CK-2 activity (Agostinis et al. 1992).

SUMMARY

The amphibian tetradecapeptide bombesin as well as the bombesin-related mammalian peptides, are potent mitogens for Swiss 3T3 cells (Rozengurt E, Sinnett-Smith J 1990). Other sole mitogens for Swiss 3T3 cells, such as PDGF and FGF, invariably signal through a tyrosine kinase receptor (Rozengurt E, Sinnett-Smith J 1990). The bombesin receptor is a member of the family of seven hydrophobic region, G-protein-linked neuropeptide receptors, whose sequence does not reveal a protein kinase domain. Since many hormones and neuropeptides that signal through similar receptors with seven membrane spanning domains, are by themselves not mitogenic for Swiss 3T3 fibroblasts, we suggest that bombesin acts through a rather special signaling pathway. Although its receptor does not feature a cytoplasmic tyrosine kinase domain, bombesin just like EGF or insulin, rapidly stimulates the activity of two MBP kinases as well as several S_6 peptide kinases. The present report furthermore implicates CK-2 in the early signal transduction pathway of this mitogen, and it is postulated that the activation of CK-2 may be an intrinsic property of "sole mitogens" like bombesin, as it may be a compulsory event leading to cell division. In that respect, CK-2 may also be the point of integration of multiple signaling pathways, initiated by several different growth factors which by their synergistic actions make cell proliferation possible.

ACKNOWLEDGEMENTS

We would like to thank H. Dewulf, L. Vanden Bosch and S. Vander Perre for expert technical assistance and V. Feytons for the synthesis of the peptides. This research was supported by grants from the "Nationaal Fonds voor Geneeskundig Wetenschappelijk Onderzoek" and the "Geconcerteerde Onderzoeksacties van het Ministerie van de Vlaamse Gemeenschap. J.V.L. is a recipient of a fellowship of "Levenslijn"; P.A. is a Senior Research Assistant and J.R.V. a Research Director of the "Nationaal Fonds voor Wetenschappelijk Onderzoek".

REFERENCES

Agostinis P, Van Lint J, Sarno S, De Witte P, Vandenheede JR, MerlevedE W (1992) Rapid stimulation of Ser/Thr protein kinases following treatment of Swiss 3T3 cells with bombesin. J. Biol. Chem. 267:9732-9737

Ahn NG, Weiel JE, Chan CP, Krebs EG (1990) Identification of multiple epidermal growth factor-stimulated protein Ser/Thr kinases from Swiss 3T3 cells. J. Biol. Chem. 265:11487-11494

Boulton TG, Nye SH, Robbins DJ, Ip NY, Radziejewska E, Morgenbesser SD, DePinho RA, Panayotatos N, Cobb MH, Yancopoulos GD (1991) ERK's: a family of protein-serine/threonine kinases that are activated and tyrosine phosphorylated in response to insulin and NGF. Cell 65:663-675

Erikson RL (1991) Structure, expression, and regulation of protein kinases involved in the phosphorylation of ribosomal protein S6. J. Biol. Chem. 266:6007-6010

Rozengurt E, Sinnett-Smith J (1990) Bombesin stimulation of fibroblast mitogenesis: specific receptors, signal transduction and early events. Phil. Trans. R. Soc. Lond. B 327:209-221

Sturgill TW, Wu J (1991) Recent progress in characterization of protein kinase cascades for phosphorylation of ribosomal protein S6. Biochem. Biophys. Acta 1092:350-357

GROWTH HORMONE (GH) INCREASES LIPOPROTEIN LIPASE mRNA IN CELLS TRANSFECTED WITH THE RAT GH RECEPTOR cDNA

S.M. Francis, S. Enerbäck#, C. Möller, B. Enberg and G. Norstedt
Center for Biotechnology
Karolinska Institute
141 57 Huddinge
Sweden

INTRODUCTION

The growth hormone (GH) receptor is phosphorylated on tyrosine residues after ligand binding (Foster *et al.* 1988). However the mechanism of intracellular signalling is unclear, since the receptor shows no apparent homology to classical tyrosine kinase receptors (Leung *et al.* 1987). One obstacle to *in vitro* studies is that few cultured cells express high levels of the GH receptor. The approach taken here was to transfect cultured cells with expression plasmids containing the rat GH receptor cDNA and the gene coding for neomycin resistance. Drug resistant cell colonies expressing high levels of GH receptors, both at the mRNA and protein level, were selected. Several different cell lines were stably transfected in this manner : Chinese hamster ovary cells showed GH-induced protein synthesis as well as GH-induced phosphorylation of MAP kinases (Emtner *et al.* 1990; Möller *et al.* 1992). The transfected Buffalo rat liver (BRL) cells in this study showed a GH-induced increase in lipoprotein lipase (LPL) mRNA indicating that the expressed receptor is functional.

MATERIALS AND METHODS

Cell culture and stable transfection

BRL cells were transfected with the rat GH receptor cDNA cloned into an expression vector containing the human metallothionein-IIa promoter and SV40 enhancer using the Lipofectin method (Felgner *et al.* 1987). Cells were co-transfected with pSVNeo which codes for resistance to the drug G418. Stably transfected clones were selected by growing cells in G418 (600 ug/ml).

Department of Molecular Biology
University of Göteborg
S-400 33 Göteborg, Sweden

NATO ASI Series, Vol. H 76
Tyrosine Phosphorylation/Dephosphorylation and Downstream Signalling
Edited by L. M. G. Heilmeyer Jr.

Measurement of mRNA by Solution Hybridization

Cells were grown to 95 % confluence and then incubated in serum free medium for 24 - 36 hours before treatment with hGH and/or cycloheximide. Total nucleic acids (TNA) were prepared from cells lysed in 1 x SET (1% SDS, 10 mM EDTA and 20 mM (pH 7.5) Tris-HCl). Samples were homogenised, digested with proteinase K (200 ug/ml), extracted with phenol-chloroform and precipitated in ethanol. TNA was dissolved in 0.2 x SET and mRNA analysed in triplicates by solution hybridization using ^{35}S-UTP-labelled RNA probes. RNase resistant radioactivity was TCA precipitated, collected on glass fibre filters and counted (Möller *et al.* 1991). Probe templates were a 50 bp oligonucleotide from mouse LPL cDNA (Kirchgessner *et al.* 1987), 560 bp of rat GH receptor cDNA (Mathews *et al.* 1989) and 1300 bp of rat glyceraldehyde-3-phosphate-dehydrogenase (GAPDH) cDNA (Fort *et al.* 1985).

RESULTS

Untransfected BRL cells contained barely detectable levels of rat GH receptor mRNA. After transfection with rat GH receptor cDNA to produce the stable BRL5 clone, a significant amount of mRNA for the GH receptor could be measured (Table 1). Binding of ^{125}I-hGH was significantly higher in BRL5 compared to untransfected BRL cells (data not shown).

mRNA		BRL	BRL5
GH Receptor		0.77 ± 0.02	3.51± 0.32
LPL	-GH	0.12 ± 0.01	0.46 ± 0.02
	+GH	0.08 ± 0.03	2.47 ± 0.13
GAPDH	-GH	47.1 ± 2.78	53.4 ± 1.08
	+GH	52.0 ± 2.59	61.6 ± 1.37

Table 1: Levels of mRNA for GH receptor (amol/ug DNA), LPL (amol/ug DNA) and GAPDH (fmol/ug DNA) in untransfected cells (BRL) and cells transfected with the rat GH receptor cDNA (BRL5) with or without hGH treatment (100 nM). Values are mean ± SEM.

Stimulation of BRL5 cells with hGH resulted in an increase in LPL mRNA levels but no effect on GAPDH mRNA (Table 1). The untransfected BRL cells did not respond to GH with increased LPL mRNA.

When protein synthesis was inhibited by cycloheximide (CHX) in GH receptor transfected BRL5 cells, basal levels of LPL mRNA were increased and there was no stimulation of LPL mRNA by GH (Figure 1).

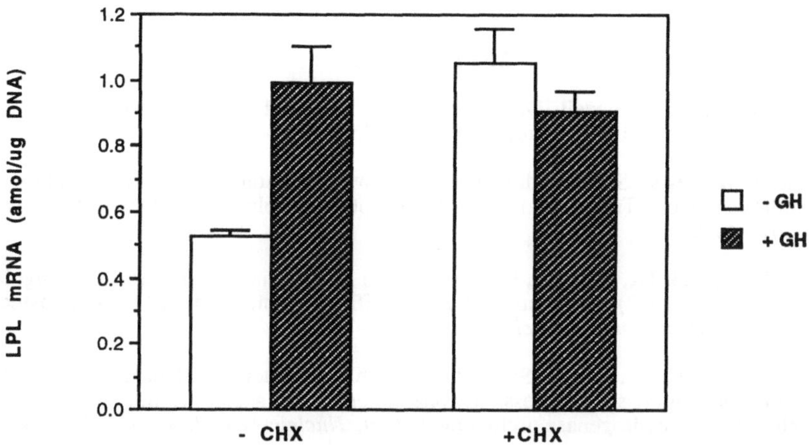

Figure 1: The effect of cycloheximide (CHX, 5 ug/ml) and hGH (50 nM) on LPL mRNA levels (fmol LPL/fmol GAPDH.) in BRL5 cells.

DISCUSSION AND CONCLUSIONS

The cloned rat GH receptor cDNA when transfected into cultured cells enabled previously non-responsive cultured Chinese hamster ovary cells to respond to GH in terms of increasing protein synthesis, protein phosphorylation and cell proliferation (Emtner *et al.* 1990; Möller *et al.* 1992). This work now demonstrates that the cloned GH receptor is functional for specific activation of gene transcription. Levels of LPL mRNA, but not GAPDH mRNA, were increased by GH in cells stably transfected with the rat GH receptor cDNA. This agrees with work in the preadipocyte cell line (Ob 1771) where GH was shown to specifically increase transcription of the LPL gene during differentiation (Pradines-Figueres *et al.* 1990). As observed in THP-1 cells (Auwerx *et al.* 1989), treatment of BRL5 cells with cycloheximide increased levels of LPL mRNA. It is thought that a labile protein is continually active on the LPL gene to suppress gene transcription. Since GH stimulation of

LPL mRNA is not seen in CHX treated cells, it is possible that the GH effect is via an interaction with this labile protein, thus reducing the suppression of LPL gene transcription.

It is concluded that the rat GH receptor cDNA when transfected into BRL cells produced a functional receptor which can be stimulated by GH to increase transcription of a specific gene. This stably transfected cell line can be used for studies of the mechanism of GH signal transduction.

REFERENCES

Auwerx JH, Deeb S, Brunzell JD, Wolfbauer G, Chait A (1989) Lipoprotein lipase gene expression in THP-1 cells. *Biochemistry* 28: 4563-4567

Emtner M, Mathews LS, Norstedt G (1990) Growth Hormone (GH) Stimulates Protein Synthesis in Cells Transfected with GH Receptor Complementary DNA. *Mol Endocrinol* 4: 2014-2020

Felgner PL, Gadek TR, Holm M, Roman R, Chan HW, Wenz M, Northrop JP, Ringold GM, Danielsen M (1987) Lipofection: A highly efficient, lipid-mediated DNA-transfection procedure. *Proc Natl Acad Sci USA* 84: 7413-7417

Fort P, Marty L, Piechaczyk M, Sabrouty SE, Dani C, Jeanteur P, Blanchard JM (1985) Various rat adult tissues express only one major mRNA species from the glyceraldehyde-3-phosphate-dehydrogenase multigenic family. *Nucleic Acids Res* 13: 1431-1442

Foster CM, Shafer JA, Rozsa FW, Wang X, Lewis SD, Renken DA, Natale JE, Schwartz J, Carter-Su C (1988) Growth Hormone Promoted Tyrosyl Phosphorylation of Growth Hormone Receptors in Murine 3T3-F442A Fibroblasts and Adipocytes. *Biochemistry* 27: 326-334

Kirchgessner TG, Svenson KL, Lusis AJ, Schotz MC (1987) The sequence of cDNA Encoding Lipoprotein Lipase. A member of a lipase gene family. *J Biol Chem* 262: 8463-8466

Leung DW, Spencer SA, Cachianes G, Hammonds RG, Collins C, Henzel WJ, Barnard R, Waters MJ, Wood WI (1987) Growth hormone receptor and serum binding protein: purification, cloning and expression. *Nature* 330: 537-543

Mathews LS, Enberg B, Norstedt G (1989) Regulation of Rat Growth Hormone Receptor Gene Expression. *J Biol Chem* 264: 9905-9910

Möller C, Arner P, Sonnenfeld T, Norstedt G (1991) Quantitative Comparison of Insulin-Like Growth Factor Messenger RNA Levels in Human and Rat Tissues Analysed by a Solution Hybridization Assay. *J Mol Endocrinol* 7: 213-222

Möller C, Hansson A, Enberg B, Lobie PE, Norstedt G (1992) Growth Hormone induction of tyrosine phosphorylation and activation of mitogen activated protein kinases in cells transfected with rat GH receptor cDNA. *J Biol Chem* In Press

Pradines FA, Barcellini CS, Dani C, Vannier C, Ailhaud G (1990) Transcriptional control of the expression of lipoprotein lipase gene by growth hormone in preadipocyte Ob1771 cells. *J Lipid Res* 31: 1283-1291

II. The Phosphatidylinositol Signalling System

ROLE OF PHOSPHATIDYLINOSITOL 3-KINASE IN GROWTH FACTOR AND ONCOGENE SIGNALING

Lewis Cantley
Dept. of Cellular and Molecular Physiology
Harvard Medical School and
Division of Signal Transduction
Beth Israel Hospital
EQRF 1st Floor
200 Longwood Ave.
Boston, MA 02115

I. Introduction

Phosphatidylinositol 3-kinase (PtdIns 3-Kinase) is an enzyme that phosphorylates the D-3 position of the inositol ring of the membrane lipid, phosphatidylinositol (PtdIns). It was discovered because of its high affinity association with the v-*src* oncogene product, $pp60^{v\text{-}src}$ and with the polyoma middle t/$pp60^{c\text{-}src}$ complex in transformed cells (Sugimoto et al., 1984; Whitman et al., 1985; Whitman et al., 1988). Prior to the discovery of this enzyme, the only phosphorylated forms of phosphatidylinositol known to exist *in vivo* were phosphatidylinositol-4-phosphate (PtdIns-4-P) and phosphatidylinositol-4,5-bisphosphate (PtdIns-4,5-P_2). These lipids are intermediates in the canonical PtdIns Turnover Cycle that leads to the well-known second messengers, inositol-1,4,5-trisphosphate and diacylglycerol (Fig. 1). *In vitro*, PtdIns 3-Kinase can phosphorylate the D-3 position of PtdIns, PtdIns-4-P or PtdIns-4,5-P_2 to produce PtdIns-3-P, PtdIns-3,4-P_2 and PtdIns-3,4,5-P_3 respectively (Auger et al., 1989; Carpenter and Cantley, 1990)). All

NATO ASI Series, Vol. H 76
Tyrosine Phosphorylation/Dephosphorylation
and Downstream Signalling
Edited by L. M. G. Heilmeyer Jr.

three of these lipids have now been detected in mammalian cells (Auger et al., 1989; Stephens et al., 1989). The lipid products of PtdIns 3-Kinase are not substrates for the phospholipases type C and are not intermediates in the canonical PtdIns Turnover Cycle (Lips et al., 1989; Serunian et al., 1989). Rather, they appear to be involved in a distinct regulatory process in the cell.

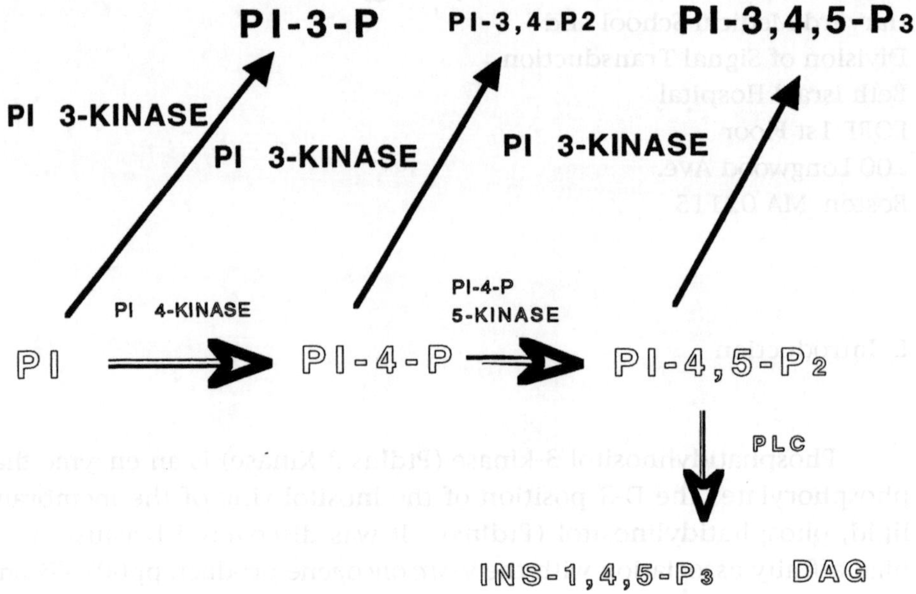

Figure 1: Synthesis of Phosphoinositides.

PtdIns 3-kinase is activated by a variety of growth factors, oncoproteins and other cellular activators (Cantley et al., 1991). Most of the factors that activate this enzyme do so by turning on a cellular protein-tyrosine kinase. PtdIns 3-kinase is activated by the PDGF, CSF-1, insulin, IGF-1, hepatocyte growth factor (HGF), NGF, stem cell growth factor (STEEL), and in some cells, EGF. The receptors for all these factors have intrinsic protein-tyrosine kinase activity. PtdIns 3-kinase is also activated by factors such as IL-2 (Remillard et al., 1991) and GM-CSF (Corey and Cantley, unpublished) whose receptors do not have intrinsic protein-tyrosine kinase activity but rather activate associated tyrosine kinases of the *src* family. PtdIns 3-kinase is also activated in cells transformed with v-*src*, polyoma middle t, v-*sis*, v-*abl* and v-*fms* (Cantley et al., 1991). Thrombin and formyl-peptides appear to activate PtdIns 3-kinase via G-protein linked pathways (Kucera and Rittenhouse, 1990; Traynor-Kaplan et al., 1988). In most of these cases the activation of PtdIns 3-kinase correlates with the recruitment of this enzyme from the cytosolic fraction to an activated protein-tyrosine kinase or to a tyrosine-phosphorylated protein (Cantley et al., 1991). Thus, PtdIns 3-kinase is a direct target of a host of protein-tyrosine kinases.

The function of the lipid products of PtdIns 3-kinase is not yet known. Examination of lipid levels in quiescent and growth factor-stimulated cells reveals that PtdIns-3-P is constitutively present (Auger et al., 1989). In contrast, PtdIns-3,4-P_2 and PtdIns-3,4,5-P_3 are nominally absent in quiescent cells and appear within seconds to minutes of growth factor stimulation (Auger et al., 1989). The levels of these lipids peak between 30 sec and 5 minutes after stimulation and then decrease with time but are still detectably elevated at more than an hour after growth factor addition. The correlation between activation of PtdIns 3-kinase and the transforming ability of a variety of mutants of polyoma middle t, v-*src*, v-*abl* and v-*fms* suggests that the lipid products have an important role in transformation in many but perhaps not all host cell lines (Cantley et al., 1991). Mutants of the PDGF receptor that fail to activate PtdIns 3-kinase fail to cause PDGF-dependent cell growth when transfected into certain cells (Fantl et al., 1992). PtdIns 3-kinase is also activated by thrombin in platelets and by formyl-peptide in neutrophils suggesting that it may play a more general role in cell signaling. The time course for appearance of PtdIns-3,4-P_2 and PtdIns-3,4,5-P_3 and the resistance of these lipids to hydrolysis by phospholipases type C suggest that the lipids themselves rather than derived inositol phosphates are acting as cellular regulators. A possible insight into the function of these lipids has come from the recent discovery that the catalytic subunit of PtdIns 3-kinase is homologous to a yeast gene involved in sorting of proteins to the vacuole (see below). Thus, these lipids may play a role in sorting of intracellular vesicles. This possibility will be discussed in more detail.

II. Structure of PtdIns 3-Kinase

PtdIns 3-kinase has been purified to homogeneity from rat liver cytosol (Carpenter et al., 1990). In unstimulated tissue, most of the enzyme appears in a soluble cytosolic fraction that migrates with an apparent size of approximately 200 kd as judged by gel filtration or sucrose gradient centrifugation. The purified enzyme is a heterodimer of a 110 kd catalytic subunit and an 85 kd regulatory subunit. Two different forms of the enzyme with distinct but homologous 110 kd subunits (110 upper and 110 lower) have been purified from rat liver cytosol (Carpenter et al., 1990). The two forms of the enzyme have similar specific activities and substrate utilization. Thus far it has not been possible to separate the 85 and 110 kd subunits without irreversible loss of enzymatic activity. However, the 110 kd subunit has PtdIns 3-kinase activity when expressed in insect cells in the absence of the 85 kd subunit (Hiles et al., 1992) Bovine thymus appears to have an active form of PtdIns 3-kinase that consists of a 110 kd subunit in the absence of the 85 kd subunit (Shibasaki et al., 1991).

A. 85 kd Regulatory Subunit

Multiple isotypes of the 85 kd subunit exist and the sequences of two distinct cDNA clones have been published (Fig. 2) (Escobedo et al., 1991; Otsu et al., 1991; Skolnik et al., 1991). These proteins have multiple domains of homology to other interesting proteins. Most notably, both have two copies of a domain found in a number of other signaling proteins called the *src*-homology 2 domain (SH2). In addition, the amino termini of these proteins have homology with another region of pp60$^{c\text{-}src}$ called the *src*-homology 3 domain (SH3). Between the SH3 domain and the first SH2 domain is another domain that has significant homology with the carboxy-terminus of the Break-Cluster Region gene product (BCR) (Otsu et al., 1991). This domain of BCR and a family of sequence-related proteins act as GTPase activating proteins (GAPs) for the *rho*/*rac*/cdc42 family of low molecular weight G-proteins (Diekmann et al., 1991). Thus, the sequence of the 85 kd regulatory subunit implies very complex regulation of PtdIns 3-kinase.

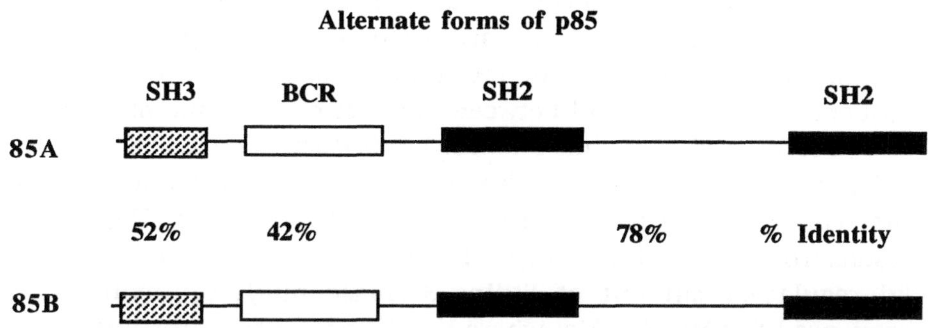

Figure 2: Domain structure of the 85 kd subunit of PtdIns 3-kinase.

1. SH2 Domains

Recent studies have elucidated the function of SH2 domains in intracellular signaling. These domains have been found in a host of protein-tyrosine kinases and intracellular targets of protein-tyrosine kinases. The function of these domains is to bind tyrosine-phosphorylated regions of other proteins (or other phosphorylated domains of the same protein) (Cantley et al., 1991; Mayer et al., 1991). Crystal structures of the pp60$^{c\text{-}src}$ SH2 domain associated with two different phosphopeptides have recently been solved (Waksman et al., 1992). An nmr solution structure of the p85 N-SH2 domain has confirmed that the folding pattern of this protein is similar to that of pp60$^{c\text{-}src}$ (Booker et al., 1992). SH2 domains form a compact globular structure of approximately 100 amino acids. The phosphotyrosine moiety of the associated peptide fits into a cleft in which arginine and lysine residues coordinate the phosphate residue. Additional contacts are made with the phenyl moiety of the tyrosine residue and with the backbone of the associated phosphopeptide. The residues that form these contacts are highly conserved in all SH2 domains described.

Individual SH2 domains demonstrate sequence-specific recognition of residues that immediately surround the phosphotyrosine moiety indicating that additional contacts are formed between the SH2 domain and the side groups of residues carboxy-terminal to the phosphotyrosine group. This sequence-specific recognition was first noticed because of sequence similarity found between the sites on polyoma middle t and PDGF receptor where PtdIns 3-kinase binds (Cantley et al., 1991). Using a degenerate phosphopeptide library it has been possible to determine the sequence specificity of a number of SH2 domains (Unpublished results from Cantley laboratory). The N-terminal SH2 domains of the 85 kd regulatory subunit of PtdIns 3-kinase binds optimally to the sequence PTyr-Met-Xxx-Met where Val, Ile or Glu can substitute for the first Met and Xxx represents any amino acid. The C-terminal SH2 domain of p85 has a similar specificity except that there is less selectivity at the first position after P-Tyr. Both SH2 domains strongly select for peptides that have a Met three residues C-terminal to the PTyr. Synthetic phosphopeptides with these motifs bind to the SH2 domains of p85 with K_D's in the range of 1 to 10 nM (Auger et al., 1991). Peptides lacking the critical Met residue or lacking the phosphate on the Tyr have affinities that are at least 1000 fold lower. In contrast, the SH2 domain of pp60$^{c\text{-}src}$ binds to the motif, P-Tyr-Glu-Glu-Ile. Thus, the primary sequence immediately surrounding the tyrosine site phosphorylated determines which SH2 domain-containing protein will associate *in vivo.*

Most of the proteins that are known to associate with PtdIns 3-kinase *in vivo* contain tyrosine phosphorylation sites with the sequence PTyr-Xxx-Xxx-Met (Cantley et al., 1991). Some examples of these are shown in Table I. In several cases, mutations of the Tyr residues of these motifs to Phe have been shown to eliminate or dramatically reduce the association of the respective proteins with PtdIns 3-kinase in intact cells. In the case of the PDGF receptor where several SH2-containing proteins are known to bind to the cytosolic domain, each protein appears to have a distinct binding site. For example, PtdIns 3-kinase binds at Tyr-740 and Tyr-751 while *ras*-GAP binds at Tyr-771 and PtdIns-specific phospholipase C-γ binds at Tyr-1021. These results are consistent with the fact that the SH2 domains of each of these

Table I: Sequences at Potential Binding Sites for PtdIns 3-Kinase in proteins known to bind this enzyme in vivo

Sequence	Protein	Site
Known Binding Sites		
ESDGG**YMDM**SKDESVD**YVPM**LD	Human PDGF Rec. β	(Y740, Y751)
EEEEE**YMPM**EDL*YLDI*LP	Mouse Polyoma MT	(Y315)
QGVDT**YVEM**RP	Mouse CSF-1 Rec.	(Y721)
DSTNE**YMDM**KP	Human c-kit	(Y721)
Possible Sites on Proteins known to bind p85		
GPGGD**YAAM**GACPASEQG**YEEM**RA	Human Erb B3	(Y1257, Y1270)
TPDEDYEYMNRQRDGGGPGGD**YAAM**GA	"	(Y1241, Y1257)
CTIDV**YMVM**VK	"	(Y922)
SPSSG**YMPM**NQ	"	(Y1,035)
DEDEE**YEYM**NR	"	(Y1,178)
LEELG**YEYM**DV	"	(Y1,203)
EELSN**YICM**GG	Rat Insulin Rec. Substrate 1	(Y460)
VSIEE**YTEM**MP	"	(Y546)
HTDDG**YMPM**SP	"	(Y608)
KGNGD**YMPM**SP	"	(Y628)
VDPNG**YMMM**SP	"	(Y658)
PCTGD**YMNM**SP	"	(Y727)
TGSEE**YMNM**DL	"	(Y939)
NSRGD**YMTM**QI	"	(Y987)
VAPVS**YADM**RT	"	(Y1,010)
ERENE**YMPM**APQIHL*YSQI*RE	Hamster polyoma MT	(298)
LSNPT**YSVM**RS	Mouse Polyoma MT	(Y250)
CPEKV**YELM**RA	Mouse V-abl oncoprotein	(Y355)

Table II Legend

Sequences in bold have similarity to the motifs that bind SH2 domains of p85. Negatively charged residues N-terminal to the Tyr are underlined. The location of the Tyr in the primary sequence is indicated in parathensis. The sequences were found using the FASTA and FIND programs to search the GenBank data base. Single letter codes for amino acids are used: A=Ala, C=Cys, D=Asp, E=Glu, F=Phe, G=Gly, H=His, I=Ile, K=Lys, L=Leu, M=Met, N=Asn, P=Pro, Q=Gln, R=Arg, S=Ser, T=Thr, V=Val, W=Trp, Y=Tyr.

proteins have distinct recognition motifs. Therefore, the primary sequence at the autophosphorylation sites of growth factor receptors determines whether they associate with PtdIns 3-kinase *in vivo*.

2. SH3 Domain

Less is known about the function of SH3 domains than SH2 domains. Many but not all proteins that contain one of these domains contain both. Recently, two proteins with high affinity for the SH3 domain of the *abl* protein-tyrosine kinase were identified by expression cloning (Cicchetti et al., 1992). The regions of these proteins that bind the SH3 domain are relatively hydrophobic and proline rich. A crystal structure of the tensin SH3 domain revealed a conserved hydrophobic surface that probably forms the binding site (Musacchio et al., 1992). It is likely that this domain at the N-terminus of p85 confers association with another protein that is relevant for cellular location of the PtdIns 3-kinase or for regulation of activity.

3. BCR Homology Domain

As mentioned above, a region of p85 between the SH3 domain and the first SH2 domain shows homology with the C-terminus of BCR. A number of proteins with homology to this domain have been recently found and most of these proteins have been shown to stimulate the GTPase activity of a *rho* family member, a *rac* family member, CDC42 or all three. Despite the homology that p85 has with these proteins, attempts to stimulate the GTPase activities of several members of the *rho*, *rac* and cdc42 families with recombinant p85 or native PtdIns 3-kinase have failed (unpublished results from L. Feig, R. Cerione and L. Cantley laboratories). On the basis of homology, It seems likely that p85 is a GAP for a low molecular weight GTP-binding proteins, but the identity of this protein remains to be determined. Interestingly, the BCR domains of p85A and p85B are relatively poorly conserved suggesting that these two proteins may interact with distinct low molecular weight GTP-binding proteins. If p85A and p85B are interacting with novel members of the rho/rac/CDC42 families then it is likely that this interaction affects actin filament rearrangement since each of these proteins has been strongly implicated in control of the actin cytoskeleton (Ridley and Hall, 1992; Ridley et al., 1992). Further work is needed to elucidate the function of the BCR domains of p85.

B. 110 kd Catalytic Subunit

The protein sequence of a110 kd catalytic subunit of PtdIns 3-kinase has been predicted based on a cDNA clone isolated from bovine brain (Hiles et al., 1992). A comparison of this sequence with partial protein sequences of the two 110 kd proteins isolated from rat liver indicates high homology with 110 lower (the higher mobility isoform). Interestingly, the bovine brain p110, as well as the rat liver 110 upper and 110 lower, have significant homology with the protein product of the *Saccharomyces cerevisiae* VPS34 gene (Herman and Emr, 1990). The VPS34 gene is required for movement of proteins from the lumen of the golgi to the lumen of the vacuole via a vesicular transport pathway. When this gene is mutated, proteins that would normally go to the vacuole are secreted from the cell. The golgi and vacuole appear to be normal in these cells, suggesting that the defect is in targeting of vesicles to the vacuole rather than in budding from the golgi or formation of the vacuole. A second defect is in recruitment of vacuoles into the budding daughter cell. The similarity between p110 and VPS34 suggest that PtdIns 3-kinase may be playing a role in vesicle sorting in mammalian cells. This possibility will be discussed below.

III. Regulation of PtdIns 3-Kinase Activity by Protein-Tyrosine Kinases

Regulation of PtdIns 3-kinase *in vivo* appears to be quite complex with multiple regulatory steps. The multidomain structure of p85 suggests that this subunit can act as a switch that integrates multiple signals to control the catalytic subunit.

The first obvious step in activation of this enzyme by growth factors such as PDGF is recruitment of the PtdIns 3-kinase from the cytosol to the inner surface of the plasma membrane via direct interaction with the cytosolic domain of the activated receptor (Fig. 3) (Coughlin et al., 1989; Kaplan et al., 1987; Whitman et al., 1987). This recruitment places the PtdIns 3-kinase close to its substrates: PtdIns-4,5-P_2 in particular is concentrated at the inner surface of the plasma membrane (Carpenter and Cantley, 1990). Thus, this change in location alone could account for a significant part of the activation observed *in vivo.*

The recruitment to the PDGF receptor also appears to change the substrate specificity of the PtdIns 3-kinase. When assayed in concentrated cell lysate from fibroblasts or liver the PtdIns 3-kinase will utilize PtdIns as a substrate but will not utilize PtdIns-4-P or PtdIns-4,5-P_2 (Carpenter et al., 1990). However, as the enzyme is further purified or merely diluted into buffer, it acquires the ability to phosphorylate all three substrates. These results and others have suggested the presence of a low affinity inhibitor of PtdIns 3-kinase in the cytosol that specifically affects utilization of PtdIns-4-P and PtdIns-4,5-P_2 as substrates (Carpenter et al., 1990). That fraction of the enzyme that immunoprecipitates with the PDGF receptor from stimulated cells will utilize all three substrates suggesting that recruitment to the receptor displaces the putative inhibitor. These results are consistent with the finding that PtdIns-3,4-P_2 and PtdIns-3,4,5-P_3 levels increase in response to PDGF while PtdIns-3-P levels do not significantly change.

The act of binding the SH2 domains of the 85 kd subunit to the phosphorylated tyrosine residues of the PDGF receptor may also cause stimulation of the catalytic subunit. A phosphopeptide based on the region of the PDGF receptor implicated in binding the PtdIns 3-kinase (including Tyr 740 and Tyr751; see Table I) was synthesized for *in vitro* studies. Forms of this peptide with phosphate at either Tyr 740 or Tyr

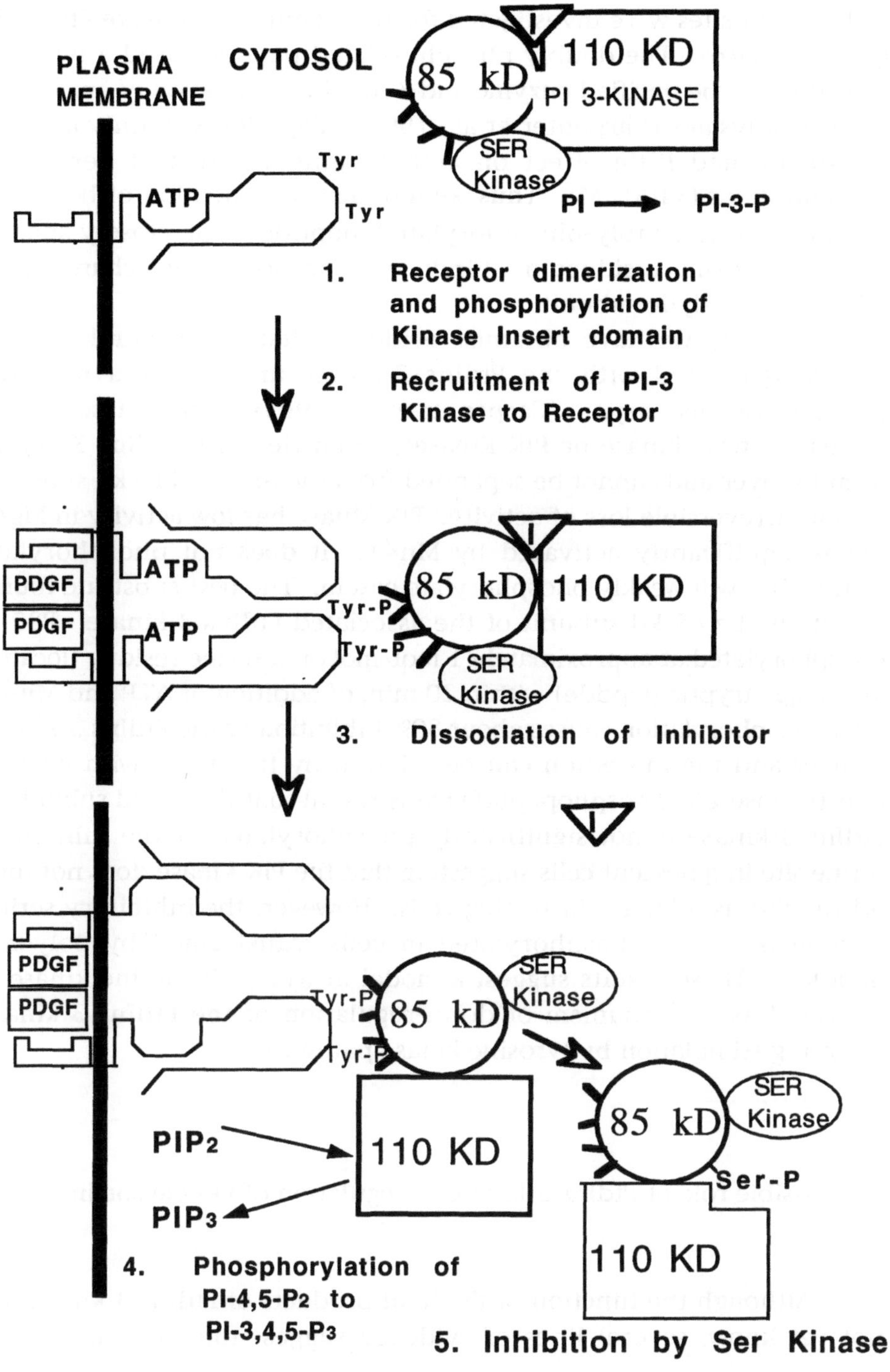
PLASMA
MEMBRANE
CYTOSOL
I
85 kD
110 KD
PI 3-KINASE
SER
Kinase
Tyr
Tyr
ATP
PI → PI-3-P
1. Receptor dimerization and phosphorylation of Kinase Insert domain
2. Recruitment of PI-3 Kinase to Receptor
PDGF
PDGF
ATP
ATP
Tyr-P
Tyr-P
3. Dissociation of Inhibitor
PIP2
PIP3
Ser-P
4. Phosphorylation of PI-4,5-P2 to PI-3,4,5-P3
5. Inhibition by Ser Kinase

751 or both sites were investigated for their ability to activate PtdIns 3-kinase *in vitro.* The doubly phosphorylated peptide cased a two fold activation of the purified enzyme and 3 to 4 fold activation of enzyme in crude cell lysates (Carpenter et al., 1993). Peptides with only a single phosphate had little effect on activity unless added at very high concentrations (>100 μM). Thus, simultaneous occupation of both SH2 domains by a doubly-phosphorylated peptide with closely-spaced phosphotyrosine residues may induce a conformational change that affects the catalytic subunit.

Recently we have obtained evidence that a Ser/Thr kinase is tightly associated with the PtdIns 3-kinase and act as a negative regulator of this enzyme (Carpenter et al., 1993). This kinase (called PtdIns 3-kinase kinase or PIK Kinase), copurifies with PtdIns 3-kinase from rat liver and cannot be separated from the 85 and 110 kd subunits without irreversible loss of activity. PIK kinase has low activity in Mg^{2+} but is significantly activated by Mn^{2+}. It does not phosphorylate histone but will weakly phosphorylate casein. The best substrate found to date is the 85 kd subunit of the associated PtdIns 3-kinase. P85 is phosphorylated at approximately 1 mol/mol on a serine residue (located in a single tryptic peptide) within 20 min. of addition of ATP and Mn^{2+}. This phosphorylation causes about 80% inhibition of the PtdIns 3-kinase activity and the inhibition can be relieved by treatment with protein phosphatase 2A. Phosphopeptide maps reveal that the 85 kd subunit of PtdIns 3-kinase is not significantly phosphorylated at the inhibitory serine site in quiescent cells suggesting that the PIK Kinase does not turn off the PtdIns 3-kinase in resting cells. However, the inhibitory serine residue is highly phosphorylated in cells transformed by polyoma middle t. These results suggest a model in which the serine kinase is activated as a mechanism of down-regulation of the PtdIns 3-kinase following stimulation by tyrosine kinase pathways.

IV. Possible role of PtdIns 3-kinase in regulation of vesicle sorting.

Although the function of the lipid products of PtdIns 3-kinase are not yet know, several lines of evidence suggest that they may have something to do with regulation of intracellular vesicle sorting. The most compelling evidence for this idea is the fact that mutants in the yeast homologue of PtdIns 3-kinase (VPS34) impair transport of luminal proteins from the golgi to the vacuole (Herman and Emr, 1990). As discussed above, this defect appears to be in the sorting (or targeting) of the vesicles since the proteins are, by default, transported to the plasma membrane and secreted. It is tempting to speculate that specific targeting to the vacuole is accomplished by binding vesicles to a specific kinesin type microtubule motor, although other possibilities exist.

In mammalian cells the equivalent of this process would be movement from the golgi to the lysosome, a mechanism mediated at least partially by microtubules (Duden et al., 1990). Thus, one might expect PtdIns 3-kinase to play a role in golgi to lysosome traffic in mammalian cells. It is possible that PtdIns-3-P has such a function in mammalian cells. This lipid is made constitutively in mammalian cells suggesting that it functions in a process that is not acutely affected by growth factors or other cellular activators. Consistent with this idea, the major fraction of PtdIns 3-kinase detectable by cytoimmunofluorescence of fixed quiescent 3t3 fibroblasts is associated with microtubules (Kapeller, Cantley and Corvera; unpublished).

In contrast to the constant levels of PtdIns-3-P, the levels of PtdIns-3,4-P_2 and PtdIns-3,4,5-P_3 dramatically increase when cells are stimulated (Auger et al., 1989). The production of these lipids presumably occurs at the location of the receptors to which PtdIns 3-kinase binds. Thus, they are initially produced at the plasma membrane and perhaps later in intracellular vesicles derived from the plasma membrane (Kelly et al., 1992). It is tempting to speculate that these lipids play a role in targeting these internalized vesicles to a specific location in the cell. Given the importance of PtdIns 3-kinase activation for growth factor and oncogene-mediated cell growth stimulation, it is unlikely that the lipid products of this enzyme merely serve to move the receptor to lysosomes for degradation. An alternative possibility is that the internalized vesicles (and their associated receptors) are moved to a location that is critical for the propagation of a mitogenic signal. In such a model, PtdIns-3,4,5-P_3 may be the agent that binds the vesicle to the appropriate motor or cytoskeletal element for proper sorting

V. Conclusion and Future Directions

Since the discovery of PtdIns 3-kinase considerable progress has been made in elucidating its structure and the extent of its involvement in signaling by a wide variety of growth factors, oncoproteins and cellular activators. The function of the lipid products of this enzyme are still obscure and progress in this area has been slow because of the difficulties in obtaining sufficient lipid for biochemical and cell biological studies. Now that clones of the genes for PtdIns 3-kinase are available, it should be possible to sort out the function of these lipids. New insight into the mechanism of action of growth factor receptors is likely to be derived from these studies.

VI References

Auger, K., Carpenter, C., Piwnica-Worms, H. and Cantley, L. (1991). Polyoma virus middle T antigen/pp60c-src complex associates with purified phosphatidylinositol 3-kinase *in vitro*. J. Biol. Chem. *267*, 5408-5415

Auger, K. R., Serunian, L. A., Soltoff, S. P., Libby, P. and Cantley, L. C. (1989). PDGF-dependent tyrosine phosphorylation stimulates production of novel polyphosphoinositides in intact cells. Cell *57*, 167-75.

Booker, G. W., Breeze, A. L., Downing, A. K., Panayotou, G., Gout, I., Waterfield, M. D. and Campbell, I. D. (1992). Structure of an SH2 domain of the p85 alpha subunit of phosphatidylinositol-3-OH kinase. Nature *358*, 684-7.

Cantley, L. C., Auger, K. R., Carpenter, C. L., Duckworth, B., Graziani, A., Kapeller, R. and Soltoff, S. (1991). Oncogenes and Signal Transduction. Cell *64*, 281-302.

Carpenter, C. L., Auger, K. A., Shoelson, S. and Cantley, L. C. (1993). Activation of Phosphatidylinositol 3-kinase by phosphotyrosine-containing peptides. (submitted)

Carpenter, C. L. and Cantley, L. C. (1990). Phosphoinositide Kinases. Biochemistry 29 11147-11156.

Carpenter, C. L., Duckworth, B., Auger, K. A., Schaffhausen, B. and Cantley, L. C. (1993). A tightly associated serine/threonine protein kinase regulates phosphoinositide 3-kinase activity. Mol. Cell. Biochem. *(in press)*,

Carpenter, C. L., Duckworth, B. C., Auger, K. R., Cohen, B., Schaffhausen, B. S. and Cantley, L. C. (1990). Purification and characterization of phosphoinositide 3-kinase from rat liver. J. Biol. Chem. *265*, 19704-19711.

Cicchetti, P., Mayer, B. J., Thiel, G. and Baltimore, D. (1992). Identification of a protein that binds to the SH3 region of Abl and is similar to Bcr and GAP-rho. Science *257*, 803-6.

Coughlin, S. R., Escobedo, J. A. and Williams, L. T. (1989). Role of phosphatidylinositol kinase in PDGF receptor signal transduction. Science *243*, 1191-4.

Diekmann, D., Brill, S., Garrett, M. D., Totty, N., Hsuan, J., Monfries, C., Hall, C., Lim, L. and Hall, A. (1991). Bcr encodes a GTPase-activating protein for p21rac. Nature *351*, 400-2.

Duden, R., Ho, W. C., Allan, V. J. and Kreis, T. E. (1990). What's new in cytoskeleton-organelle interactions? Relationship between microtubules and the Golgi-apparatus. Pathol Res Pract *186*, 535-41.

Escobedo, J. A., Navankasattusas, S., Kavanaugh, W. M., Milfay, D., Fried, V. A. and Williams, L. T. (1991). cDNA cloning of a novel 85 kd protein that has SH2 domains and regulates binding of PI3-kinase to the PDGF beta-receptor. Cell *65*, 75-82.

Fantl, W. J., Escobedo, J. A., Martin, G. A., Turck, C. W., del, R. M., McCormick, F. and Williams, L. T. (1992). Distinct phosphotyrosines on a growth factor receptor bind to specific molecules that mediate different signaling pathways. Cell *69*, 413-23.

Herman, P. K. and Emr, S. D. (1990). Characterization of VPS34, a gene required for vacuolar protein sorting and vacuole segregation in Saccharomyces cerevisiae. Mol Cell Biol *10*, 6742-54.

Hiles, I. D., Otsu, M., Volinia, S., Fry, M. J., Gout, I., Dhand, R., Panayotou, G., Ruiz, L. F., Thompson, A., Totty, N. F., Hsuan, J. J., Courtneidge, S. A., Parker, P. J. and Waterfield, M. D. (1992). Phosphatidylinositol 3-kinase: structure and expression of the 110 kd catalytic subunit. Cell *70*, 419-29.

Kaplan, D. R., Whitman, M., Schaffhausen, B., Pallas, D. C., White, M., Cantley, L. and Roberts, T. M. (1987). Common elements in growth factor stimulation and oncogenic transformation: 85 kd phosphoprotein and phosphatidylinositol kinase activity. Cell *50*, 1021-9.

Kelly, K. L., Ruderman, N. B. and Chen, K. S. (1992). Phosphatidylinositol-3-kinase in isolated rat adipocytes. Activation by insulin and subcellular distribution. J Biol Chem *267*, 3423-8.

Kucera, G. L. and Rittenhouse, S. E. (1990). Human platelets form 3-phosphorylated phosphoinositides in response to alpha-thrombin, U46619, or GTP gamma S. J Biol Chem *265*, 5345-8.

Lips, D. L., Majerus, P. W., Gorga, F. R., Young, A. T. and Benjamin, T. L. (1989). Phosphatidylinositol 3-phosphate is present in normal and transformed fibroblasts and is resistant to hydrolysis by bovine brain phospholipase C II. J. Biol. Chem. *264*, 8759-8763.

Mayer, B. J., Jackson, P. K. and Baltimore, D. (1991). The noncatalytic src homology region 2 segment of *abl* tyrosine kinase binds to tyrosine-phosphorylated cellular proteins with high affinity. Proc. Natl. Acad. Sci. *88*, 627-631.

Musacchio, A., Noble, M., Pauptit, R., Wierenga, R. and Saraste, M. (1992). Crystal Structure of a Src-homology 3 (SH3) domain. Nature *359*, 851-855.

Otsu, M., Hiles, I., Gout, I., Fry, M. J., Ruiz, L. F., Panayotou, G., Thompson, A., Dhand, R., Hsuan, J., Totty, N. and et, al. (1991). Characterization of two 85 kd proteins that associate with receptor tyrosine kinases, middle-T/pp60c-src complexes, and PI3-kinase. Cell *65*, 91-104.

Remillard, B., Petrillo, R., Maslinski, W., Tsudo, M., Strom, T. B., Cantley, L. and Varticovski, L. (1991). Interleukin-2 receptor regulates activation of phosphatidylinositol 3-kinase. J Biol Chem *266*, 14167-70.

Ridley, A. J. and Hall, A. (1992). The Small GTP-Binding Protein rho Regulates the Assembly of Focal Adhesions and Actin Stress Fibers in Response to Growth Factors. Cell *70*, 389-399.

Ridley, A. J., Paterson, H. F., Johnston, C. L., Diekmann, D. and Hall, A. (1992). The Small GTP-Binding Protein rac Regulates Growth Factor-Induced Membrane Ruffling. Cell *70*, 401-410.

Serunian, L. A., Haber, M. T., Fukui, T., Kim, J. W., Rhee, S. G., Lowenstein, J. M. and Cantley, L. C. (1989). Polyphosphoinositides produced by phosphatidylinositol 3-kinase are poor substrates for phospholipases C from rat liver and bovine brain. J Biol Chem *264*, 17809-15.

Shibasaki, F., Homma, Y. and Takenawa, T. (1991). Two types of phosphatidylinositol 3-kinase from bovine thymus. Monomer and heterodimer form. J Biol Chem *266*, 8108-14.

Skolnik, E. Y., Margolis, B., Mohammadi, M., Lowenstein, E., Fischer, R., Drepps, A., Ullrich, A. and Schlessinger, J. (1991). Cloning of PI3 kinase-associated p85 utilizing a novel method for expression/cloning of target proteins for receptor tyrosine kinases. Cell *65*, 83-90.

Stephens, L., Hawkins, P. T. and Downes, C. P. (1989). Metabolic and structural evidence for the existence of a third species of polyphosphoinositide in cells: D-phosphatidyl-myo-inositol 3-phosphate. Biochem. J. *259*, 267-276.

Sugimoto, Y., Whitman, M., Cantley, L. C. and Erikson, R. L. (1984). Evidence that the Rous sarcoma virus transforming gene product phosphorylates phosphatidylinositol and diacylglycerol. Proc Natl Acad Sci U S A *81*, 2117-21.

Traynor-Kaplan, A. E., Harris, A. L., Thompson, B. L., Taylor, P. and Sklar, L. A. (1988). An inositol tetrakisphosphate-containing phospholipid in activated neutrophils. Nature *334*, 353-356.

Waksman, G., Kominos, D., Robertson, S. C., Pant, N., Baltimore, D., Birge, R. B., Cowburn, D., Hanafusa, H., Mayer, B. J., Overduin, M. and et, al. (1992). Crystal structure of the phosphotyrosine recognition domain SH2 of v-src complexed with tyrosine-phosphorylated peptides. Nature *358*, 646-53.

Whitman, M., Downes, C.P., Keeler, M., Keller, T. and Cantley, L. (1988) Type I Phosphatidylinositol Kinase Makes a Novel Inositol Phospholipid, Phosphatidylinositol-3-phosphate Nature *332*, 644-646.

Whitman, M., Kaplan, D., Roberts, T. and Cantley, L. (1987). Evidence for two distinct phosphatidylinositol kinases in fibroblasts: Implications for cellular regulation. Biochem J *247*, 165-74.

Whitman, M., Kaplan, D. R., Schaffhausen, B., Cantley, L. and Roberts, T. M. (1985). Association of phosphatidylinositol kinase activity with polyoma middle-T competent for transformation. Nature *315*, 239-42.

ISOLATION AND CHARACTERIZATION OF PHOSPHATIDYLINOSITOL 4-KINASE ISOENZYMES FROM BOVINE BRAIN

G. Vereb*, M. Schmidt, D. Klix, M. Varsányi , L.M.G. Heilmeyer , Jr.
Institut für Physiologische Chemie
Ruhr-Universität Bochum
Universitätsstr. 150
4630 Bochum 1
Germany

INTRODUCTION

The existence of two membrane-bound isoforms of phosphatidylinositol (PtdIns) 4-kinase has been demonstrated in brain (Endemann et al., 1987; Yamakawa and Takenawa, 1988). The type 2 isoform (Mr of 55 kDa) exhibits a high affinity for ATP and a high sensitivity to adenosine inhibition; it does not seem to be organ-specific. The type 3 isoform has little affinity for ATP and adenosine and exists predominantly in brain.

Here we describe a new procedure to separate type 2 and type 3 isoforms from bovine brain. They were identified on the basis of their kinetic constants, Mr values and sensitivity to SDS-denaturation. Furthermore, their reaction with 5'-p-fluorosulfonylbenzoyladenosine (FSBA) was studied.

METHODS

The determination of PtdIns-kinase activity has been described by Varsányi et al. (1989).

130 g bovine brain was homogenized with 0.32 M sucrose, 10 mM Tris.HCl pH 7.5, 1 mM EDTA, 1 mM EGTA, 1 mM DTE and protease inhibitors (PMSF, aprotinin, leupeptin, aminohexanoic acid). After removal of cell debris by centrifugation at 1000 x g the supernatant was ultracentrifuged at 100,000 x g. The resulting pellet was washed with 0.15 M NaCl, 10 mM Tris.HCl pH 7.5, 1 mM DTE, 0.1 mM EDTA, recentrifuged and solubilized with 0.1 % Triton X-114, 10 mM Tris.HCl pH 7.5, 1 mM DTE and protease inhibitors. After phase separation at 30°C the PtdIns 4-kinase was precipitated from the waterphase by ammonium sulfate at 30 % saturation. The precipitate was dissolved and dialysed against 0.1 % Triton X-100, 5 mM Tris.HCl pH 7.5, 0.5 mM DTE, 0.05 mM EDTA. The solubilisate was loaded onto a 5 x 5 cm hydroxyapatite column, equilibrated with 5 mM K-phosphate pH 7.0, 0.5 M NaCl, 1 mM DTE and washed with about 4 L of this buffer. The type 3 isoform was eluted by increasing K-phosphate concentration up to 50

* Department of Medical Chemistry, University School of Medicine, 4026 Debrecen, Hungary

NATO ASI Series, Vol. H 76
Tyrosine Phosphorylation/Dephosphorylation and Downstream Signalling
Edited by L. M. G. Heilmeyer Jr.

mM. Following washing the column with 300 mM K-phosphate, 0.5 M KCl, 1 mM DTE, pH 7.0 the type 2 isoform was eluted with this buffer containing additionally 0.2 % Triton X-100.

Type 3 isoform was loaded onto a 1.5 x 12 cm Heparin Sepharose CL-6B column equilibrated with 5 mM $MgCl_2$, 10 mM Tris.HCl pH 7.5, 1 mM DTE, 0.1 mM EDTA and was eluted with a NaCl gradient. Active fractions were rechromatographed on hydroxyapatite using a phosphate gradient of 70-350 mM, pH 7.0.

RESULTS AND DISCUSSION

Chromatography of the solubilized membrane protein on hydroxyapatite results in two enzyme fractions which proved to be approximately equal amounts of the type 3 and type 2 isoforms. After chromatography on Heparin-Sepharose and rechromatography on hydroxyapatite type 3 can be obtained with a specific activity of 1250 nmol x min^{-1} x mg^{-1}. The type 2 isoform has not been purified further and its properties are investigated after separation on hydroxyapatite. Characteristics of the separated isoforms are listed in Table I.

Table 1: Characteristics of PtdIns-Kinase Isoforms

Type	Type 3	Type 2
Binding to hydroxyapatite	weak	strong
K_M (PtdIns)	730 μM	200 μM
K_M (ATP)	700 μM	312 μM 133 μM
K_i (adenosine)	500 μM	260 μM 63 μM
Inhibition by 0.005 % SDS	99 %	7 %
Activation by Triton X-100	+	+
Mr by SDS-PAGE	200 kDa	55 kDa
Mr by gelfiltration	470 kDa	not determined

The type 3 enzyme eluted as the first peak from the hydroxyapatite by 50 mM K-phosphate shows kinetic constants similar to those described by Endemann et al. (1987). A striking feature of this isoform is the high sensitivity to SDS-denaturation.

The type 2 enzyme fraction eluted from hydroxyapatite as a second peak exhibits a ca. 4-fold lower Mr than type 3. Furthermore, the type 2 isoform is more resistant to SDS-inhibition than type 3. However, this enzyme preparation does not seem to be homogeneous; it shows two K_M values for ATP and two K_i values for adenosine which might be explained by the presence of further isoform(s) with binding sites of different affinity.

Assays for enzyme activity in gel slices after SDS-PAGE (see Schmidt et al. in this volume) yield an apparent molecular weight of 200 kDa for the type 3 isoform. A major activity band at 55 kDa has been found for the type 2 isoform. Gelfiltration of the purified type 3 on Superdex 200 column results in an apparent molecular weight of 470 kDa for the native enzyme. However, SDS-PAGE of the fractions reveals a major polypeptide at 200 kDa coeluted with the enzymic activity. The deviation from the native Mr value suggests a high degree of asymmetry and/or association of type 3 PtdIns 4-kinase under conditions of gelfiltration.

The reaction of PtdIns 4-kinases with FSBA, an ATP-like affinity labelling reagent, is demonstrated in Fig.1. Part A shows that FSBA inactivates both isoforms in a time-dependent manner without significant difference. Competition experiments, presented in Part B, show that ATP can not prevent this inactivation whereas adenosine is a powerful protective agent. The protective effect of adenosine is higher for type 2 than for type 3, correlating with the inhibitory potency of adenosine on the enzymic activity of these isoforms. The fact that neither ATP nor Mg-ATP exert any significant protection against FSBA inactivation, while adenosine proved to be protective, might be explained by the hydrophobicity of the ATP binding domain to which less polar ATP-analogues like adenosine or FSBA bind preferentially.

Affinity labelling of the type 3 isoform with 400 μM [^{14}C]FSBA and subsequent separation by SDS-PAGE show a preferential labelling of a 200 kDa protein supporting its identity with the type 3 isoform.

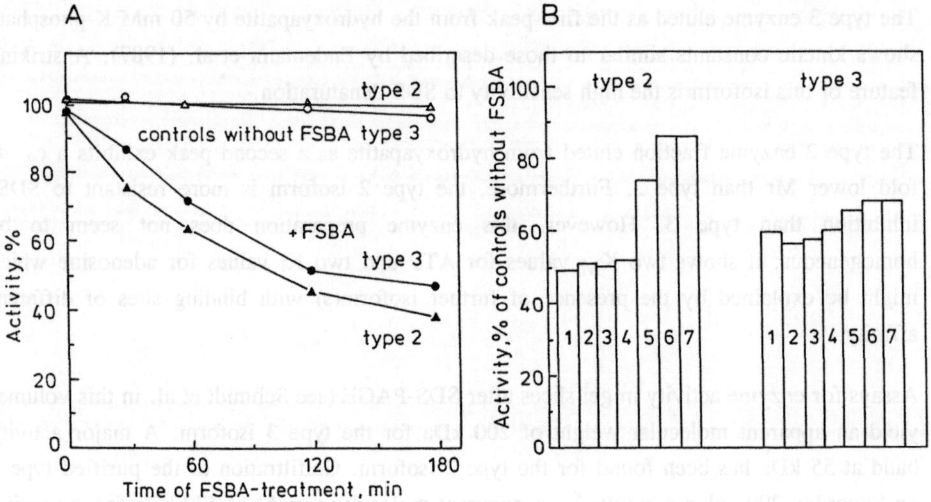

Fig. 1. Inactivation of PtdIns 4-kinase isoforms by FSBA

Part A: type 2 or type 3 was incubated at 25°C with 208 μM FSBA in 0.8 % dimethylformamide (DMF), 150 mM or 220 mM K-phosphate pH 7.0, 72 mM NaCl or 360 mM KCl, 1 mM DTE, 0.05 % CHAPS (type 3) or 0.2 % Triton X-100 (type 2). Controls without FSBA contained 0.8 % DMF. Activities were expressed in percent of initial activity measured without FSBA and DMF.

Part B: the following additions were included prior to starting the incubation by FSBA or DMF: 1: none; 2: 400 μM ATP; 3: 1.2 mM $MgCl_2$; 4: 400 μM ATP + 1.2 mM $MgCl_2$; 5: 400 μM adenosine; 6: 1 mM adenosine; 7: 1 mM adenosine + 1.2 mM $MgCl_2$. After 60 min incubation at 25°C the remaining activities were determined and expressed in percent of controls run with the appropriate additions but without FSBA.

REFERENCES

Endemann, G., Dunn, S.N., Cantley, L.C. (1987) Bovine brain contains two types of phosphatidylinositol kinase. Biochemistry 26, 6845-6852

Varsányi, M., Messer, M., Brandt, N.R. (1989) Intracellular localization of inositol-phospholipid-metabolizing enzymes in rabbit fast-twitch skeletal muscle. Eur.J.Biochem. 179, 473-479

Yamakawa, A., Takenawa, T. (1987) Purification and characterization of membrane-bound phosphatidylinositol kinase from rat brain. J.Biol.Chem. 263, 17555-17560

DETERMINATION OF PHOSPHOINOSITIDES AND INOSITOL PHOSPHATES. PRINCIPLES AND RULES.

G. W. Mayr
Institut für Physiologische Chemie
Ruhr-Universität Bochum
W 4630 Bochum 1
F.R.G.

Only cellular PLC# activity can be determined presently without a HPLC or LC instrumentation. However, a state-of-art analysis of the complex metabolism of inositol phosphates and phosphoinositides requires sophisticated sample workup and HPLC analysis protocols. These are laborious and still mostly based on radioactive prelabeling of cells or tissues. The novel HPLC-MDD technique allows a fully automatized non-radioactive analysis (i) of masses of all inositol phosphates and polyphosphoinositides from any biological source and (ii) of binding or interconversion of such compounds by proteins *in vitro*. This and other frequently used techniques are compared.

PRINCIPLES OF PHOSPHOINOSITIDE AND INOSITOL PHOSPHATE ANALYSIS

Hundreds of papers describing the changes and effects of InsP's and PtdInsP's in response to extracellular stimuli have been published over the past years with steadily increasing frequency. Ins(1,4,5)P_3 is as a calcium mobilizing second messenger in animal cells (Berridge & Irvine 1989, Putney 1991) and seems to play a similar role in plant cells (Alexandre et al. 1990). Complex routes of phosphorylation and dephosphorylation of InsP's released from the PtdInsP's and of others starting from *de novo* synthesized Ins(3)P (Stephens & Irvine 1990) have been described (reviewed by Berridge & Irvine & 1989, Downes and MacPhee 1990, Shears 1991). Not only in plant cells, but also in most animal

---

Abbreviations: MDD, metal-dye detection; PLC, phospholipase C/ phosphoinositidase C; PI3-kinase, phosphoinositide 3-kinase; PAR, 4-(2-pyridylazo)resorcinol; HPLC, high performance liquid chromatography; LC, liquid chromatography; TLC, thin layer chromatography; HPTLC, high performance thin layer chromatography; TCA, trichloroacetic acid; DVB, divinylbenzene; PCA, perchloric acid; TCA trichloroacetic acid: All D-*myo*-inositol phosphates and corresponding *myo*-inositol phospholipids (phosphoinositides) as well as glycerophospho-*myo*-inositol phosphates are abbreviated according to the 1988 relaxed NC-IUB recommendations e.g: Ins(1,4,5)P_3, D-*myo*-inositol 1,4,5-trisphosphate; PtdIns(4,5)P_2, phosphatidyl(1)-D-*myo*-inositol 4,5-bisphosphate; $InsP_4$-$(PP)_2$, bisdiphosphoinositol tetrakisphosphate; GroPIns(4,5)P_2, α-glycero-phospho(1)-D-*myo*-inositol 4,5-bisphosphate.

cells anabolic routes are leading to the fully phosphorylated $InsP_6$ (Stephens & Irvine 1990, reviewed by Berridge and Irvine 1989, Downes and MacPhee 1990, Shears 1991). Specific functions have been postulated for this compound in plants (Raboy 1990) and are now being also postulated in animal systems (Sawada et al. 1989, Beck and Keen 1991, Teibert et al. 1991, review by Mayr 1988, Downes and MacPhee 1990, Shears 1991). Furthermore, recent evidence indicates the existence *in vivo* of pyrophosphorylated InsP's derived from $InsP_5$ and $InsP_6$ (Mayr et al. 1992, Stephens et al 1993). It is completely unknown what functions these most densely phoshorylated carbohydrates may have in cells. Beside the key role of Ins(1,4,5)P_3, also some $InsP_2$, $InsP_4$, and $InsP_5$ intermediates of animal cells appear to exert specific functions on enzymes, ion channels and structural proteins (e.g. Berridge & Irvine & 1989, Koppitz et al. 1986, Mayr 1988, Mayr 1989, Pittet et al. 1989, Thieleczek et al. 1989, Gärtner et al. 1991, Ivorra et al. 1991, Beck & Keen 1991, Sawada et al. 1989, Theibert et al. 1991). A recent discovery is also that certain agonists, beside activating PLC to release Ins(1,4,5)P_3, activate a novel PI3-kinase which generates 3-phosphorylated PtdInsP's, most rapidly PtdIns(3,4,5)P_3 from PtdIns(4,5)P_2 (Traynor-Kaplan et al. 1988, Stephens et al. 1991). These putative phosphoinositide messengers are presumably activating or targeting membrane associated key proteins (Traynor-Kaplan 1988, Carpenter & Cantley 1990, chapter by Cantley in this volume).

The complexity of the inositide system in animal and plant cells requires sophisticated analyses techniques for both the water soluble and insoluble messengers and intermediates of the system. In most studies, cell cultures were kept for hours to days in [^{3}H]*myo*-inositol or for shorter periods in [^{32}P]phosphate and changes in the radioactivities of extracted PtdInsP's and InsP's were determined after their separation by TLC and anion exchange chromatography or HPLC, respectively. However, many primary cell preparations, intact tissues of animal and plant origin, plant cell cultures, food products, an human biopsy material are not amenable to such an analysis due to insufficient or impossible uptake of label. For the key compound Ins(1,4,5)P_3, and recently also for Ins(1,3,4,5)P_4, sensitive enzymatic (Tarver et al. 1987) or radioligand binding assays have been developed (Palmer & Wakelam 1989, Donié & Reiser 1989). Picomol amounts of these isomers can be determined now with acceptable isomer specificity and without the need for a HPLC system. Similar isomer specific assays might be useful for other InsP's and PtdInsP's.

Organic chemists have begun to synthesize all theoretical isomers of *myo*-InsP's and inositol lipids as well as structurally modified, non-hydrolysable, and membrane permeant analogues thereof, affinity labels, fluorescent or radioactive derivatives and immobilized compounds (e.g. Henne et al. 1988, Reiser et al. 1991, Dreef et al. 1991, see also papers in Carbohydrate Res., Vol. 234, pp 1- 246). In order to assess the purity and stability of these compounds as well as to monitor their purification simple but structure-selective analysis techniques are also desirable. NMR-spectroscopy requires amounts in the micromol or higher range (Scholz et al. 1990) and is not suited to detect small amounts of contaminates. Since some of the synthetic analogues are presently investigated for their potential therapeutic use (Irth et al. 1990, Shamsuddin et al. 1989) analytical techniques should also allow for a determination of plasma or tissue levels of such compounds.

In our laboratory has been developed a novel and extremely sensitive complexometric post-column derivatization system for HPLC separated InsP's based on a ligand exchange reaction and allowing an on-line monitoring at visible wavelength (Mayr 1988b, Mayr 1990). This detection principle termed *metal-dye detection* (MDD) employs yttrium as a 'detector metal' and the metal specific dye PAR as a visible indicator of ligand-exchange. Ligand exchange, i.e. release of free dye is taking place when an analyte forms a stronger complex with the 'detector metal' than the dye. In this case the metal-specific absorbance of the dye at a wavelength of about 520 nm is linearly decreasing with the concentration of the analyte present in the eluent. Conditions for a ligand exchange reaction with InsP analytes are brought about by bringing the pH of the eluent either originally or after elution from the column into the alkaline range (7.5 to 9.5). The detection system is very versatile concerning the solvent conditions and the columns employed for HPLC analysis. It fulfils most of the requirements for a direct isomer selective analysis of practically all physiological InsP's, synthetic analogues, and PtdInsP's (Mayr 1990). The latter compounds are well resolved into isomeric forms by anion exchange HPLC and susceptible to metal-dye detection when made water soluble by mild deacylation (Clarke & Dawson 1981, Traynor-Kaplan et al. 1988). Also other substances containing at least one phosphomonoester group, e.g. carbohydrate phosphates can be determined by this fully non-destructive derivatization technique. Substances containing only phosphodiesters are not sensitively enough determined. In contrast to the above described isotopic techniques, metal-dye detection allows a direct analysis of masses of these compounds without the necessity for a radioactive prelabeling. Thus any type of specimen, cells, tissues, extracts, and food products can be analysed for these compounds. A comparison of our technique with other published techniques (Palmer & Wakelam 1989, Dean & Beaven 1989, Mayr 1990, Coté et al. 1990) shows that the detection sensitivity of our technique is higher than that of most other non-isotopic techniques. With a standard HPLC set-up, the minimal amounts detected are in the range of 5 to 20 pmols (4 to 10 ng) depending on the number of phosphates and the eluent and derivatization conditions (see fig. 1, below). When a

Table I. Isomers of *myo*-InsP's in theory and *in vivo*.

Isomer	*pairs of enantiomers*	*meso forms*	*D-isomers detected in animals and plants*
InsP	2	2	1; 2; 3; 4; 5; c1:2
$InsP_2$	6	3	1,2[a]; 1,3; 1,4; 1,6[a]; 2,4; 3,4; 3,6; 4,5; c1:2,4
$InsP_3$	8	4	1,2,3; 1,2,6[a]; 1,3,4; 1,4,5; 1,5,6[a]; 2,4,5; 3,4,5; 3,4,6; c1:2,4,5, 1,4,6[a]
$InsP_4$	6	3	1,2,3,4[a]; 1,3,4,5; 1,3,4,6; 1,4,5,6; 3,4,5,6; 1,2,3,4
$InsP_5$	2	2	1,2,3,4,5[a]; 1,3,4,5,6; 1,2,4,5,6[a]; 1,2,3,4,6
$InsP_6$		1	
$InsP_4$-PP			?
$InsP_4$-$(PP)_2$			1,4[a]
$InsP_5$-PP			1[a], 4[a]
total	63 (39[b])		47[c]

[a] Isomers which were found in our extracts from animal and plant tissues or cell cultures and fungi. Their enantiomeric form, i.e. whether they are D or L enantiomers or mixtures thereof was not determined.
[b] The number of isomers separable by anion HPLC.
[c] The number is based on the assumption the isomers marked [a] are present in both enantiomeric forms.

Table II. Isomers of glycerol containing head groups derived from inositides in animal cells.

phosphate content	*isomerism of phosphates*
1	GroPIns
2	GroPIns(3)P, GroPIns(4)P
3	GroPIns(3,4)P_2; GroPIns(4,5)P_2
4	GroPIns(3,4,5)P_2

specimen has micromolar concentrations of the InsP's or inositides of interest, specimen wet weights between 20 and 100 mg are sufficient for an analysis. In a microbore system (presently developed in our laboratory) the sensitivities can be increased further by about an order of magnitude.

As discussed (Mayr 1990), alternative dyes and metals might be used in order to modify the technique and adapt it to special requirements. However, the sensitivities reported are not higher than those obtained by our system and fluorescence detection is subject to many more physical problems than absorbance detection. One of the latter methods was combined with ion-pairing chromatography (Irth et al. 1990). The older technique monitoring the absorbance decrease of the red Fe^{III}-sulfosalicylic acid complex at low pH when inorganic or organic phosphate-containing analytes are present (Cilliers & VanNiekerk 1986) is analogous to our technique in its principle, but by orders of magnitude less sensitive.

The about 50! *myo*-InsP isomers hitherto detected *in vivo* (compare with theoretically possible isomers in table I) and the 6 known phosphoinositide head groups encountered in animal cells (table II) make it clear that highly resolving HPLC separation techniques are required for physiological analyses. Although InsP isomer separations are possible with modern anion exchange HPLC techniques, pairs of enantiomorphs are not resolved. HPLC techniques separating enantiomorphs of InsP's are still completely missing. But also a complete separation of all 24 pairs of enantiomorphs and the 15 *meso* forms from one another are hardly achieved by one HPLC separation . In the upper chromatogram depicted in fig. 1 a mixture of InsP's generated from partially hydrolysed phytic acid was mixed with defined amounts of Fru(1,6)P_2, PPi, and 2,3 bisphosphoglycerate. The InsP's produced from the phytic acid are separated into 18 individual peaks (about half of the expected isomers). Under the strongly acidic elution conditions, inositol mono- and bisphosphates are only poorly separated into individual isomers whereas InsP's containing three and more phosphates exhibit excellent separation. In the lower chromatogram, very low amounts of the mixture of InsP's were applied in order to demonstrate the detection limits of this system. It is mainly the signal-to-noise ratio and not the specific absorbance which is limiting the detection sensitivity. In fig. 2 an analysis of deacylated phospholipids (fig. 2a) and a corresponding standard chromatogram (fig. 2b) is shown. By analysing extracts or mixtures of isomers under two or more different chromatographic conditions we have meanwhile succeeded in separating most $InsP_2$ and $InsP_3$ isomers theoretically separable (Radenberg et al. 1989, Pittet et al. 1989, Nehls-Sahabandu et al. 1989, Schäfer et al. 1990, Reiser et al. 1991, Mayr & Thieleczek 1991, Freund et al. 1992, Mayr et al. 1992, Li et al. 1992, Stephens et al. 1993).

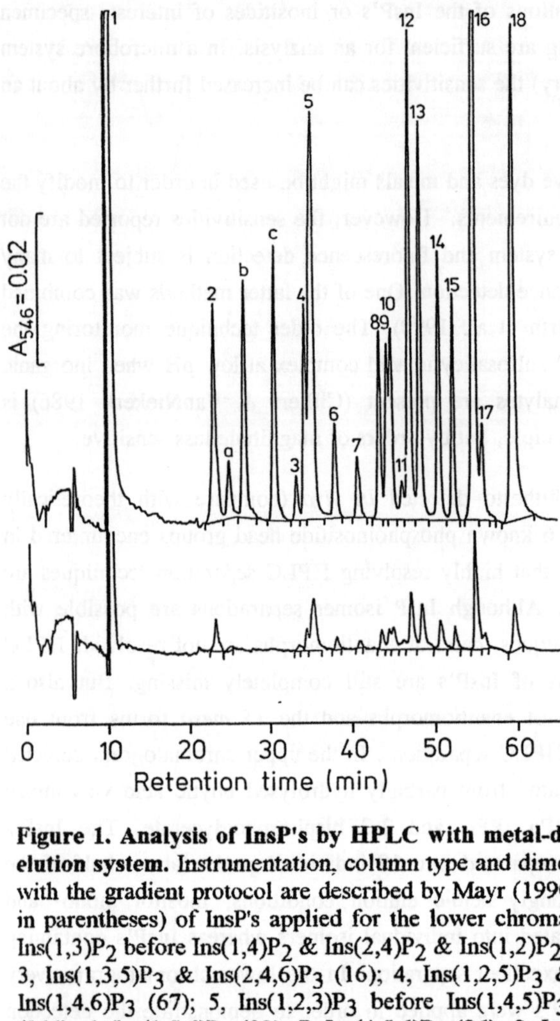

Figure 1. Analysis of InsP's by HPLC with metal-dye detection using a strongly acidic elution system. Instrumentation, column type and dimension and the elution system together with the gradient protocol are described by Mayr (1990). The isomers and quantities (pmols in parentheses) of InsP's applied for the lower chromatogram are: 1, InsP & Pi (2,600); 2, Ins(1,3)P_2 before Ins(1,4)P_2 & Ins(2,4)P_2 & Ins(1,2)P_2 & Ins(1,6)P_2 before Ins(4,5)P_2 (284); 3, Ins(1,3,5)P_3 & Ins(2,4,6)P_3 (16); 4, Ins(1,2,5)P_3 & Ins(1,2,4)P_3 before Ins(1,3,4)P_3 & Ins(1,4,6)P_3 (67); 5, Ins(1,2,3)P_3 before Ins(1,4,5)P_3 before Ins(1,2,6)P_3 & Ins(2,4,5)P_3 (266); 6, Ins(1,5,6)P_3 (33); 7, Ins(4,5,6)P_3 (26); 8, Ins(1,2,3,5)P_4 & Ins(1,2,4,6)P_4 (50); 9, Ins(1,2,3,4)P_4 & Ins(1,3,4,6)P_4 (46); 10, Ins(1,2,4,5)P_4 (52); 11, Ins(1,3,4,5)P_4 (8); 12, Ins(1,2,5,6)P_4 (154); 13, Ins(2,4,5,6)P_4 (127); 14, Ins(1,4,5,6)P_4 & Ins(1,2,3,4,6)P_5 (85); 15, Ins(1,2,3,4,5)P_5 (61); 16, Ins(1,2,4,5,6)P_5 (394); 17, Ins(1,3,4,5,6)P_5 (31); 18, InsP_6 (300). The order of elution of these isomers was assessed by isolated compounds co-chromatographed with the whole mixture of standards (Mayr 1990). The erroneous order of elution of Ins(1,4,6)P_3 given in table 1 of (Mayr 1990) has been corrected. Incompletely separated species are always assigned to one peak number. The precise order of elution of individual isomers in the constantly separated peaks 1 through 18 seems to be subject to slight variations according to sample pH and the presence of other anions. E.g. Ins(1,4,5)P_3 and Ins(2,4,5)P_3 are separated in certain samples, in other they are not. In the upper chromatogram, a tenfold amount of the mixture of InsP's was applied together with : a, Fru(1,6)P_2 (2,000); b, PPi (2,000); c, 2,3-bisphosphoglycerate (2,000).

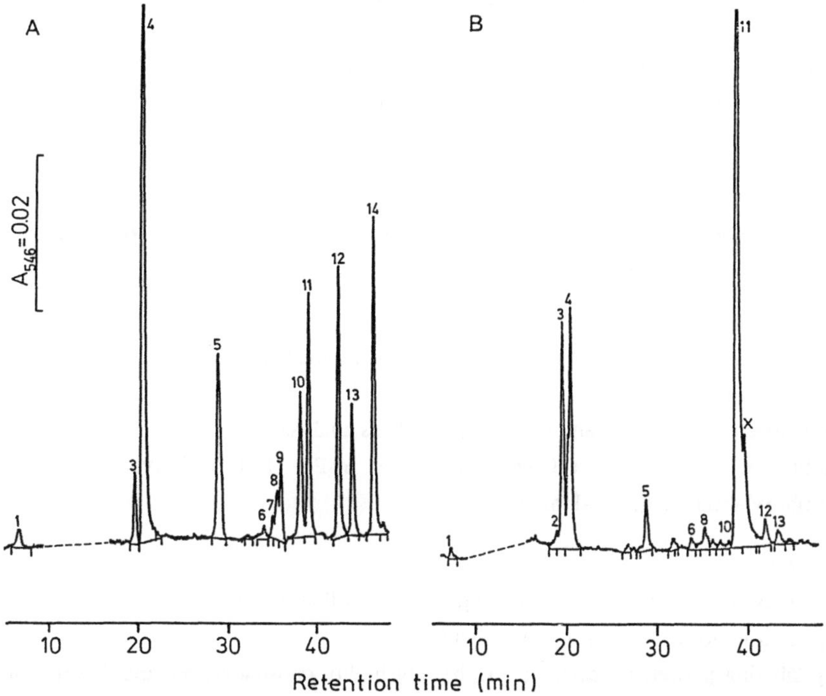

Figure 2. Analysis of GroPInsP's by HPLC with metal-dye detection using a pH of 7.5 for the eluent. Instrumentation, column type and details of the controller protocol are described by Mayr (1990). The eluents and the postcolumn dye reagent are described in the text (above). The following gradient (flow rate 1 ml/min for eluent A/B, 0.5 ml for reagent C) was applied 5 min after injection: linear increase from 4% to 17% B in 10 min; linear increase from 17% B to 35% B in 10 min; linear increase from 35% B to 63% B in 10 min; linear increase from 63% B to 100% B in 12 min; 100% B for 4 min: 0% B for 10 min (for regeneration). The following isomers and quantities (nmols in parentheses) of InsP's and GroPIns(P)'s were applied for the chromatogram A: 1, GroPIns (50); 3, _-GroP (5); 4, Pi (10); 5, GroPIns4P (10); 8, Ins(2,4)P_2 (0.3): 9, Ins(1,4)P_2 (0.4); 10, GroPIns(3,4)P_2 (0.5); 11, GroPIns(4,5)P_2 (0.9); 12, Ins(1,4,5)P_3 (0.5); 13, GroPIns(3,4,5)P_3 (0.3); 14, Ins(1,3,4,5)P_4 (0.5). Further identified compounds in the standard mixture are: 6, Ins(4,5)P_2; 7, Ins(3,4)P_2 & Ins(1,3)P_2. For chromatogram B 55 mg (wet weight) of white matter from rabbit brain cortex have been extracted and deacylated as described in the text (above). Further compounds not contained in the standard mixture were: 2, Ins1P; X, unknown compound of putative structure GroPXP_2; X could be inositol or an unidentified sugar or polyol. The region between retention time 9 and 17 contains a number of "system peaks" corresponding to detected monovalent anions, e.g. F^- and HCO_3^-. These are not depicted. GroPInsP standards were prepared as described by Koppitz et al. (1986). The 3-phosphorylated standards were kindly provided by Lew Cantley and coworkers. As deduced from the large amount of GroPIns applied in the standard mixture and the very small peak obtained, the detection sensitivity for this compound containing no phoshomonoester is only 1-2% of that of GroPIns4P. Unless large amounts are present, this technique is insufficient for a determination of GroPIns.

VALIDATION OF TECHNIQUES OF PTDINSP AND INSP ANALYSIS

In the subsequent section advantages and disadvantages of the individual "standard" techniques used to analyse InsP's and inositides are compared.

A. Radioactive Prelabeling Techniques.

1. Prelabeling of cultured cells with [^{3}H]Ins and determination of labeled InsP's and PtdIns(P)'s by LC/HPLC/TLC.

Advantages:

- high sensitivity (<2 pmol; analyses from <5 Million cells possible),
- specificity for Ins-containing compounds,
- whole spectrum of InsP's and PtdIns(P)'s and Ins analysable,
- suitable for measuring flux from PtdInsP's to InsP's (PLC activity) and inositide phosphorylations (e.g. PI3-kinase).

Disadvantages:

- high costs due to high amounts of [^{3}H]Ins and scintillation counting,
- long-term radioactive handling of cell cultures,
- long labeling periods necessary (>12 h), but highly phosphorylated InsP's often still not labeled to equilibrium
- HPLC separation for isomer specificity required,
- mass determinations in many cases impossible,
- labeling often insufficient in primary cell cultures and tissue specimens.

This technique is the most frequently employed radioactive prelabeling technique. It is normally employed in permanently cultured cells which can be prelabeled for up to 48 h.

2. Prelabeling of cells with [^{32}P]Pi and determination of labeled InsP's and PtdIns(P)'s by LC/HPLC/TLC.

Advantages:

- lower costs as with [^{3}H]Ins labeling,
- very high sensitivity (<<1 pmol; analyses from <1 Million cells),
- whole spectrum of PtdIns(P)'s and most InsP's analysable,
- brief labeling periods (pulse labeling) possible (1-2h),
- possible in primary and permanent cell cultures,
- easy detection of labeled compounds by Serene radiation,

- rapid labeling of PtdIns(P)'s,
- suitable for measurements of flux from PtdInsP's to InsP's and of PtdIns(P) and InsP phosphorylations.
- easier than 1. fully automatized when using on-line Serene radiation monitoring

Disadvantages:

- no specificity for Ins containing compounds (sugar phosphates and nucleotides also labeled),
- [^{32}P]Pi implicit problems (long range radiation, $t_{1/2}$ 14 days),
- sample pretreatment for removal of nucleotides necessary,
- HPLC separation for isomer specificity required.
- mass determinations always impossible due to heterogeneous precursor (ATP) pools.

This technique is a frequently employed alternative to technique 1. when cells cannot be held for long prelabeling periods or when only special enzyme activities, e.g. PLC or PI-kinase activities are of interest.

B. Mass Determination Techniques.

1. Radioligand displacement assays of Ins(1,4,5)P_3 and Ins(1,3,4,5)P_4.

Advantages:

- commercial availability of Ins(1,4,5)P_3 assay,
- high specificity for Ins(1,4,5)P_3 and Ins(1,3,4,5)P_4, respectively,
- high sensitivity (about 10 pmol),
- determinations possible from non-radiolabeled cells and tissues.

Disadvantages:

- only one single isomer determined,
- high costs of commercial Ins(1,4,5)P_3 assay,
- interferences possible by phosphocompounds (see below),
- no simple working protocol,
- no commercial availability of Ins(1,3,4,5)P_4 assay.

The commercial Ins(1,4,5)P_3 assay is increasingly applied for InsP_3 determinations and PLC assays in cell type or tissue. It poses specificity problems in presence of high concentrations of ATP (>10 mM), Pi (>50 mM), InsP_5, InsP_6 (> 0.5 mM) and polyphosphates (>1 mM P).

2. Enzymatic determination of Ins(1,4,5)P_3 and GroPIns(4,5)P_2 with InsP_3 3-kinase.

Advantages:

- very high specificity for these isomers,
- high sensitivity (< 5 pmol),
- determinations possible from non-radiolabeled cells and tissues.

Disadvantages:

- no commercial availability of enzyme,
- need of [gamma^{32}P]ATP as co-substrate,
- complex working protocol, LC/HPLC for product identification necessary,
- losses during deacylation of PtdIns(4,5)P_2 to form GroPIns(4,5)P_2.

Enzymic assays for these compounds by use of InsP_3 3-kinase are extremely specific due to the selectivity of conversion of bound ligands. Up to now not commercially available. Comparably selective other inositol kinases might be useful for enzymatic assays as well.

3. HPLC/TLC separation of InsP's/PtdIns(P)'s with off-line determination of masses by enzymatic Ins or total P assays after hydrolysis.

Advantages:

- whole spectrum of PtdIns(P)'s and most InsP's analysable,
- analyses possible from non-radiolabeled cells and tissues,
- many synthetic derivatives amenable for assay.
- determinations possible from non-radiolabeled cells and tissues.

Disadvantages:

- limited specificity of enzymatic Ins or P assays,
- complex working protocols for off-line mass determinations,
- limited sensitivity (>100 pmol),
- hydrolysis losses and artifacts,
- no automatization possible.

4. Anion exchange HPLC separation of InsP's/GroPInsP's with on-line determination of masses by metal-dye detection (HPLC-MDD)

Advantages:

- whole spectrum of deacylated PtdInsP's and of InsP's analysable,
- low running costs,

- can be fully automatized and thus made highly reproducible,
- high sensitivity of metal-dye detection (<10 pmol),
- determinations possible from non-radiolabeled cells and tissues,
- other derivatives with phosphomonoester groups amenable for assay,
- rapid HPLC-MDD assays of novel enzymes and binding proteins specific for any InsP realized.

Disadvantages:
- no specificity for Ins containing compounds (sugar phosphates and nucleoside phosphates also detected with InsP's),
- sample pretreatment for removal of nucleotides (sugar phosphates and creatine-P) necessary,
- deacylation losses of PtdIns(P)'s,
- special HPLC equipment with extra HPLC pump necessary.

Alternative HPLC techniques are limited in isomer selectivity (anion exchange HPLC with pulsed amperometric detection, isocratic ion-pairing HPLC with absorbance detection) or sensitivity (anion-exchange HPLC with suppressed conductivity detection).

Check-list of adequate techniques for PLC and PI3-kinase assays.

The following lists facilitate a selection of the adequate technique in pharmacological or endocrinological studies of PLC and PI3-kinase activities in intact cells.

Study of PLC activation

1. *in permanent cell cultures: use*

- radioligand binding assay of Ins(1,4,5)P_3 if sufficient cells available and [^{3}H]Ins handling or HPLC impossible,
- [^{3}H]Ins prelabeling and HPLC/LC/TLC of InsP's (and PtdIns(P)'s) if only low amounts of cells available or more detailed metabolic and kinetic information required,
- HPLC-MDD of unlabeled cells if sufficient cells available and more detailed metabolic and kinetic information required;

2. *in short-lived primary cells or tissue slices: use*

- radioligand binding assay if sufficient cells available and HPLC impossible,

- [^{32}P]Pi prelabeling and HPLC analysis of InsP's (and PtdIns(P)'s) if only low amounts of cells available or more detailed metabolic and kinetic information required,
- HPLC-MDD of unlabeled cells if sufficient cells available and more detailed metabolic and kinetic information required;

3. *in tissue specimens from intact organs or ex vivo: use*

- radioligand binding assay if sufficient material available,
- HPLC-MDD if more detailed kinetic and metabolic information required.

Study of PI3-kinase activation

1. *in long-term cell cultures: use*

- [^{3}H]Ins prelabeling and HPLC of deacylated PtdInsP's if only low amounts of cells available,
- HPLC-MDD of unlabeled cells for masses of deacylated PtdInsP's if sufficient cells available;

2. *in short-lived primary cells or tissue slices: use*

- [^{32}P]Pi prelabeling and HPLC of deacylated PtdInsP's if only low amounts of cells available,
- HPLC-MDD of unlabeled cells for masses of deacylated PtdInsP's if sufficient cells available;

3. *in tissue specimens from intact organs or ex vivo: use*

- HPLC-MDD for masses of deacylated PtdInsP's,

RULES FOR PTDINSP AND INSP DETERMINATION

The appended protocols represent state-of-art working instructions for the most frequently used isomer specific InsP and PtdIns(P) determination procedures. For all of them certain rules are identical.

It is essential that upon acidic extraction of InsP's the cellular precipitate containing membranes is removed from the extract as rapidly as possible and all steps are performed on ice. If not, an acid hydrolysis of the inositides contained in the precipitates may give rise to artificially increased InsP levels. For the same reason, the final TCA concentration should not be higher than 8% (w/v). Since in some cells and tissues the concentration of inositides is several hundred μM, 1% artificial hydrolysis may already give rise to InsP's in their maximal physiological concentration. Likewise, if inositides are extracted from these pellets or from the original cell or tissue material with acidic chloroform/methanol, this procedure has to be completed as rapid as possible in order to prevent losses due to acid hydrolysis of inositides. Re-extractions of pellets both for InsP's and inositides are not recommended since they only prolong the times of exposure to acid. In order to increase the extraction yields, large extraction volumes (> 10 volumes per volume of cells) should rather be used. When doubt exists whether acid hydrolysis of any labile compounds of interest has rally been avoided, cells or tissues can be extracted alternatively by adding or homogenizing in icecold methanol (65% final concentration) instead of TCA. All InsP's (except those containing more than 5 phosphates) are quantitatively extracted into the methanol/water phase and an acidic chloroform/ methanol extraction of pellets can be performed as normal. In the acidic extracts of InsP's the acid has to be removed again in order to prevent acid catalysed phosphate migration and acid hydrolysis in these compounds. Since PCA is easily removed by precipitation as the potassium salt, this technique was frequently employed in the past. However, co-precipitation losses of InsP's, in particular when the final pH is alkaline and Me^{2+} is present, are occurring in a relatively uncontroled manner. Also when TCA is extracted with freon/n-octylamine, losses of InsP's by complexation with octylamine are occurring. Therefore, the preferred technique in any instance is TCA extraction with subsequent removal of the acid by ether extraction. An addition of EDTA to prevent precipitation losses of highly phosphorylated InsP's as Me^{2+} complexes after TCA removal and of NaF to do the same and to inhibit phosphatases have turned out to be advantageous. In the case of [^{32}P]phosphate prelabeling and of HPLC-MDD a charcoal treatment of InsP extracts is obligate to remove analytical interferences from nucleotides. The presence of EDTA and NaF previously added as well as of high salt concentrations during charcoal extraction have turned out to avoid adsorption losses of InsP's to the charcoal particles. When the amounts of salt included in individual samples are so large that they give rise to earlier, non-reproducible retention times upon anion exchange HPLC, a solid phase extraction on a small column or cartridge of Q-Sepharose can be performed in order to remove the carrier ions. This procedure gives >98% recoveries of all InsP's when performed with freshly recycled lots of Q-Sepharose as described for the HPLC-MDD technique (but leads to a complete loss of Ins). Reasons for losses during deacylation are incompletely dissolved samples, an incomplete deacylation

(Clarke & Dawson 1981) due to a lower as the nominal reaction temperature, unavoidable side reactions during deacylation, and extraction losses after deacylation.

Due to all these potential losses the recoveries of InsP's and PtdInsP's are often far below 100% depending on the type of experiment and degree of optimization of extraction and sample workup. Only by routinely added radioactive internal standards of InsP's and PtdInsP's to control samples the recovery problems can be adequately met. These additions are done immediately before or after addition of stopping agent since at earlier times dephosphorylation by secreted or ekto-phosphatases might occur. With internal standards present in control samples, the extraction and sample treatment can be optimized and the incomplete recoveries of certain compounds can be adequately corrected for.

PROTOCOLS FOR PTDINSP AND INSP DETERMINATION

1. *Prelabeling of continuously cultured cells with [^{3}H]Ins*

Cell culture:

- Plan all stimulation/control conditions twice or in triplicate.
- Use culture medium with low Ins content (100 μM).
- Adjust cell number/assay (5 to 50 million/assay, 1 dish per assay for adherently growing cells.
- Add 2 to 10 μCi/ml [^{3}H]Ins of high specific radioactivity to medium for >12 h (normally 2-4 days, 1 or 2 medium changes).
- Remove radioactive medium completely and wash at least once with non- radioactive medium or physiological buffer suitable for stimulation.
- Add a defined volume of this medium/buffer for stimulation.
- Let cells adapt for 30 min.

Stimulation and stopping:

- Add a small volume of stimulant in same medium/buffer and stop before and up to 30 min (include 5",10", 30",60" values) after stimulation by adding 2 ml of icecold 8% TCA (with prior removal of medium/buffer if possible. If impossible, add 1 medium/buffer volume of 15% TCA).
- Do not forget some control incubations without stimulant for up to 30 min.
- For an internal standardization and a validation of recoveries, defined amounts of [^{3}H]InsP's and/or [^{3}H]PtdIns(P)'s (also analysed in identical amounts as pure

untreated mixtures) are added to dishes not prelabeled with [^{3}H]Ins but otherwise treated likewise.

- Put tubes on ice. In case of dishes rapidly (within 5 min) scratch and pipette out the whole extract and precipitated cells into a cooled Falcon or glass tube.
- Centrifuge immediately for 5-10 min at 5,000xg at 4° C.
- Completely remove the supernatant (containing InsP's) and put it on ice.

Alternative non-acidic stopping:

- Do not remove medium, add 2 medium volumes of icecold methanol and continue as described above.

Inositide extraction:

- Immediately after centrifugation (< 20 min after stopping) resuspend pellets in 4 ml of water:methanol = 1:1 and rapidly perform the acidic chloroform/methanol extractions in order to extract PtdIns(P)'s into the organic phase without their hydrolysis.
- To 4 ml water:methanol add 3 ml/of chloroform and 50 μl 4 M HCl, shake vigorously for 2 min, centrifuge for phase separation, aspirate upper water:methanol phase without removing interphase (add this phase to the methanol:buffer supernatant obtained with non-acidic extractions).
- Add 3 ml of chloroform:methanol:HCl(0.6N) = 3:48:47 to the lower (chloroform:methanol) phase, re-extract by vigorous shaking for 1 min, phase-separate by centrifugation and remove upper phase.
- Repeat the same procedure with 2 ml of chloroform:methanol:HCl and transfer chloroform:methanol phase into a glass tube if deacylation is planned.
- Dry the organic phases completely down with a stream of nitrogen or in a vacuum centrifuge and prepare material for PtdIns(P) analysis (by TLC/ HPTLC on oxalate impregnated silica or by HPLC after deacylation).

InsP extract treatment:

- Add EDTA to supernatant (adjusted to pH 7 with NaOH) over the Mg^{2+}- plus Ca^{2+} content of the sample ($\geq$ 2 μmol, check medium/buffer Me^{2+} content).
- Add 2 μmol NaF (pH 7).
- 4x extract with ether (two volumes) to remove TCA and freeze or, if volumes are large, freeze-dry water phases until analysed.

InsP Analysis:

- Analyse for [^{3}H]InsP's by HPLC on a >25 cm x 0.4 cm column of SAX resin (silica or polystyrene/DVB copolymer beads, 10 μm) eluted with a gradient of ammonium formate (0.1 to 2.5 M, about 100 ml in 100 min), adjusted with phosphoric acid to pH 3.7.
- Collect 0.5 or 0.75 ml fractions into 5 ml scintillation vials, add 3 ml of scintillator suitable for high salt plus water content and count for >1 min.

Inositide analysis by TLC:

- Dissolve inositide extracts in small volumes of chloroform (or chloro-form:methanol = 2:1) and spot defined amounts onto (pre-developed) oxalate impregnated TLC or HPTLC silica plates (spray with 1% p.a. K-oxalate in methanol:water = 2:3, heat for 15 min at 110 ^{o}C) which are developed with solvent chloroform:acetone:methanol: acetic acid:water = 40:15:13:12:8.
- After drying, autoradiograph plates with high sensitivity X-ray film.
- For quantitation, scratch out radioactive spots, transfer into a scintillation vial, extract with 0.5 ml chloroform:methanol = 2:1, supplement with scintillation cocktail for organic solvents and count.

Inositide analysis by HPLC after deacylation (for PtdIns3P's):

- Dissolve dried inositides in 0.1 ml n-butanol by sonifying for 10 s.
- Add 0.3 ml monomethylamine reagent (33 % monomethylamine in ethanol: water = 7:3).
- Slowly warm up solution under swirling to 53 ^{o}C and keep at this temperature for 40 min.
- Cool down on ice and add 0.15 ml ice-cold n-propanol.
- Dry by vacuum centrifugation.
- Suspend in 0.2 ml water and vigorously vortex.
- Add 0.25 ml n-butanol:petroleum:ethyl formate = 20:4:1, shake vigorously for 1 min, centrifuge for phase separation and remove upper organic phase containing amidated fatty acids.
- Add 0.25 ml of same organic mixture to lower (water) phase and re-extract as before.
- Analyse water phase containing GroPIns(P)'s by HPLC under similar conditions as employed for InsP's (modified eluent gradient up to 1.5 M ammonium formate).
- Collect fractions and count as described for InsP's.

2. *After prelabeling of cells with [^{32}P]Pi*

Cell culture:

- Use final culture medium with low phosphate content ($\leq$100 μM).
- Plan all stimulation/control conditions twice or in triplicate.
- Adjust cell number/assay (1 to 20 Million/assay, 1 dish per assay for adherently growing cells).
- Add 5 to 20 μCi/ml [^{32}P]Pi ($\leq$1 μmol/ml) to medium for 1-4 h.
- Remove radioactive medium completely and wash at least once with non-radioactive medium or physiological buffer suitable for stimulation.
- Add a defined volume of this medium/buffer for stimulation
- Let cells adapt for 30 min.

Stimulation and stopping:

- Proceed as with [^{3}H]Ins labeled cells. Be very fast.
- For an internal standardization and a validation of recoveries, defined amounts of [^{3}H]InsP's and/or [^{3}H]PtdIns(P)'s (also analysed as pure untreated mixtures) are added to some dishes/samples.

Inositide extraction and extract treatment:

- Proceed as with [^{3}H]Ins labeled cells.

InsP analysis:

- Resuspend freeze-dried samples in $\leq$0.5 ml water and treat twice with 20 μl of a 20 % (w/v) suspension of acid-treated charcoal to remove nucleotides.
- Analyse for [^{32}P]InsP's by HPLC as with [^{3}H]Ins labeled cells.
- Do either on-line Serene radiation monitoring or collect 0.5 or 0.75 ml fractions into 5 ml scintillation vials. Count for Cerenkov radiation for 1 min. Or count both for [^{3}H] and [^{32}P] scintillation after addition of scintillant (as described for [^{3}H]Ins labeling) when internal standard was added.

Inositide analysis by TLC:

- Proceed as with [^{3}H]Ins labeling. Autoradiograph for shorter periods.

Inositide analysis by HPLC after deacylation (for PtdIns3P's):

- Proceed as with [^{3}H]Ins labeling and count as described for [^{32}P]InsP's.

3. *Direct mass determination of InsP's and PtdIns(P)'s by HPLC-MDD*

Cell culture:

- Chose cell number/assay (50 to 200 million/assay, 1 dish/flask per assay for adherently growing cells).
- Plan all stimulation/control conditions twice or in triplicate.
- Replace culture medium by a minimal buffer with $\leq$ 1 mM Pi, $\leq$ 1 mM sulphate.
- Wash once with this buffer and add a defined volume before stimulation.
- Let cells adapt for 30-60 min.

Stimulation and stopping:

- As with [^{3}H]Ins labeled cells. Extraction should be $\geq$2 ml/50 million cells. Keep final TCA concentration at 6-8%. Proceed very rapid.

Tissue extracts:

- Stop and extract 50 to 500 mg wet weight with 2 ml/100 mg of 8% TCA.
- Proceed as with [^{3}H]Ins labeled cells.

Internal standardization and validation of recoveries:

- Defined amounts of [^{3}H]InsP's and/or [^{3}H]PtdIns(P)'s (also analysed in identical amounts as pure untreated mixtures) are added to some dishes/ samples immediately before extraction.

Inositide extraction:

- Immediately after removal of supernatant resuspend pellets in 4 ml/100 mg wet weight of water:methanol = 1:1 and rapidly perform the acidic chloroform/ methanol extractions with appropriately increased volume in order to extract PtdIns(P)'s into the organic phase.
- Proceed rapidly as with [^{3}H]Ins labeled cells

InsP extract treatment:

- Add EDTA to supernatant (adjusted to pH 7 with NaOH) over the Mg^{2+}- plus Ca^{2+} content of the sample ($\geq$ 4 μmol, check tissue/cell plus buffer Me^{2+} content).
- Add 4 μmol of NaF (pH 7).
- 4x extract with ether (1-2 volumes) to remove TCA and freeze. If water volume is >1 ml, freeze-dry.

- Resuspend samples in 0.5-1 ml water and treat twice (test in pilot experiments) with 20-50 μl/100 mg wet weight of a 20 % (w/v) suspension of acid-treated charcoal to remove nucleotides.
- Solid phase-extract samples if containing much salt (check buffer volume for salt) as follows: dilute with water to 50 ml, apply to a disposable 1 ml column of Q-Sepharose (acetate), wash with 2 x 3 ml 2 mM HCl to remove carbonic acids, finally elute InsP's with 2 x 3 ml 1.5 M ammonium acetate (pH 5) and lyophilise until all ammonium acetate is removed.

InsP analysis:

- Analyse for InsP's by HPLC with postcolumn metal-dye detection (HPLC-MDD) on a 25 cm x 0.4 cm column of Mono Q (10 μm) eluted with a gradient of HCl (0.2 mM to 0.5 M, about 60 ml in 60 min) containing 15 μM YCl_3 or a gradient of KCl, adjusted to pH 8.5 with 10 mM triethanolamine/HCl (0 to 0.4 M, 60 ml in 60 min) containing no YCl_3. In the former case, the postcolumn derivatization reagent contains 1.6 M triethanolamine pH 9.2 and 300 μM PAR, in the latter case 2 mM ammonium acetate, 300 μM PAR, and 45 μM YCl_3.
- In case of simultaneous labeling or added internal standards, either do on-line radioactivity detection by Cerenkov radiation or collect 0.5 or 0.75 ml fractions into 5 ml scintillation vials, and count for Cerenkov radiation or [^{3}H]Ins radioactivity (see above).

Inositide analysis by TLC:

- Proceed as with [^{3}H]Ins labeling, spray with Hanes-Isherwood reagent for phosphate detection or with sulphuric acid for ashing, or scratch out unstained radioactively labeled or iodine stained spots for total P determination.

Inositide analysis by HPLC after deacylation (for PtdIns3P's):

- Proceed as with [^{3}H]Ins labeling but duplicate volumes for deacylation procedure and extraction of GroPIns(P)'s.
- Analyse GroPInsP's by HPLC-MDD (same column as above) employing a gradient of KCl adjusted to pH 7.5 with 10 mM triethanolamine/HCl (0 to 0.4 M, 60 ml in 60 min). The postcolumn derivatization reagent is as for the pH 8.5 system described above.
- Collect fractions and count as described for [^{3}H]InsP's or for [^{32}P]InsP's in case of radioactive labeling or added internal standards.

Acknowledgement

This work was supported by the Deutsche Forschungsgemeinschaft (grants Ma 989 and SFB 354) and the Bundesministerium für Forschung und Technologie. Thanks are expressed to all collaborators who by contributing many samples for InsP or PtdInsP analysis helped to optimize many of the procedures described here.

REFERENCES

Alexandre J, Lassalles JP, Kado RT (1990) Opening of Ca^{2+} channels in isolated red beet vacuole membrane by inositol 1,4,5-trisphosphate. Nature 343: 567-570

Beck KA, Keen JH (1991) Self association of the plasma membrane-associated assembly protein AP-2. J Biol Chem 266: 4437-4441.

Berridge MJ, Irvine RF (1989) Inositol phosphates and cell signalling. Nature 341: 197-205

Carpenter CL, Cantley LC (1990) Phosphoinositide kinases. Biochemistry 29: 11147-11156

Cilliers JJL, VanNiekerk PJ (1986) LC determination of phytic acid in food by postcolumn colorimetric detection. J Agric Food Chem 34: 680-683

Clark NG, Dawson RMC (1981) Alkaline O-N-transacylation. A new method for the quantitative deacylation of phospholipids. Biochem J 165: 301-306

Coté GG, Quarmby LM, Satter RL, Morse MJ, Crain RC (1990) Extraction, separation, and characterization of metabolites of the inositol phospholipid cycle. Inositol metabolism in plants. Wiley-Liss New York pp 113-137

Dean NM, Beaven MA (1989) Methods for the analysis of inositol phoshates. Anal Biochem 183: 199-209.

Donié F, Reiser G (1989) A novel, specific binding protein assay for quantitation of intracellular inositol 1,3,4,5-tetrakisphosphate ($InsP_4$) using a high-affinity $InsP_4$ receptor from cerebellum. FEBS Lett 254: 155-158

Downes CP, MacPhee CH (1990) *myo*-Inositol metabolites as cellular signals. Eur J Biochem 193: 1-18

Dreef CE, Mayr GW, Roelen HCPF, van der Marel GA, van Boom JH (1991) An expeditious synthesis of *myo*-inositol phosphorothioates. Bioorg Med Chem Lett 1: 239-242

Freund WD, Mayr GW, Tietz C, Schultz JE (1992) Metabolism of inositol phosphates in the protozoan *Paramecium*. Characterization of a novel inositol hexakisphosphate dephosphorylating enzyme. Eur J Biochem 207: 359-367

Gaertner A, Mayr GW, Wegner A (1991) Binding of sugar phosphates, inositol phosphates and phosphorylated amino acids to actin. Eur J Biochem 198: 67-91

Henne V, Mayr GW, Grabowsky B, Koppitz B, Söling HD (1988) Semisynthetic derivatives of inositol(1,4,5)trisphosphate substituted at the 1-phosphate group. Effects on calcium release from permeabilized guinea pig parotid acinar cells and comparison with binding to aldolase A. Eur J Biochem 174: 95-101

Irth H, Lamoree M, DeJong GJ, Brinkman, UAT, Frei RW, Kornfeldt RA, Persson L (1990) Determination of D-*myo*-1,2,6-inositol trisphosphate by ion-pair reverse phase liquid chromatography with postcolumn ligand exchange and fluorescence detection. J Chromatogr 499: 617-625

Ivorra I, Gigg R, Irvine RF, Parker I (1991) Inositol 1,3,4,6-tetrakisphosphate mobilizes calcium in *Xenopus* Oocytes with high potency. Biochem J 273: 317-321

Koppitz B, Vogel F, Mayr GW (1986) Mammalian aldolases are isomer-selective high-affinity inositol polyphosphate binders. Eur J Biochem 161: 421-433

Li G, Pralong WF, Pittet D, Mayr GW, Schlegel W, Wollheim CB (1992) Inositol tetrakisphosphate isomers and elevation of cytosolic Ca2+ in vasopressin-stimulated insulin-secreting RINm5F cells. J Biol Chem 267: 4349-4356

Martin TF (1991) Receptor regulation of phosphoinositidase C. Pharmac Ther 49: 329-345

Mayr GW (1988a) Inositol phosphates. Structural components, regulators, and signal transducers of the cell - a review. Topics in Biochemistry vol 7, Boehringer Mannheim

Mayr GW (1988b) A novel metal-dye detection permits picomolar range h.p.l.c. analysis of inositol polyphosphates from non-radioactively labelled cell or tissue specimens. Biochem J 254: 585-591

Mayr GW (1989) Inositol 1,4-bisphosphate is an allosteric activator of muscle-type 6-phosphofructo-1-kinase. Biochem J 259: 463-470

Mayr GW (1990) Mass determination of inositol phosphates by high-performance liquid chromatography with postcolumn complexometry (metal-dye detection). Methods in inositide research (ed: Irvine RF) Raven Press London pp 83-108

Mayr GW, Koppitz B (1986) Inositol 1,4,5-P_3 and inositol 1,4-P_2 are effectors of phosphofructokinase, aldolase, and phosphogluconate dehydrogenase. Biol Chem Hoppe-Seyler 367: 362

Mayr GW, Radenberg T, Thiel U, Vogel G, Stephens LR (1992) Phosphoinositol diphosphates. non-enzymic formation *in vitro* and occurrence *in vivo* in the slime mold *Dictyostelium*. Carbohyd Res 234: 247-262

Mayr GW, Thieleczek R (1991) Masses of inositol phosphates in resting and tetanically stimulated vertebrate skeletal muscles. Biochem J 280: 631-640

Nehls-Sahabandu M, Falke R, Mayr GW (1989) A rapid non-radioactive HPLC-with-column-switching assay for identification of proteins involved in the inositol phosphate metabolism. Biol Chem Hoppe-Seyler 370: 938

Palmer S, Wakelam MJO (1989) Mass measurements of inositol phosphates. Biochim Biophys Acta 1014: 239-246

Pittet D, Lew DP, Mayr GW, Monod A, Schlegel W (1989) Chemoattractant receptor promotion of Ca2+-influx across the plasma membrane of HL60 cells. A role for cytosolic free calcium elevations and Ins(1,3,4,5)P_4 production. J Biol Chem 264: 7251-7261

Pittet D, Schlegel W, Lew DP, Monod A, Mayr GW (1989) Mass changes of inositol tetrakis- and pentakisphosphate isomers induced by chemotactic peptide stimulation in HL-60 cells. J Biol Chem 264: 18489-18493

Putney J W (1991) Receptor-regulated calcium entry. Pharmc Ther 48: 427-434

Raboy V (1990) Biochemistry and genetics of phytic acid synthesis. Inositol metabolism in plants. Wiley-Liss New York pp 55-76

Radenberg T, Scholz P, Mayr GW (1989) The quantitative spectrum of inositol phosphate metabolites in avian erythrocytes, analysed by proton n.m.r. and h.p.l.c. with direct isomer detection. Biochem J 264: 323-333

Reiser G, Schäfer R, Donie F, Hülser E, Nehls-Sahabandu M, Mayr GW (1991) A high affinity inositol 1,3,4,5-tetrakisphosphate receptor from brain is specifically labelled by a newly synthesized photoaffinity analogue, N-(4-azidosalicyl)aminoethanol(1)-1-phospho-D-myo-inositol 3,4,5-trisphosphate. Biochem J 280: 533-539

Sawada M, Ichinose M, Maeno T (1989) Intracellularly injected inositol hexakisphosphate induces a biphasic current in identified neurons of *Aplysia*. Neurosci Lett 106: 328-333.

Schäfer R, Nehls-Sahabandu M, Grabowsky B, Dehlinger-Kremer M, Schulz I, Mayr GW (1990) Synthesis and application of photoaffinity analogues of inositol 1,4,5-

trisphosphate selectively substituted at the 1-phosphate group. Biochem J 272: 817-825

Scholz P, Bergmann G, Mayr GW (1990) Nuclear magnetic resonance spectroscopy of myo-inositol phosphates. Methods in Inositide Research (ed: Irvine RF) Raven Press London pp 65-82

Shamsuddin AM, Ullah A, Chakravarthy AK (1989) Inositol and inositol hexakisphosphate suppress cell proliferation and tumor formation in CD-1 mice. Carcinogenesis 10: 1461-1463

Shears SB (1991) Regulation of the metabolism of 1,2-diacylglycerols and inositol phosphates that respond to receptor activation. Pharmac Ther 49: 79-104

Skolnik EY, Margolis B, Mohammadi M, Lowenstein E, Fischer R, Drepps A, Ullrich A, Schlessinger J (1991) Cloning of PI3 kinase- associated p85 utilizing a novel method for expression/cloning of target proteins for receptor tyrosine kinases. Cell 65: 83-90

Stephens L, Radenberg T, Thiel U, Vogel G, Khoo KH, Dell A, Jackson TR, Hawkins PT, Mayr GW (1993) The detection, purification, structural characterzation, and metabolism of diphosphoinositol pentakisphosphate(s) and bisdiphosphoinositol tetrakisphosphate(s). J Biol Chem 268: in press

Stephens LR, Hughes KT, Irvine RF (1991) Pathway of phosphatidylinositol(3,4,5)-trisphosphate synthesis in activated neutrophils. Nature 351: 33-39

Stephens LR, Irvine RF (1990) Stepwise phosphorylation of *myo*-inositol leading to *myo*inositol hexakisphosphate in *Dictyostelium*. Nature 346: 580-583

Tarver AP, King WG, Rittenhouse SE (1987) Inositol 1,4,5-trisphosphate and inositol 1,2-cyclic 4,5-trisphosphate are minor components of total mass of inositol trisphosphate in thrombin-stimulated platelets. J Biol Chem 262: 17268-17271

Theibert AB, Estevez VA, Ferris CD, Danoff SK, Barrow RK, Prestwich GD, Snyder SH (1991) Inositol 1,3,4,5-tetrakisphosphate and inositol hexakisphosphate receptor proteins: isolation and characterization from brain. Proc Natl Acad Sci USA 88: 3165-3169

Thieleczek R, Mayr GW, Brandt NR (1989) Inositol polyphosphate-mediated repartitioning of aldolase in skeletal muscle triades and myofibrils. J Biol Chem 264: 7349-7356

Traynor-Kaplan AE, Harris AL, Thompson BL, Taylor P, Sklar LA (1988) An inositol tetrakisphoshate-containing phosholipid in activated neutrophils. Nature 334: 353-356

PHOSPHOINOSITIDES IN RABBIT SKELETAL MUSCLE MEMBRANES

H. Milting, R. Thieleczek, L.M.G. Heilmeyer jr.
Institut für Physiologische Chemie
Abteilung für Biochemie Supramolekularer Systeme
Ruhr-Universität
W-4630 Bochum
Germany

Introduction

The phosphoinositides phosphatidylinositol (PtdIns), phosphatidylinositol 4-phosphate (PtdIns(4)P) and phosphatidylinositol 4,5-bisphosphate (PtdIns(4,5)P_2) establish a membrane localized signaling pathway, which ultimately leads to the generation of the second messengers inositol 1,4,5-trisphosphate (Ins(1,4,5)P_3) and diacylglycerol by phospholipase C (PLC).

Ins(1,4,5)P_3 is involved in intracellular calcium release in many cell systems. In skeletal muscle Ins(1,4,5)P_3 is mobilized as a consequence of tetanic stimulation (Mayr & Thieleczek, 1991). Different authors report calcium release by Ins(1,4,5)P_3 in skeletal muscle and postulate a central role of this compound in the process of excitation contraction coupling (ECC) (Valdivia et al., 1992; Hidalgo & Jaimovich, 1989). However, the strong calcium dependence of PLC activity suggests that Ins(1,4,5)P_3 generation is a consequence rather than a cause of ECC (Varsányi et al., 1989). The role of inositolphosphates in ECC is still a matter of debate (Hannon et al., 1992). On the other side, there are effects of Ins(1,4,5)P_3 and derivatives on glycolytic enzymes located in the triadic gap where ECC takes place (Thieleczek et al., 1989). In this study we analyzed by non-radioactive mass determinations the amounts and locations of PtdIns, PtdIns(4)P and PtdIns(4,5)P_2 in the different membranes of the skeletal muscle triad. These data are compared with available data on the enzymes involved in the formation of the phosphoinositides to give a comprehensive picture of this signaling pathway.

NATO ASI Series, Vol. H 76
Tyrosine Phosphorylation/Dephosphorylation
and Downstream Signalling
Edited by L. M. G. Heilmeyer Jr.

Material and Methods

Membrane preparation

Membranes of rabbit skeletal muscle were isolated as described previously (Caswell et al., 1976) with modifications. The distribution of triads, terminal cisternae (TC) and t-tubular (TT) membranes in a sucrose gradient after centrifugation was examined using specific radioligands (TC: [^{3}H]-ryanodine, TT: [^{3}H]-PN200/110). In case of TT and TC, fractions containing only one specific membrane type were combined and used for phosphoinositide analysis.

Extraction and Deacylation of Phospholipids

Lipids from characterized membrane fractions were isolated according to Schacht (1976). The isolated lipids were deacylated as described by Clarke and Dawson (1981) resulting in the glycero-derivatives of the phosphoinositides and other phospholipids. The overall yield of extraction and deacylation was determined with [^{3}H]-PtdIns(4)P and [^{3}H]-PtdIns(4,5)P_2.

Analysis of Glyceroinositol Derivatives

The glycero-derivatives of PtdIns(4)P and PtdIns(4,5)P_2 were analysed according to the method of Mayr (1988). Identification and quantification of these glycero-derivatives was done by calibration with standard compounds (Boehringer, Mannheim). The measured data were corrected by the yield coefficient obtained for the extraction and deacylation of phosphoinositides. Mass determinations were related to total phospholipid phosphate determined in parallel (Lanzetta et al., 1979).

High Performance Thin Layer Chromatography (HPTLC)

PtdIns of triads, TT and TC membranes were analysed according to the method of Sun and Lin (1990). Quantification was performed by determination of phospholipid phosphate content in characterized spots.

Results and Discussion

Irrespective of a number of studies on the major phospholipid composition in skeletal muscle membranes there are no quantitative data on the amounts of the different phosphoinositides in triads, TT and TC membranes. We have determined the localization and the amounts of phosphoinositides in characterized membranes by HPLC-metal dye detection and HPTLC. The phosphoinositide contents obtained with the different skeletal muscle membranes are listed in table 1.

Table 1: Phosphoinositide content of different rabbit skeletal muscle membranes [mol/(mol phospholipid phosphate) $\pm$ SE, n experiments]

Membrane type	PtdIns	PtdIns(4)P	PtdIns(4,5)P_2
Triads	$(74 \pm 15) \times 10^{-3}$ (n = 2)	$(3.8 \pm 0.2) \times 10^{-5}$ (n = 2)	$(4.8 \pm 3.6) \times 10^{-6}$ (n = 4)
T-Tubules (TT)	$(32 \pm 3.7) \times 10^{-3}$ (n = 2)	$(4.8 \pm 1.0) \times 10^{-3}$ (n = 5)	$(2.4 \pm 0.6) \times 10^{-4}$ (n = 9)
Terminal Cisternae (TC)	$(16 \pm 1.5) \times 10^{-2}$ (n = 3)	$(6.2 \pm 2.0) \times 10^{-3}$ (n = 5)	not detectable (n = 6)

These results indicate that the phosphoinositide composition of TT membranes is significantly different from that of TC membranes. Triadic TT contain all the phosphoinositides which establish the complete signaling system whereas TC are apparently devoid of PtdIns(4,5)P_2.

In TT and TC membranes the detected phosphoinositides agree well with the distribution of the corresponding kinases, PtdIns-kinase and PtdIns4P 5'-kinase (Varsanyi et al., 1989). The restriction of PtdIns(4,5)P_2 to T-tubular membranes was also shown for frog skeletal muscle by means of radioactive methods (Lagos & Vergara, 1990). The lack of PtdIns(4,5)P_2 in TC membranes is probably due to the lack of PtdIns4P 5'-kinase in TC membranes (Varsanyi et al., 1989).

When the mass determinations were carried out with washed muscle homogenate (microsomes) a PtdIns(4,5)P_2 content of 22-56 nmol/g muscle was obtained. This value is much higher than the cellular PtdIns(4,5)P_2 content which can be extrapolated from the data obtained with purified TT membranes (0.29-0.41 nmol/g muscle). The results suggest that PtdIns(4,5)P_2 is not restricted to the membranes of the triadic structure and that a major PtdIns(4,5)P_2 pool is located in the plasma membrane away from the triadic junction.

During tetanic activation of skeletal muscle the total myoplasmic content of Ins(1,4)P_2 and Ins(1,4,5)P_3 increase by about 1.2 nmol/g muscle on a time scale of seconds (Mayr & Thieleczek, 1991). This increase would comprise about 2-5 % of the total PtdIns(4,5)P_2 of skeletal muscle membranes and could not be supplied by the triadic T-tubular PtdIns(4,5)P_2 pool alone.

References

Caswell, A.H., Lau, Y.H. and Brunschwig, J.P. (1976) Ouabain-binding vesicles from skeletal muscle. Arch. Biochem. Biophys. 176, 417-430

Clarke, N.G. and Dawson, M.C. (1981) Alkaline O-N transacylation. A new method for the quantitative deacylation of phospholipids. Biochem. J. 195, 301-306

Hidalgo, C. and Jaimovich, E. (1989), Inositol trisphosphate and excitation contractioncoupling. J. Bioenerg. Biomembr. 21, 267-281

Hannon, J.D., Lee, N.K.M, Yandong. C. and Blinks, J.R. (1992) Inositol trisphosphate (InsP_3) causes contraction in skeletal muscle only under artificial conditions, evidence that Ca^{2+} release can result from depolarization of T-tubules. J. Musc. Res. Cell Mot. 13, 447-456

Lagos, N. and Vergara, J. (1990) Phosphoinositides in frog skeletal muscle: a quantitative analysis. Biochim. Biophys. Acta 1043, 235-244

Lanzetta, P.A., Alvarez, L.J., Reinach, P.S. and Candia, O.A. (1979) An improved assay for nanomolar amounts of inorganic phosphate. Anal. Biochem. 100, 95-97

Mayr, G.W. (1988) A novel metal-dye detection system permits picomolar-range h.p.l.c. analysis of inositol polyphosphates from non-radioactively labelled cell or tissue specimens. Biochem. J. 254, 585-591

Mayr, G.W. and Thieleczek, R. (1991) Masses of inositol phosphates in resting and tetanically stimulated vertebrate skeletal muscles. Biochem. J. 280, 631-640

Schacht, J. (1976) Extraction and purification of polyphosphoinositides. Meth. Enzymol. 72, 626-631

Sun, G.Y., Lin, T.N. and Irvine, R.F. (eds) (1990) Separation of phosphoinositides and other phospholipids by high-performance thin-layer chromatography. Methods in Inositide Research, Raven Press New York, pp 153-158

Thieleczek, R., Mayr, G.W. and Brandt, N.R. (1989) Inositol polyphosphate-mediated repartitioning of aldolase in skeletal muscle. J. Biol. Chem.264, 7349-7356

Valdivia, C., Vaughan, D., Potter, B.V.L. and Coronado, R. (1992) Fast Release of $^{45}Ca^{2+}$ induced by inositol 1,4,5-trisphosphate and Ca^{2+} in the sarcoplasmic reticulum of rabbit skeletal muscle: evidence for two types of Ca^{2+} release channels. Biophys. J. 61, 1184-1193

Varsanyi, M., Messer, M. and Brandt, N.R. (1989) Intracellular localization of inositol-phospholipid-metabolizing enzymes in rabbit fast-twitch skeletal muscle. Eur. J. Biochem. 179, 473-479

RENATURATION OF PHOSPHATIDYLINOSITOL 4- AND PHOSPHATIDYLINOSITOL 4-PHOSPHATE 5'-KINASES FOLLOWING POLYACRYLAMIDE GELELECTROPHORESIS IN PRESENCE OF SDS. STUDIES ON THEIR SUBSTRATE BINDING REQUIREMENTS USING SYNTHETIC SUBSTRATE ANALOGUES.

M. Schmidt, G. Vereb[1], M. Varsányi, D. Klix, C.E. Dreefs[2], and L.M.G. Heilmeyer, Jr.
Institut für Physiologische Chemie
Abteilung für Biochemie Supramolekularer Systeme
Ruhr-Universität Bochum
Universitätsstr. 150
4630 Bochum 1
F.R.G.

INTRODUCTION

Phosphatidylinositol (PtdIns) 4-kinase and phosphatidylinositol 4 phosphate (PtdIns4P) 5'-kinase catalyze the phosphorylation of PtdIns in two subsequent steps leading to the formation of PtdIns(4,5)P_2. From bovine brain, two membrane bound PtdIns 4-kinases were separated by hydroxyapatite chromatography and identified as type 2 and type 3 isoforms (see Vereb et al., p. 167 in this volume). In parallel, also the PtdIns4P 5'-kinase was eluted from the same column with three activity maxima. A new renaturation procedure has been developed which allows the determination of the apparent molecular weight of both lipid kinases following polyacrylamide gelelectrophoresis in presence of SDS (SDS-PAGE). Furthermore, we explored substrate binding requirements using synthetic substrate analogues with the aim to determine structural features on PtdIns and PtdIns4P which are essential for the function as substrate for the respective kinases.

METHODS

Details for preparation of PtdIns 4-kinase, separation of its isoforms and further enrichment of type 3 were described by Vereb et al. in this volume. Determination of PtdIns 4-kinase and PtdIns4P 5'-kinase activities was carried out as described by Varsányi et al. (1989).

Phosphorylation specificity of PtdIns 4- and PtdIns4P 5'-kinases. Following extraction of the phosphorylated lipids into the chloroform phase they were deacylated with monomethylamine according to the method of Clarke and Dawson (1981). The resulting

1 Dept. of Medical Chemistry, University School of Medicine, 4026 Debrecen, Hungary
2 Dept. of Chemistry, University of Leiden, 2333 Leiden, The Netherlands

NATO ASI Series, Vol. H 76
Tyrosine Phosphorylation/Dephosphorylation and Downstream Signalling
Edited by L. M. G. Heilmeyer Jr.

glycerophosphoinositol derivatives were separated on a HPLC Mono Q HR 5/5 column and then detected with the online metal-dye detection method of Mayr (1988).

Renaturation of PtdIns 4- and PtdIns4P 5'-kinase activity following SDS-PAGE. 7.5 % SDS-PAGE was carried out according to Laemmli (1970) with the modification that the SDS concentration has been reduced to 0.05 % in the sample and chamber buffers. Type 3 PtdIns 4-kinase (first activity peak after hydroxyapatite chromatography I) was concentrated by AMICON ultrafiltration and ca. 80 μg protein in the modified SDS-sample buffer was applied onto the polyacrylamide gel. Type 2 PtdIns 4-kinase (second activity peak after hydroxyapatite chromatography I) was diluted (without concentration) with three volumes of modified sample buffer. PtdIns4P 5'-kinases (membrane fraction) were solubilized in sample buffer containing 3 % (w/v) SDS. Electrophoresis was carried out at +4°C for 5 h. The slots with the samples were cut in 2 mm segments which were incubated overnight at +4°C by gently shaking in 100 μl and/or 500 μl renaturation buffer consisting of 10 mM Tris-HCl, 1 mM DTE, 0.1 mM EDTA, 0.1 % CHAPS, 0.2 % Triton X-100, 0.03 % sodiumazid, pH 7.5, for the type 3 and 10 mM Tris-HCl, 1 mM DTE, 0.1 mM EDTA, 0.1 % DOC, 0.2 % Triton X-100, 0.03 % sodiumazid, pH 7.5, for the type 2 PtdIns 4-kinase and PtdIns4P 5'-kinases, respectively. Aliquots were tested for lipid kinase activities.

RESULTS AND DISCUSSION

Table 1 shows that 0.005 % SDS completely abolishes the activity of type 3 PtdIns 4-kinase. However, 90 % and 80 % of the original activity can be recovered, respectively, when 0.1 % CHAPS and 0.2 % Triton X-100 or alternatively 0.1 % desoxycholate (DOC) and 0.2 % Triton X-100 are added. In presence of SDS the activity of the type 2 kinase is much more stable than that of the type 3 kinase. 0.005 % SDS causes an inhibition of only ca. 7 %. Additional presence of 0.1 % DOC and 0.2 % Triton X-100 enhances the activity to 120 %.

To achieve an SDS concentration of 0.005 % in the renaturation assay during electrophoresis we have reduced the detergent concentration in the sample- and chamber-buffers to 0.05 %. Fig. 1 shows the renaturation of the type 3 isoform after SDS-PAGE.

After renaturation one main peak can be detected coinciding with an apparent molecular weight of 200 kDa. Following separation and renaturation of polypeptides of the type 2 PtdIns 4-kinase containing fraction in SDS-PAGE results in one main activity peak coinciding with an apparent molecular weight of 55 kDa (not shown).

The renaturation of a polypeptide representing PtdIns4P 5'-kinase following SDS-PAGE is successful in presence of 0.1 % DOC and 0.2 % Triton X-100, the apparent molecular weight is 80 kDa.

Table 1: Influence of SDS, CHAPS, DOC and Triton X-100 on type 2 and type 3 PtdIns 4-kinase activities.

Type 2 and type 3 kinase were fractions obtained after hydroxyapatite chromatography I exhibiting activities of 12.5 nmol x min^{-1} x ml^{-1} (= 100 %) and 3.8 nmol x min^{-1} x ml^{-1} (= 100 %), respectively. Preincubation of these enzymes was carried out for 30 min at +4°C with SDS and afterwards for 5 min at +24°C with the detergent mixture as indicated. Activities were measured from aliquots and expressed in percent of the initial activity determined without any detergent.

Detergent (%; w/v)				Type 3 PtdIns 4-kinase activity (%)	Type 2 PtdIns 4-kinase activity (%)
SDS	Triton X-100	CHAPS	DOC		
-	-	-	-	100	100
0.005	-	-	-	2	93
0.005	0.2	0.1	-	90	not determined
0.005	0.2	-	0.1	80	120

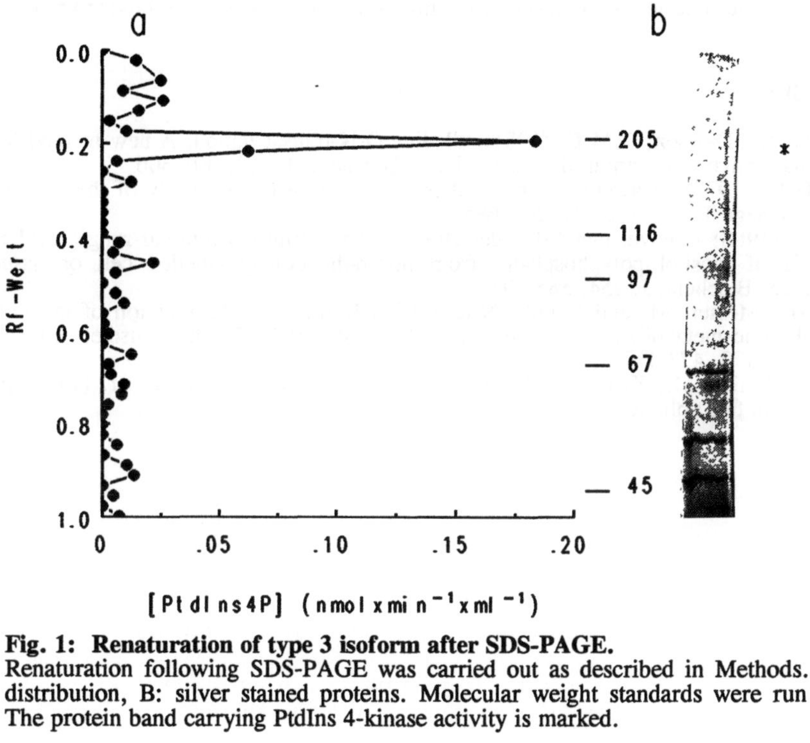

Fig. 1: Renaturation of type 3 isoform after SDS-PAGE.
Renaturation following SDS-PAGE was carried out as described in Methods. A: activity distribution, B: silver stained proteins. Molecular weight standards were run in parallel. The protein band carrying PtdIns 4-kinase activity is marked.

To study substrate binding requirements of both lipid kinases the following synthetic substrate analogues have been used:

[1]: D-myo-inositol (1-((S)-3,4-dipalmitoyloxybutyl-1-phosphonate), [2]: D-myo-inositol 1-(1,2-di-O-palmitoyl-sn-glycer-3-yl methylphosphonate) high running isomer, [3]: the same as [2], but low running isomer, [4] D/L-myo-inositol (1-(S)-3,4 dipalmitoyloxybutyl-1-sulfonate), [5]: D-myo-inositol (1-((S)-3,4-dipalmytoyloxybutyl-1-phosphonate)-4-phosphate. Analoga [3] and [4] were neither substrates nor inhibitors of the PtdIns 4-kinase isoenzymes. The type 3 isoform accepts [1] as substrate. Under identical conditions the rate is 5 % of that measured with PtdIns. [2] is no substrate. Thus, for expression of the enzymic activity the type 3 isoform requires the intact phosphodiester bond for recognition. In contrast, the type 2 isoform accepts [1] as substrate reaching maximally ca. 74 % of the rate obtained with PtdIns. Furthermore, analogon [2] is employed as substrate to ca. 5 % in comparison to PtdIns by the type 2 isoform.

Analogon [5] has been phosphorylated in a partially enriched membrane fraction by the PtdIns4-P kinase but absolutely not by the renaturated kinase following SDS-PAGE. These observations might indicate the existence of multiple PtdIns4P 5'-kinases in bovine brain.

REFERENCES

Clarke, N.G. and Dawson, R.M.C. (1981) Alkaline O-N-transacylation. A new method for the quantitative deacylation of phospholipids. Biochem. J. 195, 301-306

Laemmli, U.K. (1970) Cleavage of structural proteins during the assembly of the head of bacteriophage T4. Nature 227, 680-685

Mayr, G.W. (1988) A novel metal-dye detection system permits picomolar-range h.p.l.c. analysis of inositol polyphosphates from non-radioactively labelled cell or tissue specimen. Biochem. J. 254, 585-591

Varsányi, M., Messer, M. and Brandt, N.R. (1989) Intracellular localization of inositol-phospholipid-metabolizing enzymes in rabbit fast-twitch skeletal muscle. Eur. J. Biochem. 179, 473-479

Vereb, G., Schmidt, M., Klix, D., Varsányi, M. and Heilmeyer, L.M.G., Jr. (1993) pp 167-170 in this volume.

III. Serine/Threonine Protein Kinases

THE CATALYTIC SUBUNIT OF cAMP-DEPENDENT PROTEIN KINASE

Friedrich W. Herberg, Wes Yonemoto, and Susan S. Taylor
Department of Chemistry, 0654
University of California, San Diego
9500 Gilman Drive
La Jolla, CA 92093-0654

Introduction

Protein phosphorylation is a common posttranslational modification associated with regulatory processes in eukaryotic cells. Protein kinases represent a large and diverse family of enzymes that catalyze the transfer from the γ–phosphate of ATP to a target protein or peptide. These enzymes regulate most known metabolic pathways by phosphorylating either a serine, threonine or tyrosine residue (Krebs & Beavo, 1979). The first protein kinase to be discovered was phosphorylase kinase (Krebs et al., 1959), but the best understood biochemically is the second to be described, cAMP-dependent protein kinase (cAPK) (Walsh et al., 1968). cAPK is one of the simplest and smallest in the large family of protein kinases, which overall show tremendous diversity in terms of size, subunit structure, subcellular localization, and activation mechanism. Despite these differences, however, all eukaryotic protein kinases, including the catalytic subunit of the cAPK, share a conserved catalytic core (Hanks et al., 1988). For this reason, the C-subunit can serve as a framework for the entire protein kinase family (Taylor et al., 1992).

The activation mechanism of cAPK is unique amongst the protein kinases and accounts, in large part, for its simplicity. cAPK is composed of both regulatory (R) and catalytic (C) subunits. In the absence of cAMP the subunits associate with a high affinity to form an inactive tetrameric (R_2C_2) holoenzyme. cAMP, serving as a second messenger *in vivo*, binds at nanomolar concentrations to the R-subunit promoting dissociation of the complex into a dimeric R-subunit and two monomeric, active C-subunits. This mechanism of activation contributes to the simplicity of cAPK, because, unlike other kinases, the regulatory component can be easily removed and both

NATO ASI Series, Vol. H 76
Tyrosine Phosphorylation/Dephosphorylation
and Downstream Signalling
Edited by L. M. G. Heilmeyer Jr.

subunits can be studied independently (Beebe & Corbin, 1986; Taylor et al., 1992).

Two classes of physiological inhibitors are known to interact with the C-subunit. The R-subunits constitute one class. Several forms of the R-subunit exist and are classified as type I and type II (Beebe & Corbin, 1986; Taylor et al., 1990). The other class of inhibitors are the heat stable protein kinase inhibitors (PKI's) (Thomas et al., 1991; Van Patten et al., 1991; Walsh et al., 1990; Walsh et al., 1971). The PKI's are relatively small proteins that bind with high affinity to the C-subunit, typically in the low nanomolar range. The mechanism of inhibition for both classes of inhibitors is the same. Each contains a substrate-like sequence, Arg-Arg-X-Ser/Thr-hydrophobic residue, and this segment occupies the substrate recognition site in the complex. The R-subunit and PKI's are thus competitive inhibitors of protein substrates. The primary difference between the two inhibitors is that the inhibition of the C-subunit is reversible in the case of the R-subunit by the binding of cAMP, whereas no physiological mechanism for releasing the C-subunit from PKI is known so far.

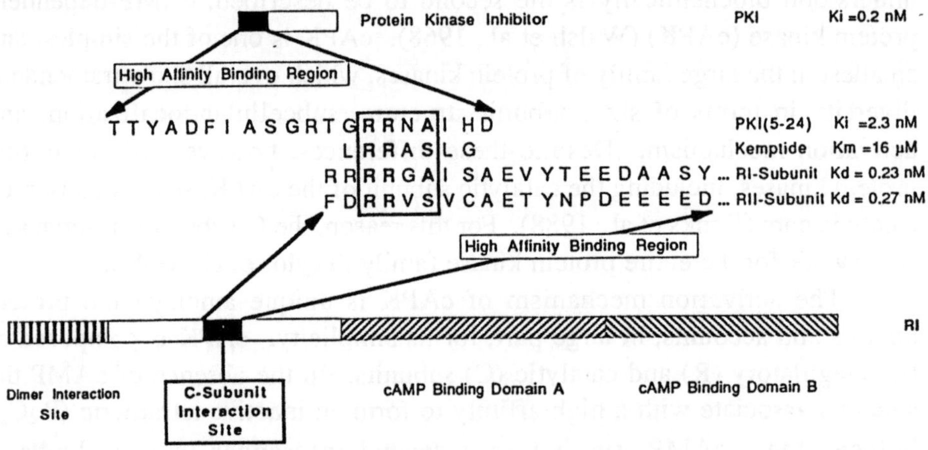

Figure 1: Sequence comparison of physiological inhibitors of the C-subunit. The consensus recognition sequence is boxed displaying the common motif Arg-Arg-X-Ser/Thr as a substrate or Arg-Arg-X-Ala as an inhibitor. Residues N-terminal from the consensus sequence determine high affinity binding for the Heat Stable Protein Kinase Inhibitor (PKI) where the important residues for the interaction with the R-subunit are C-terminal from the sequence.

The common consensus site shared by both the R-subunits and PKI is shown in Figure 1. This consensus site is sufficient to convey low-affinity binding, however, in order to achieve high affinity binding, additional interactions are required. Surprisingly, the interaction sites for high affinity binding seem to differ for the R-subunits and PKI. In the case of PKI, it is the region that lies N-terminal to the consensus site that conveys high affinity binding (Walsh et al., 1990). In the case of the R-subunit, recent evidence suggests that the region C-terminal to the consensus site is important. This is based on proteolytic fragments (Weldon et al., 1983) and on the analysis of mutants in both the yeast and mammalian C-subunits (Gibbs & Zoller, 1991; Orellana & McKnight, 1992). The recently solved crystal structure of a complex of a peptide derived from PKI, PKI(5-24), and the C-subunit in the presence (ternary complex) and absence (binary complex) of MgATP has given us a first insight into understanding the molecular basis for the binding of PKI (Knighton et al., 1991b; Zheng et al., Submitted 1992).

Methodology and Strategy

A variety of techniques were used to investigate structure/function relationships in the C-subunit (Bramson et al., 1982; Zoller et al., 1981). Protein chemistry, both affinity labeling and group-specific labeling, were used to identify specific residues important for substrate binding and catalysis (Buechler & Taylor, 1989; Buechler & Taylor, 1990; Buechler et al., 1989). This is combined with kinetic studies (Adams & Taylor, 1992; Cook, 1982) and analog mapping (Flockhart et al., 1980). The development of an expression system in *E. coli* has enabled the use of site-directed mutagenesis as well (Slice & Taylor, 1989). Finally, we have the large family of protein kinases where sequences in the catalytic core can be compared. Those residues that are conserved throughout the family also allow us to identify essential residues that are conserved throughout evolution.

The recently solved crystal structure allows for the localization of essential residues and, in addition, provides an overall topology for the folding of the polypeptide chain (Knighton et al., 1991a; Knighton et al., 1991b). The structure also displays a unique nucleotide binding motif where the orientation of MgATP in the molecule, and its interaction with conserved residues is clearly defined. This binding site is dominated by a 5-stranded antiparallel β sheet with the adenine ring buried in a hydrophobic pocket between the two lobes of the C-subunit. The small lobe of the C-subunit constitutes the nucleotide binding domain. The ATP is held in place by a glycine-rich loop between β strands 1 and 2 and is further fixed by ion pairing with Lys72 and Glu91. This small lobe fixes the nontransferable α and γ-phosphates, while the γ-phosphate interacts with several residues in the large lobe, specifically Asp184 which binds to the γ-phosphate, which chelates the magnesium and Lys168 (Zheng et al., Submitted 1992). Another invariant residue, Asp166, may serve as a catalytic base (Yoon & Cook, 1987).

The hydrodynamic properties of the C-subunit provide information about the interaction of the enzyme with inhibitory proteins and peptides. Determination of the Stokes' radius via analytical gel filtration and determination of the radius of gyration by neutron scattering (Parello et al., In press 1992) show interesting results considering the overall shape of the molecule in interaction with inhibitors. Where the radius of gyration is reduced by 1.5 Å due to binding of the PKI(5-24) peptide in the presence of MgATP, the Stokes' radius is increased by 1 Å (unpublished data). This suggest that structural changes inside of the molecule are not necessarily reflected in overall conformational changes

Characterization of Isozymes of C-subunit

The C-subunits purified from skeletal muscle (type I and II), liver (type II), and heart (type II) have similar enzymatic and biochemical properties and correspond to the C_{α} gene product (Bechtel et al., 1977; Kinzel & Kubler, 1976; Peters et al., 1977; Sugden et al., 1976; Zoller et al., 1979). Each can be separated into two or three different isoelectric forms (Kinzel et al., 1987; Peters et al., 1977; Sugden & Corbin, 1976); however, the physical basis for these differences in the isozymes of the catalytic subunit was never determined. At least two posttranslational modifications are known to be associated with the catalytic subunit, phosphorylation and myristylation. Two phosphorylation sites were identified, Thr(P)197 and Ser(P)338, based on amino acid sequencing (Shoji et al., 1983; Shoji et al., 1979). These were termed "silent" phosphorylation sites because the phosphates could not be removed readily with phosphatases (Chiu & Tao, 1978). There is evidence that Ser10 also can be autophosphorylated slowly (Toner-Webb & Taylor, In press 1992). In addition, the mammalian C-subunit is myristylated at its N-terminus (Carr et al., 1982). The recombinant protein contains additional phosphorylation sites at Ser 10 and Ser 139 (Yonemoto et al., Manuscript in Preparation).

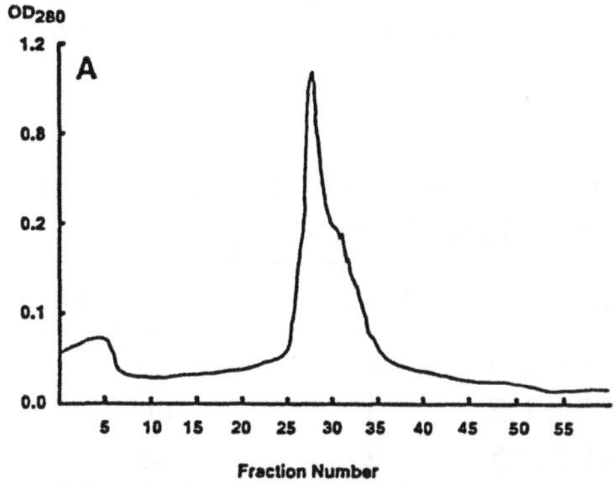

Figure 2A: Elution profile of recombinant C-subunit from a P11 phosphocellulose column (Whatman). *E. coli* BL21. DE3 cells were lysed in a French press as described (Slice & Taylor, 1989) batchbound to the P11 resin and eluted with KPO_4 gradient from 0 - 300 mM according to (Yonemoto et al., 1991). Only one homogeneous protein with an apparent molecular weight of 38,000 on SDS-gel electrophoresis was eluted.

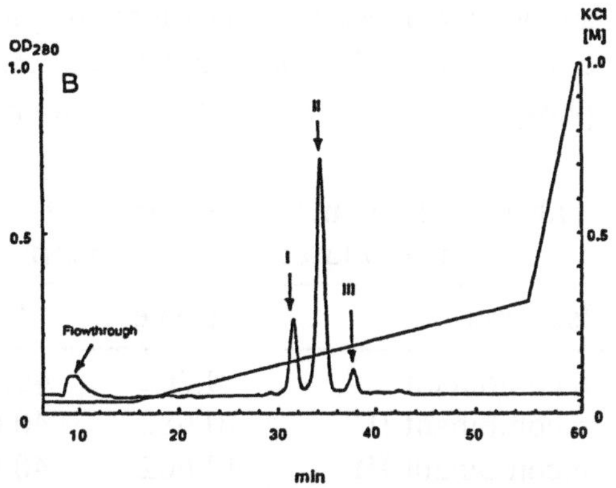

Figure 2B: Protein as purified in Figure A (30 mg) was loaded on a Mono S 10/10 column (Pharmacia/LKB) and eluted with a salt gradient from 0 - 1 M KCl yielding three homogeneous isozymes (I-III) eluting at 101 mM KCl, 127 mM KCl and 158 mM KCl, respectively.

The recombinant C-subunit ($C\alpha$) of cAPK can be separated into three distinct forms using Mono-S ion exchange chromatography (Figure 2) (Herberg et al., Manuscript submitted). These forms (isozyme I-III) correspond to three isoelectric variants with pI's of 6.4 (30 % of the total protein), 7.2 (60 %) and 8.2 (5-15 %). The Stokes' radius of each form measured by analytical gel filtration is 27.7 Å, 27.1 Å and 26.3 Å, respectively. Using electrospray mass spectrometry, the differences between the isozymes were shown to be due to phosphorylation with each form differing by 80 mass units corresponding to a single phosphate (see Table). The fully phosphorylated recombinant enzyme (isozyme I) contains four phosphates displaying an absolute molecular weight of 40,762. Comparison with the calculated molecular weight according to the amino acid sequence (Chrivia et al., 1988) gives a value of 40,761. Isozyme II is not phosphorylated at Ser 139 and isozyme III lacks an additional phosphate at Ser 10.

Comparison with the mammalian enzyme, porcine heart Cα, shows some significant differences in posttranslational modifications. The mammalian enzyme also can be separated into two isozymes. However. each isozyme has an identical mass of 40,840 according to electrospray mass spectroscopy. This correlates with posttranslational modifications corresponding to two phosphates and an amino-terminal myristyl moiety.

Table: Molecular Weight of Isozymes of the Cα-subunit Calculated and Determined by Mass Spectroscopy

Isozyme	Calculated	Observed	# PO_3	Myristylation
recombinant I	40,762	40,763 ± 2.0	4	no
recombinant II	40,682	40,682 ± 1.5	3	no
recombinant III	40,602	40,601 ± 2.5	2	no
mammalian I	40,839	40,840 ± 4.0	2	yes
mammalian II	40,839	40,840 ± 5.0	2	yes

The theoretical atomic mass of the murine alpha C-subunit was calculated according to (Chrivia et al., 1988) with a MacProMass 1.05 software. The electrospray mass spec data were obtained on PE Sciex Biomolecular Mass Analyzer API III from isozymes separated by Mono S chromatography (Herberg et al., Manuscript submitted). Proteins were additionally purified by Superdex 75 gel chromatography and 500 pmoles were analyzed by on line reverse phase HPLC mass spectroscopy. Data were analyzed using a Macspec 3,11b25 software. The mass spec was operated in positive mode showing multiple charged molecules with an median charge of approximately +50.

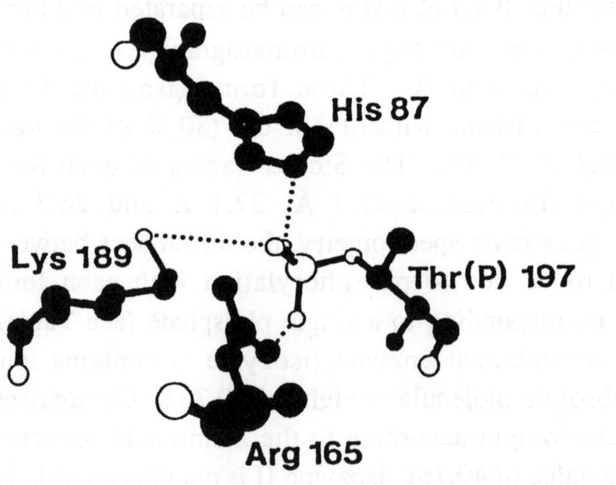

Figure 3: Interaction sides of the phosphate of Thr197 with essential residues in the C-subunit. The interaction to this sides accounts probably for the high resistance of Thr 97 against phosphatases. Arg165 is adjacent to Asp166, the proposed catalytic base for the phosphotransfer eventually so affecting activity of the cAPK. (Taken from (Taylor et al., 1992).)

The stability of the phosphorylation sites in the C-subunit can be explained on the basis of the crystal structure. Ser338 lies at a turn near the C-terminus: it forms an ion pair with Lys342 and also interacts with the backbone amide of residue 339. Thr197, likewise, interacts with multiple residues. These include His87, Lys189 and Arg165 (see Figure 3). These interactions explain clearly why especially this phosphate is so resistant to removal by phosphatases. Removing this phosphate could have a direct effect on the catalytic loop by weakening the interaction with Arg165. It could also be important for proper folding of the C-subunit and the overall structural integrity of the protein. When Thr197 is not phosphorylated it does not bind the R-subunit and PKI properly.

The stability of the mammalian C-subunit is increased by N-terminal myristylation as shown by heat denaturation experiments (Yonemoto et al., In press 1992).

The Importance of MgATP for the Interaction of the Subunits and with PKI

In the interaction of the type I R-subunit and PKI with the C-subunit MgATP plays an essential role in stabilizing both complexes. By developing an FPLC method for measuring directly the association of both of the physiological inhibitors to the C-subunit, we have been able to quantitative this effect under saturating MgATP and under MgATP-free conditions (Herberg, in preparation). PKI, for example, binds to the C-subunit with an apparent K_d of less than 1 nM in the presence of MgATP, but this is altered to 2.3 μM in the absence of MgATP. The apparent K_d of the RI-subunit is also less than 1 nM in the presence of MgATP, but 125 nM without MgATP. We are now using these techniques to analyze mutant forms of the C-subunit and of the RI-subunit that are defective in binding to the corresponding subunit. The effect of other nucleotides and other divalent cations on the stability of the complexes formed has to be determined as well.

Acknowledgments

We thank following individuals and resources for their contributions: Dr. Steven Kent at The Scripps Research Institute, La Jolla, for use of the PE Sciex Mass Analyzer and, in particular Sean Bell for performing the isoelectric focusing experiments and Jianhua Zheng for helpful discussion concerning the topology of MgATP in the crystal structure. We also thank Gene Hasegawa for assistance in preparing this manuscript.

References

Adams JA & Taylor SS (1992) The Energetic Limits of Phosphotransfer in the Catalytic Subunit of cAMP-dependent Protein Kinase as Measured by Viscosity Experiments. Biochemistry. 31: 8516-8522

Bechtel PJ, Beavo JA & Krebs EG (1977) Purification and Characterization of Catalytic Subunit of Skeletal MUscle Adenosine 3':5'- Monophosphate-dependent Protein Kinase. J. Biol. Chem. 252(8): 2691-2697

Beebe SJ & Corbin JD (1986) Cyclic Nucleotide-Dependent Protein Kinases. In *The Enzymes: Control by Phosphorylation Part A* (Krebs EG and Boyer PD, eds) pp 43-111, Academic Press, Inc. New York

Bramson HN, Thomas N, Matsueda R, Nelson NC, Taylor SS & Kaiser ET (1982) Modification of the Catalytic Subunit of Bovine Heart cAMP-Dependent Protein Kinase with Affinity Labels Related to Peptide Substrates. J. Biol. Chem. 257: 10575-10581

Buechler JA & Taylor SS (1989) Dicyclohexyl carbodiimide crosslinks two essential residues at the active site of the catalytic subunit of cAMP-dependent protein kinase: Asp 184 and Lys 72. Biochemistry. 28: 2065-2070

Buechler JA & Taylor SS (1990) Differential Labeling of the Catalytic Subunit of cAMP-Dependent Protein Kinase with a Water-Soluble Carbodiimide: Identification of Carboxyl Groups Protected by MgATP and Inhibitor Peptides. Biochemistry. 29: 1937-1943

Buechler JA, Vedvick TA & Taylor SS (1989) Differential Labeling of the Catalytic Subunit of cAMP-dependent Protein Kinase with Acetic Anhydride. Biochemistry. 28: 3018-3024

Carr SA, Biemann K, Shuji S, Parmalee DC & Titani K (1982) n-Tetradecanoyl in the NH2 Terminal Blocking Group of the Catalytic Subunit of the Cyclic AMP-Dependent Protein Kinase from Bovine Cardiac Muscle. Proc. Natl. Acad. Sci. USA. 79: 6128-6131

Chiu YS & Tao M (1978) Autophosphorylation of Rabbit Skeletal Muscle cAMP-Dependent Protein Kinase Catalytic Subunit. J. Biol. Chem. 253: 7145-7148

Chrivia JC, Uhler MD & McKnight GS (1988) Characterization of Genomic Clones Coding for the Cα and Cβ Subunits of Mouse cAMP-dependent Protein Kinase. J. Biol. Chem. 263: 5739-5744

Cook PF (1982) Kinetic Studies To Determine the Mechanism of Regulation of Bovine Liver Glutamate Dehydrogenase by Nucleotide Effectors. Biocemistry. 21: 113-116

Flockhart DA, Freist W, Hoppe J & Corbin JD (1980) Use of ATP Analogues to Map the ATP Binding Sites of the Cyclic Nucleotide-Dependent Protein Kinases. Fed. Proc. 39(Abst.): 3165

Gibbs CS & Zoller MJ (1991) Rational scanning mutagenesis of a protein kinase identifies functional regions involved in catalysis and substrate interactions. J. Biol. Chem. 266: 8923-31

Hanks SK, Quinn AM & Hunter T (1988) The Protein Kinase Family: Conserved Features and Deduced Phylogeny of the Catalytic Domains. Science. 241: 42-52

Herberg FW, Bell S & Taylor SS (Manuscript submitted) Separation and Characterization of Isozymes of the Catalytic Subunit of cAMP-dependent Protein Kinase. J. Biol. Chem.

Kinzel V, Hotz A, Konig N, Gagelmann M, Pyerin W, Reed J, Kubler D, Hofmann F, Obst C, Gensheimer HP, Goldblatt D & Shaltiel S (1987) Chromatographic Separation of Two Heterogeneous Forms of the Catalytic Subunit of Cyclic AMP-Dependent Protein Kinase Holoenzyme Type I and Type II from striated Muscle of Different mammalian Species. Arch. Biochem. Biophys. 253: 341-349

Kinzel V & Kubler D (1976) Single Step Purification of the Catalytic Subunit(s) of Cyclic 3',5'-Adenosine Monophosphate-Dependent Protein Kinase(s) from Rat Muscle. Biochem. Biophys. Res. Commun. 71: 257-264

Knighton DR, Zheng J, Ten Eyck LF, Ashford VA, Xuong N-h, Taylor SS & Sowadski JM (1991a) Crystal Structure of the Catalytic Subunit of cAMP-dependent Protein Kinase. Science. 253: 407-414

Knighton DR, Zheng J, Ten Eyck LF, Xuong N-h, Taylor SS & Sowadski JM (1991b) Structure of a Peptide Inhibitor Bound to the Catalytic Subunit of Cyclic Adenosine Monophosphate-Dependent Protein Kinase. Science. 253: 414-420

Krebs EG & Beavo JA (1979) Phosphorylation-Dephosphorylation of Enzymes. Annu. Rev. Biochem. 48: 923-959

Krebs EG, Graves DJ & Fischer EH (1959) Factors affecting the activity of muscle phosphorylase b kinase. J. Biol. Chem. 234: 2867-2873

Orellana SA & McKnight GS (1992) Mutations in the Catalytic Subunit of cAMP-Dependent Protein Kinase Result in Unregulated Biological Activity. Proc. Natl. Acad. Sci. USA. 89: 4726-4730

Parello J, Timmins PA, Sowadski JM & Taylor SS (In press 1992) Major Conformational Changes in the Catalytic Subunit of cAMP-Dependent

Protein Kinase Induced by the Binding of a Peptide Inhibitor: A Small-Angle Neutron Scattering Study in Solution. J. Mol. Biol.

Peters KA, Demaille JG & Fischer EH (1977) Adenosine 3':5'-Monophosphate-Dependent Protein Kinase from Bovine Heart: Characterization of the Catalytic Subunit. Biochemistry. 16: 5691-5697

Shoji S, Ericsson LH, Walsh DA, Fischer EH & Titani K (1983) Amino Acid Sequence of the Catalytic Subunit of Bovine Type II Adenosine Cyclic 3',5'-Phosphate Dependent Protein Kinase. Biochemistry. 22: 3702-3709

Shoji S, Titani K, Demaille JG & Fischer EH (1979) Sequence of Two Phosphorylated Sites in the Catalytic Subunit of Bovine Cardiac Muscle Adenosine 3':5'-Monophosphate-Dependent Protein Kinase. J. Biol. Chem. 254: 6211-6214

Slice LW & Taylor SS (1989) Expression of the Catalytic Subunit of cAMP-dependent Protein Kinase In *Escherichia coli*. J. Biol. Chem. 264: 20940-20946

Sugden PH & Corbin JD (1976) Adenosine 3':5'-Cyclic Monophosphate-Binding Proteins in Bovine and Rat Tissues. Biochem. J. 159: 423-437

Sugden PH, Holladay LA, Reimann EH & Corbin JD (1976) Purification and Characterization of the Catalytic Subunit of Adenosine 3':5'-Cyclic Monophosphate-Dependent Protein Kinase. Biochem. J. 149: 409-422

Taylor SS, Buechler JA & Yonemoto W (1990) cAMP-dependent Protein Kinase: Framework for a Diverse Family of Regulatory Enzymes. Annu. Rev. Biochem. 59: 971-1005

Taylor SS, Knighton DR, Zheng J, Ten Eyck LF & Sowadski JM (1992) Structural Framework for the Protein Kinase Family. Annu. Rev. Cell Biol. 8: 429-462

Thomas J, Van Patten SM, Howard P, Day KH, Mitchell RD, Sosnick T, Trewhella J, Walsh DA & Maurer RA (1991) Expression in Escherichia coli and characterization of the heat-stable inhibitor of the cAMP-dependent protein kinase. J Biol Chem. 266: 10906-11

Toner-Webb J & Taylor SS (In press 1992) Autophosphorylation of the Catalytic Subunit of cAMP-Dependent Protein Kinase. J. Biol. Chem.

Van Patten SM, Ng DC, Th'ng JP, Angelos KL, Smith AJ & Walsh DA (1991) Molecular Cloning of a Rat Testis Form of the Inhibitor Protein of cAMP-dependent Protein Kinase. Proc. Natl. Acad. Sci. USA. 88: 5383-5387

Walsh DA, Angelos KL, Van Patten SM, Glass DB & Garetto LP (1990) The Inhibitor Protein of the cAMP-Dependent Protein Kinase. In *Peptides and Protein Phosphorylation* (Kemp BE, eds) pp 43-84, CRC Press, Inc. Boca Raton

Walsh DA, Ashby CD, Gonzalez C, Calkins D, Fischer EH & Krebs EG (1971) Purification and Characterization of a Protein Inhibitor of Adenosine 3',5'-Monophosphate-Dependent Protein Kinases. J. Biol. Chem. 246: 1977

Walsh DA, Perkins JP & Krebs EG (1968) An Adenosine 3',5'-Mono-Phosphate-Dependent Protein Kinase from Rabbit Skeletal Muscle. J. Biol. Chem. 243: 3763-3765

Weldon SL, Mumby MC, Beavo JA & Taylor SS (1983) Monoclonal Antibodies as Structural Probes of Surface Residues in the Regulatory Subunit of cAMP-Dependent Protein Kinase II from Porcine Heart. J. Biol. Chem. 258: 1129-1135

Yonemoto W, McGlone ML, Slice LW & Taylor SS (1991) Prokaryotic Expression of the Catalytic Subunit of cAMP-dependent Protein Kinase. In *Protein Phosphorylation (Part A)* (Hunter T and Sefton BM, eds) pp 581-596, Academic Press, Inc. San Diego

Yonemoto W, McGlone ML & Taylor SS (In press 1992) N-Myristylation of the Catalytic Subunit of cAMP-dependent Protein Kinase Conveys Structural Stability. J. Biol. Chem.

Yonemoto W, McGlone ML & Taylor SS (Manuscript in Preparation) Functional Importance of Phosphorylation Sites in the Recombinant Catalytic Subunit of cAMP-dependent Protein Kinase.

Yoon M-Y & Cook PF (1987) Chemical Mechanism of the Adenosine Cyclic 3',5'-Monophosphate Dependent Protein Kinase from pH Studies. Biochemistry. 26: 4118-4125

Zheng J, Knighton DR, Ten Eyck LF, Karlsson R, Xuong N-h, Taylor SS & Sowadski JM (Submitted 1992) Crystal Structure of the Catalytic Subunit of cAMP-dependent Protein Kinase Complexed with MgATP and Peptide Inhibitor.

Zoller MJ, Kerlavage AR & Taylor SS (1979) Structural Comparisons of cAMP-Dependent Protein Kinases I and II from Porcine Skeletal Muscle. J. Biol. Chem. 254: 2408-2412

Zoller MJ, Nelson NC & Taylor SS (1981) Affinity Labeling of cAMP-dependent Protein Kinase with *p*-Fluorosulfonylbenzoyl Adenosine. J. Biol. Chem. 256: 10837-10842

CASEIN KINASE I AND THE CLASSIFICATION OF PROTEIN KINASES: WHAT'S IN A NAME?

Peter J Roach

Department of Biochemistry and Molecular Biology, Indiana University School of Medicine, Indianapolis, IN 46223, USA

INTRODUCTION.

The discovery of new protein kinases, whether through analysis of proteins or the characterization of genes, has reached a remarkable rate that will only abate once the supply of genes in the genome becomes limiting. Probably a couple of hundred protein kinases are identified at present (Hunter, 1991). For the yeast *Saccharomyces cerevisiae*, 55 of 1700 entries in the database maintained by Mark Goebl, Indiana University, are protein kinase catalytic subunits (as of October 22, 1992; M. Goebl, personal communication). Extrapolated from these numbers, a yeast genome of say 6,000 genes would contain ~200 genes encoding protein kinases[1]. Phrased another way, approximately 3.2% of the genome would code for protein kinases. Note that this count does not include any regulatory subunits. For a human genome ten times as large, figures of ~2000 protein kinases could be projected. This type of prediction is not novel and, of course, the forerunner was the noted "1001 kinases" paper of Tony Hunter, Salk Institute (Hunter, 1987). What looked to some like a fairly high estimate aimed mainly at a good title may well turn out to be low. The principal point to be made here is simply that the protein kinases constitute one of the larger gene families known. Interestingly, several other signaling molecules such as the low molecular weight GTPases, trimeric G-protein α-subunits, protein phosphatases, and the G-protein coupled transmembrane receptors are members of rapidly expanding gene families. While attesting to the importance of regulatory processes and the signal transduction molecules involved, this

[1] This extrapolation does not account for any bias in the discovery of genes to date. Also, the fraction of protein kinases in the genome may well vary from organism to organism.

NATO ASI Series, Vol. H 76
Tyrosine Phosphorylation/Dephosphorylation and Downstream Signalling
Edited by L. M. G. Heilmeyer Jr.

high degree of diversity poses some interesting problems of taxonomy, nomenclature and classification. This article takes one example, that of casein kinase I, to highlight the problems of classifiying protein kinases in an expanding market.

ENZYMOLOGICAL DEFINITION OF PROTEIN KINASES.

Protein kinases were first defined biochemically as enzymes catalyzing the stable transfer of phosphate to proteins. Some of the first protein kinases characterized, cyclic AMP-dependent protein kinase (Walsh et al., 1968), phosphorylase kinase (Krebs and Fischer, 1956), and pyruvate dehydrogenase kinase (Linn et al., 1969), were distinguishable by their regulatory ligands and/or a high degree of selectivity in terms of substrate. The operational definition of the activity was meaningful, fitted well with the given name, and still does so today[2]. So long as the number of phosphoproteins was easily countable and the number of protein kinases was limited, standard enzymological separations and analyses were sufficient to demonstrate the existence of many different protein kinases.

In the early 80s, as the number of distinct protein kinases was just beginning its upward surge, the approach to the characterization of a potentially new enzyme would typically include purification, an examination of substrate specificity and assay of a small battery of ligands. For substrate specificity, if some specialized substrate was not already the basis for identification of the activity, favourites to test were histones, caseins and phosvitin. Protein kinases could usually be distinguished by preference for basic versus acidic protein substrates. ATP versus GTP use was usually tested even if relatively few enzymes are effective with GTP. There were not too many effectors to test, but cyclic nucleotides and Ca^{2+} were obvious ones. Polyanions and polycations such as heparin and polylysine were frequently analyzed for effects on activity, and inhibition by heparin became a discriminator for casein kinase II. One of the first truly specific probes was the inhibitor protein of cyclic AMP-dependent protein kinase identified by Walsh et al. (1971). Thus, so long as there were relatively few kinases, it was often possible, by a combination of positive and negative biochemical criteria, to define a given protein kinase as distinct or not. This is not to say there were not cases of confusion and dispute, but these examples most often arose from comparison of relatively impure enzyme preparations.

The first two interconvertible enzymes identified, phosphorylase and phosphorylase kinase, resulted in the identification of only one kinase each, phosphorylase kinase and cyclic AMP-dependent protein kinase. The third example of an enzyme controlled by reversible

---

[2] True, the acronym PKA is used frequently but in my mind this is an abbreviation of the real name.

phosphorylation, glycogen synthase, also yielded cyclic AMP-dependent protein kinase (Huijing and Larner, 1966) but subsequently proved to be substrate for numerous protein kinases (Roach, 1990). In stark contrast to phosphorylase, this enzyme was modified at multiple sites and, *in vitro*, by multiple protein kinases. At the time, study of glycogen synthase phosphorylation contributed significantly to the growing diversity of protein kinases and was one of the first instances where enzymes that started out as glycogen synthase kinases turned out to be enzymes already known in other contexts. Two examples are casein kinase I and casein kinase II.

Since casein had been a frequent choice as an available protein substrate, even in the early work of Burnett and Kennedy (1954), "casein kinases" had been known for many years. Two major activities were identified in many studies that, after having been accorded a variety of names, are now termed casein kinase I and casein kinase II by almost everyone (see Pinna, 1989, and Tuazon and Traugh, 1991, for reviews). Casein kinase II was a tetramer, $\alpha_2\beta_2$, could effectively utilize GTP as well as ATP, and was extremely sensitive to inhibition by heparin. Casein kinase I was usually defined as monomeric, although estimates of M_r ranged from 25 to 55 kDa. The enzyme was unable to utilize GTP. Both enzymes had been reported to be present in many cell types and in various intracellular fractions, including cytosol, nucleus and membranes. Though the question of physiological function remained largely unanswered, casein kinase I and II, though perhaps poorly named, seemed quite well defined in the late 1980's.

SUBSTRATE RECOGNITION AND THE SPECIFICITY OF PROTEIN KINASES.

The most fundamental classification of protein kinases by substrate specificity is by the nature of the amino acid side chain modified in the protein substrate. The protein kinases, upon which this article is focused, were first shown to modify serine and threonine and then in 1980, tyrosine was added to this list (Hunter and Sefton, 1980). As discussed later, these enzymes are defined primarily by a single gene family[3]. Note that some enzymes, often called dual specificity kinases, have begun to emerge that have the capacity to modify Ser, Thr or Tyr (Posada and Cooper, 1992). Classification of enzymes by substrate specificity became more sophisticated as the sequences around more and more phosphorylation sites became available (Pearson and Kemp, 1991). It was apparent that some enzymes, with cyclic

[3] Enzymes capable of phosphorylating several other residues are known (Hunter, 1991) but this article addresses only the major eukaryotic Ser/Thr/Tyr kinase family. One confounding recent discovery (Popov et al., 1992), however, is that the mitochondrial branched chain ketoacid dehydrogenase kinase, which is thought to modify Ser residues in its substrate, belongs to a different gene family and bears more resemblance to the bacterial so called histidine kinases.

AMP-dependent protein kinase again leading the way, recognized certain primary sequence features in the vicinity of the modified residue. The use of synthetic peptides as protein kinase substrates allowed hypotheses about local recognition determinants to be tested. Thus, for several protein kinases, consensus recognition motifs have been proposed (Pearson and Kemp, 1991). These motifs are certainly not foolproof but have been useful. However, for many protein kinases it has proven difficult to describe recognition in terms of local sequence determinants. Some enzymes, like phosphorylase kinase, are very specific whereas for others, including many tyrosine kinases, it has been hard to define recognition determinants in terms of primary structures surrounding the phosphorylation sites. Perhaps in these cases, other types of macromolecular interactions, such as those mediated by SH2 domains (Koch et al., 1991), are involved in modulating specificity.

The casein kinases fitted in the category of acidotropic enzymes (Roach, 1991) or kinases for which acidic residues were important in recognition. The laboratories of Pinna (Marin et al., 1986; Marchiori et al., 1988) and Krebs (Kuenzel et al., 1987), for example, showed that Asp and Glu residues were critical to substrate recognition by casein kinase II. It was later found that phosphoamino acids could also fulfill this requirement (Meggio and Pinna, 1988; Meggio et al., 1988; Litchfield et al., 1990). In my laboratory, we investigated the substrate determinants of casein kinase I, initially in the context of its phosphorylation of glycogen synthase (Flotow and Roach, 1989,1991; Flotow et al., 1990). Briefly, we found that optimal recognition required prior substrate phosphorylation, just as we had earlier reported for the GSK-3 enzyme (Fiol et al., 1987). The preferred motif was -S(P)-X-X-S-, and the phosphoserine could not be substituted by either an Asp or a Glu residue. This conclusion was confirmed using peptides based on other protein substrate sequences (Meggio et al., 1991; Umphress et al., 1992). We did find that a block of acidic residues with the same spacing from the target serine, eg. -D-D-D-D-X-X-S-, generated a substrate that was a couple of orders of magnitude worse (Flotow and Roach, 1991). Synthetic peptides with this sequence, though poor substrates, were nonetheless specific for casein kinase I; that is, no other kinases that we had to test could phosphorylate the peptide except casein kinase I. Similar conclusions were reached by Agostinis et al. (1989). Thus, casein kinase I seemed quite well documented as the monomeric, low M_r casein kinase that had substrate specificity as defined above.

PRIMARY SEQUENCE COMPARISONS AND THE PROTEIN KINASE STRUCTURE

Overlapping with these studies of substrate recognition, sequence information began to emerge for the protein kinases. Hanks et al. (1988) compiled some 65 sequences and were able to identify highly conserved regions or residues of the protein kinase family. An updated version has appeared in 1991 (Hanks and Quinn, 1991). Phylogenetic trees allowed the relatedness of protein kinases to be visualized. Some clear distinctions were obvious, and it was, for example, easy to view the protein Tyr kinases as a separate branch of the family as compared with the protein Ser/Thr kinases. Some residues characteristic of these branches can be identified. Sub-families composed of cyclic nucleotide-dependent protein kinases or protein kinase C, for example, are also apparent. The alignment of Hanks et al. (1988) has proven invaluable for those identifying or cloning new enzymes and was given even more meaning once the three dimensional structure of the catalytic subunit of cyclic AMP-dependent protein kinase was determined (Knighton et al., 1991a,b). From this work, the highly conserved sequences can be seen to correlate with essential features of the protein structure. Variable regions or inserts generally fall such that the fundamental structure is not disrupted. Since the structure was also obtained with an inhibitor peptide bound, it was possible to map residues involved in binding substrate and, in this case, linking with the known consensus recognition sequence -R-R-X-S-. This is an exciting time and there is a frenzy of modeling, with one goal of trying to rationalize substrate recognition for other protein kinases. Of course, the protein kinases distinguish themselves not only by variations in sequence in the catalytic domain but also by specific sequences that are either inserts or are located at the NH_2- and COOH-termini. These regions are not covered by the existing crystal structure but can obviously be involved in substrate recognition. Other protein kinase structures are now being analyzed and we can expect in the next decade to understand a great deal more about protein kinase interactions with substrates.

ISOFORMS AND SUB-FAMILIES

To return to the saga of casein kinase I, in 1991 the first success in molecular cloning was achieved. Cobb and colleagues (Rowles et al., 1991) reported sequences for several cDNAs and a small PCR product. The two full length cDNAs were termed CKIα and CKIβ. A partial cDNA was designated CKIγ and the small PCR product CKIδ. Work in this laboratory identified a full length CKIδ cDNA (Robinson et al., 1992; Graves, P.R., Haas, D.W., Hagedorn, C.H., DePaoli-Roach and Roach, P.J., unpublished results) and very recently three distinct forms of CKIγ (Zhai, L., Graves, P.R. and Roach, P.J., unpublished). Alternative splicing of the CKIα has also been noted (Rowles et al., 1991). Four distinct *Saccharomyces cerevisiae* homologues have been identified, termed YCK1 and YCK2 (Robinson et al., 1992; Wang et al., 1992), HRR25 (Hoekstra et al., 1991), and NUF4 (Hoekstra, M., unpublished results). Thus, in the space of about a year this seemingly innocuous protein kinase has yielded over 10 distinct forms based on DNA sequences. Initial efforts are underway to analyze the enzymes biochemically, using overexpression in bacteria or yeast to produce recombinant protein. With the limited information available so far, the enzymes tested seem to behave like casein kinase I in terms of substrate specificity and other of the original biochemical characterizations. This tale is told to exemplify how a situation of apparent relative simplicity can rapidly become complex. The multiplicity of casein kinase I isoforms can, with benefit of hindsight, help to explain some of the inconsistencies in the biochemical literature but now poses critical questions as to what the functions of all these enzymes are. Students of several other protein kinases will recognize a similar problem.

Casein kinase I has scarcely been an ideal name, although it has had the merit of general acceptance and usage. Now, we hardly know what the term casein kinase I defines. Is it descriptive of a kinase bearing the conserved signature sequences from alignment of the different forms? Does it describe substrate recognition characteristics? Probably, yes but, in the end, we need to define the actual isoform. It is specificity of function that is ultimately important and this is likely to be dictated not by the common features but by the differences, most probably in the non-catalytic domains.

CONCLUSION

Considering the more general issue, how do we deal with such sub-families of biochemically related protein kinases? Some level of redundancy in function can be postulated, but one almost has to presume some multiplicity of distinct tasks. These could be separated in time and space, in different cells or subcellular localizations, and by expression under different physiological or developmental conditions. Such knowledge will help define and distinguish different isoforms and a major challenge of the future will be to address the multiplicity of functions of these enzymes. A mundane but practically important issue will simply be to keep track of and to name all the individual members. Since any meaningful nomenclature must allow for operational detection of different enzymes, we can see that many of the old biochemical criteria will run the risk of being ambiguous. It will not always be sufficient to use a particular well-defined substrate, observe stimulation by a particular ligand, or even inhibition by some of the more specific inhibitors. Exact definition of the enzymic species present will require more precise molecular strategies. One approach is the use of anti-peptide or monoclonal isoform-specific antibodies - even with such probes, caution must be exercised until all members of a sub-family are known to have been identified! The other approach that will have to assume more importance is the genetic one so that individual genes can be modified. If only every enzyme could be named as easily and lastingly as cyclic AMP-dependent protein kinase.

ACKNOWLEDGMENTS

Research from the author's laboratory was supported in part by research grants DK27221 and DK42576 from the National Institutes of Health.

REFERENCES.

Agostinis, P., Pinna, L. A., Meggio, F., Marin, O., Goris, J., Vandenheede, J. R., and Merlevede, W. (1989) FEBS Lett. 259, 75-78
Burnett, G. and Kennedy, E.P. (1954) J. Biol. Chem. 211, 969-980
Fiol, C.J., Mahrenholz, A. M., Wang, Y., Roeske, R. W., and Roach, P. J. (1987) J. Biol. Chem. 262,14042-14048.
Flotow, H. and Roach, P. J. (1991) J. Biol. Chem. 266, 3724-3727
Flotow, H., and Roach, P. J. (1989) J. Biol. Chem. 264, 9126-9128
Flotow, H., Graves, P. R., Wang, A., Fiol, C. J., Roeske, R. W. and Roach, P. J. (1990) J. Biol. Chem. 265, 14264-14269
Hanks, S.K., and Quinn, A.M. (1991) Methods Enzymol. 200, 38-62
Hanks, S.K., Quinn, A.M., Hunter, T. (1988) Science 241, 42-52
Hoekstra, M. F., Liskay, R. M., Ou, A. C., DeMaggio, A. J., Burbee, D. G. and Heffron, F. (1991) Science 253, 1031-1034
Huijing, F. and Larner, J. (1966) Biochem. Biophys. Res. Commun. 23, 259-263
Hunter, T. (1987) Cell 50, 823-829
Hunter, T. (1991) in Methods Enzymol 200, 3-37
Hunter, T. and Sefton, B.M. (1980) Proc. Natl. Acad. Sci. USA 77, 1311-1315
Knighton, D. R., Zheng, J., Ten Eyck, L.F., Ashford, V.A., Xuong, N., Taylor, S.S., Sowadski, J.M. (1991a) Science 253, 407-414
Knighton, D. R., Zheng, J., Ten Eyck, L.F., Ashford, V.A., Xuong, N., Taylor, S.S., Sowadski, J.M. (1991b) Science 253, 414-420.
Koch, C.A., Anderson, D., Moran, M.F., Ellis, C., and Pawson, T. (1991) Science 252, 668-718
Krebs, E.G. and Fischer, E.H. (1956) Biochim. Biophys. Acta 20, 150-157
Kuenzel, E. A., Mulligan, J. A., Sommercorn, J., and Krebs, E. G. (1987) J. Biol. Chem. 262, 9136-9140.
Linn, T.C., Pettit, F.H., Hucho, F., and Reed, L.J. (1969) Proc. Natl. Acad. Sci. USA 64, 227-234
Litchfield, D. W., Arendt, A., Lozeman, F. J., Krebs, E. G., Hargrave, P. A., and Palczewski, K. (1990) FEBS Lett. 261, 117-120.
Marchiori, F., Meggio, F., Marin, O., Borin, G., Calderan, A., Ruzza, P. and Pinna, L. A. (1988) Biochim. Biophys. Acta 971, 332-338.
Marin, O., Meggio, F., Marchiori, F., Borin, G., and Pinna, L. A. (1986) Eur. J. Biochem. 160, 239-244.
Meggio, F. and Pinna, L. A. (1988) Biochem. Biophys. Acta 971, 227-231.
Meggio, F., Perich, J. W., Johns, R. B., and Pinna, L. A. (1988) FEBS Lett. 237, 225-228.
Meggio, F., Perich, J. W., Reynolds, E. C., and Pinna, L. A. (1991) FEBS Lett. 283, 303-306
Pearson, R.B. and Kemp, B.E. (1991) Methods Enzymol. 200, 62-81
Pinna, L. A. (1989) in "Peptides and Protein Phosphorylation." (B. Kemp, ed.). pp 145-169
Popov, K.M., Zhao, Y., Shimomura, Y., Kuntz, M.J., and Harris, R.A. (1992) J. Biol. Chem. 267, 13127-13130
Posada, J. and Cooper, J.A. (1992) Mol. Biol. Cell 3, 583-592
Roach, P.J. (1990) FASEB J. 4, 2961-2968
Roach, P.J. (1991) J. Biol. Chem.266, 14139-14142

Robinson, L.C., Hubbard, E.J.A., Graves, P.R., DePaoli-Roach, A.A., Roach, P.J., Kung, C., Haas, D.W., Hagedorn, C.H., Goebl, M.G., Culbertson, M.R., and Carlson, M. (1992) Proc. Natl. Acad. Sci. (USA) 89, 28-32

Rowles, J., Slaughter, C., Moomaw, C., Hsu, J., and Cobb, M. H. (1991) *Proc. Natl. Acad. Sci.* 88, 9548-9552

Tuazon, P. T. and Traugh, J. A. (1991) Adv. Second Messenger Phosphoprotein Res. 23, 123-164

Umphress, J. L., Tuazon, P. T., Chen, C-J. and Traugh, J. A. (1992) Eur. J. Biochem. 203, 239-243

Walsh, D.A., Ashby, C.D., Gonzalez, C., Calkins, D., Fischer, E.H., and Krebs, E.G. (1971) J. Biol. Chem. 246, 1977-1985

Walsh, D.A., Perkins, J.P., and Krebs, E.G. (1968) J. Biol. Chem. 243, 3763-3765

Wang, P.-C., Vancura, A., Mitcheson, T. G. M., and Kuret, J. (1992) Mol. Biol. Cell. 3, 275-286

Downregulation of protein kinase C is associated with phenotypic changes and enhanced proliferation of vascular smooth muscle cells

C. Lindschau, H. Haller, P. Quass, A. Distler

Dept. Nephrology, Klinikum Steglitz, Freie Universität Berlin, Germany

Introduction

Vascular smooth muscle cell hypertrophy and proliferation play an important role in the pathogenesis of hypertension and atherosclerosis. Dedifferentiation of VSMC seems to be a prerequisite for proliferation and is accompanied by typical phenotypic changes. However, the underlying cellular mechanisms are poorly understood.

As protein kinase C (PKC) is a key enzyme in cellular differentiation and proliferation we investigated enzyme activity and immunoreactivity of the PKC isoforms α and β in isolated vascular smooth muscle cells (VSMC) from spontaneous hypertensive rats (SHR) and, as a control, from Whistar Kyoto rats (WKY) during the first passages (1-4) to analyse the relationship between VSMC differentiation and PKC expression *in vitro*.

Methods

Smooth muscle cells were isolated from 12-14 weeks old SH- and WKY-rats by an combined collagenase/elastase treatment of the aorta after removal of the endothelial cells. PKC activity was measured by *in vitro* phosphorylation of histone after purification of the cell fractions on DEAE-sepharose column by a step gradient. PKC immunoreactivity was assessed by western blot using commercial available isozyme specific polyclonal antibodies raised against specific oligopeptides in rabbits (Gibco). Cell proliferation was measured by cell count and a non-radioactive proliferation assay (Promega). VSMC phenotype was characterised by immunostaining and western blot analysis for actin and desmin.

NATO ASI Series, Vol. H 76
Tyrosine Phosphorylation/Dephosphorylation and Downstream Signalling
Edited by L. M. G. Heilmeyer Jr.

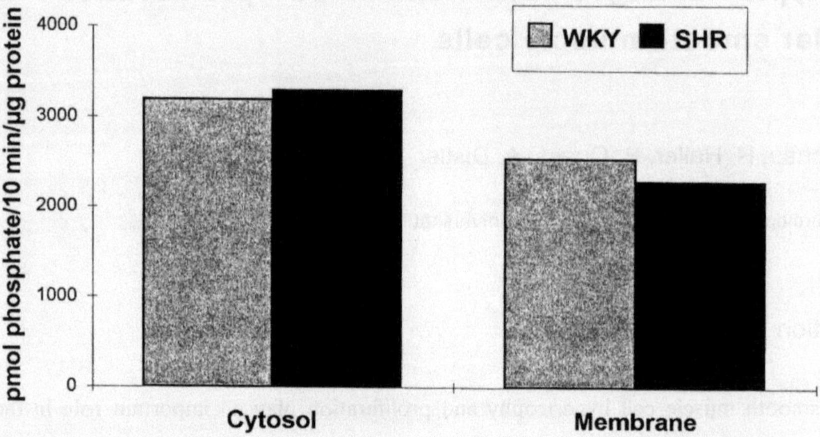

Figure 1: PKC activity was measured in the cytosolic and membrane fraction of vascular smooth tissue (aorta) from WKY- and SH- rats after homogenisation and cell fractionation. Enzyme activity was not different between the two strains (n=7).

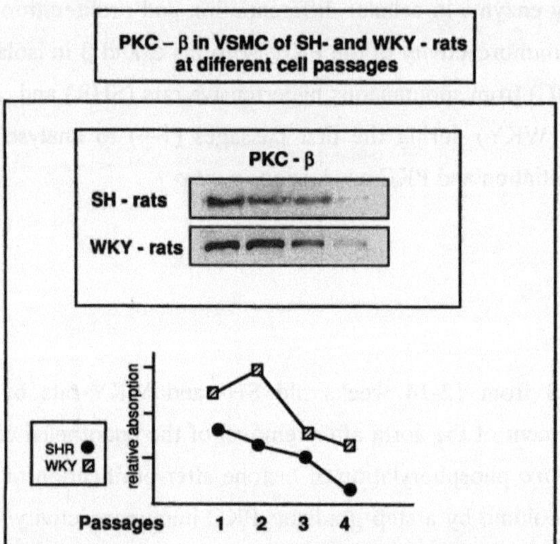

Figure 2: Western blot analysis of PKC isoforms β in subcultured rat aortic vascular smooth muscle cells from SHR and age-matched WKY- rats during passages 1 -4. Cells were grown on culture dishes and harvested at confluency. Cells from SHR showed a decreased expression of β-PKC (representative of 4 experiments).

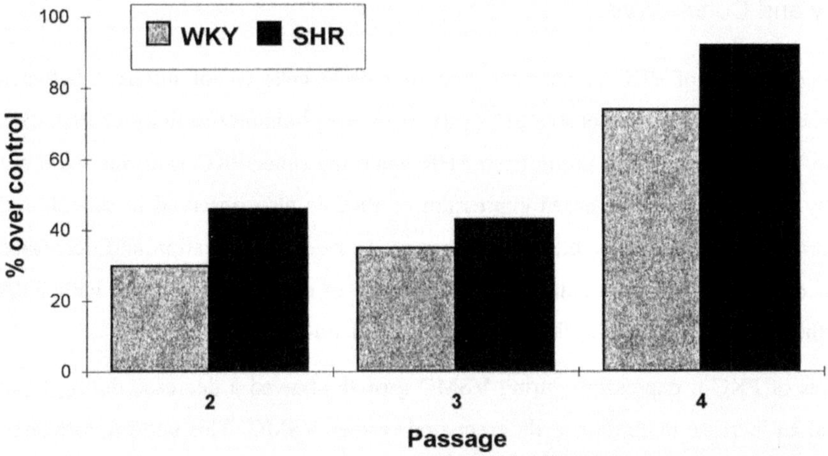

Figure 3: PDGF-induced proliferation of subcultured VSMC from SH- and WKY rats. Cell proliferation of VSMC was measured by dye incorporation (Promega) two days after passaging. VSMC from both strains showed an increased rate of proliferation from passage 2 to 4, which corresponded to the decreased PKC expression.

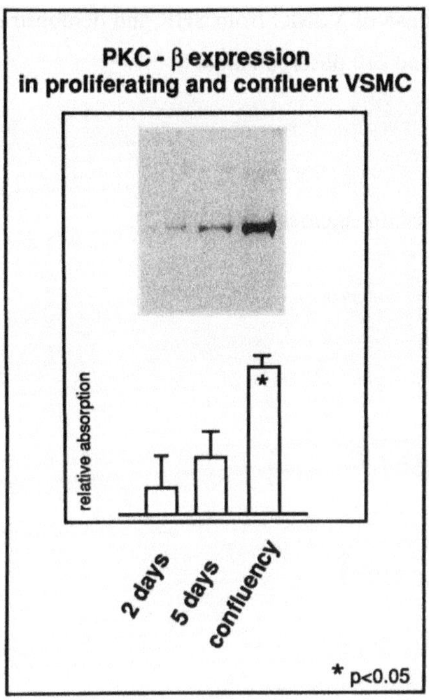

Figure 4: Western blot analysis of β-PKC from VSMC (WKY rat aorta, passage 2) during *in vitro* proliferation on day 2, day 5 and two days after reaching confluency. Cells were seeded at a cell density of 2.5×10^5 cells/cm^2.

Summary and Conclusions

The enzyme activity of PKC in vascular smooth muscle cells is not different between normotensive WKY- and hypertensive SH - rats. However, immunoreactivity of PKC-β is significantly decreased in aortic tissue from SHR while the other PKC isoforms were not significantly different.. The decreased expression of PKC is also observed in subcultured VSMC from SHR and is associated with an increased rate of proliferation and decreased expression of α-actin. Down-regulation of PKC by pretreatment of VSMC with TPA abolished the difference in proliferation between SH- and WKY-rats.

The analysis of PKC-β expression during VSMC growth showed a decrease during rapid growth and an increase in confluent, differentiated resting VSMC. This pattern correlated well with the decrease in actin and desmin i.e. the change from the secretory to the contractile phenotype of the cells.

We conclude that PKC is down-regulated in rapidly growing VSMC. Our finding suggest a role of PKC in the increased proliferative response of VSMC from SHR and demonstrate a negative correlation between PKC expression and cell differentiation.

Acknowledgement

This study was supported by the Deutsche Forschungsgemeinschaft (DFG).

INSULIN-STIMULATED SERYL/THREONYL PROTEIN KINASES IN DIABETIC RATS AND THE EFFECTS OF VANADIUM TREATMENT

Yong-jiang Hei, Steven L. Pelech*, Jack Diamond and John H. McNeill
Faculty of Pharmaceutical Sciences and Medicine*,
The University of British Columbia
2146 East Mall, Vancouver, B. C.
Canada V6T 1Z3

Insulin plays an essential role in the regulation of cellular metabolism and growth. The exact mechanisms of action for insulin are not completely understood, although recent studies suggest the involvement of a sequentially activating protein phosphorylation cascade which includes p74*raf*, mitogen-activated protein (MAP) kinase kinase, MAP kinase, and ribosomal S6 protein kinase (Pelech and Sanghera, 1992). Another kinase, casein kinase II (CK-II), has also been implicated in insulin signalling (Sommercorn *et al.*, 1987; Krebs *et al.*, 1988). However, very little is known about the insulin signalling pathways in diabetes mellitus, a pathological condition of absolute or relative insulin deficiency. Previous studies have shown that Type-I diabetes was associated with an increase in insulin receptor number (Burant *et al.*, 1986), a decrease in insulin receptor tyrosine kinase activity and its autophosphorylation (Okamato *et al.*, 1986; Thies *et al.*, 1990), and an enhanced phosphorylation of IRS-1 (Giorgino *et al.*, 1992). There have been no studies to date, however, on the seryl/threonyl protein kinases in the insulin signalling cascade in diabetes. Therefore, we have used skeletal muscles from chronic diabetic rats to examine the activities of MAP kinases, CK-II and S6 kinases. Additionally, we investigated whether vanadyl sulphate, an insulin-mimetic, has any effects on these kinases.

Materials and Methods

1. Animals. Diabetes was induced in Wistar rats of about 200 grams by a bolus intravenous injection of streptozotocin (60 mg/kg). Age-matched control rats were injected with saline. The rats were kept for either 1 month or 6 months before preparing samples for biochemical analysis. Vanadyl sulphate treatment was administered in the one-month groups at a dose of 0.75-1 mg/ml in drinking water.

2. Sample preparation and protein kinase assays. Skeletal muscles were removed from the hind legs of the rats and crude extracts were prepared by homogenization and ultracentrifugation. Some rats from the one-month groups were injected with insulin (10 U/kg, iv) 20 min before removing the skeletal muscles. Kinases in the skeletal muscle extracts were assayed as the phosphotransferase activities of the samples with myelin basic protein (MBP), phosvitin and ribosomal S6 protein or an S6 peptide (RRLSSLRA) as substrates, as described previously (Pelech and Krebs, 1987; Ahn and Krebs, 1990).

NATO ASI Series, Vol. H 76
Tyrosine Phosphorylation/Dephosphorylation
and Downstream Signalling
Edited by L. M. G. Heilmeyer Jr.

Results and Discussion

1. General features of the animals. The diabetic rats in both the one-month groups and the 6-month groups had significantly elevated plasma glucose levels and decreased body weight. One week after the treatment with vanadyl sulphate in the one-month diabetic rats, the plasma glucose levels were significantly lowered and by four weeks, the glucose levels were completely normalized by the treatment (Figure 1), indicating that vanadium compounds can produce insulin-like effects on glucose metabolism in diabetic rats. The glucose levels in the control rats were not altered by the treatment. Similar effects of vanadium compounds have been previously reported (Heyliger *et al.*, 1985; Ramanadham *et al.*, 1989; also reviewed in Shechter 1990 and Brichard *et al.*, 1991).

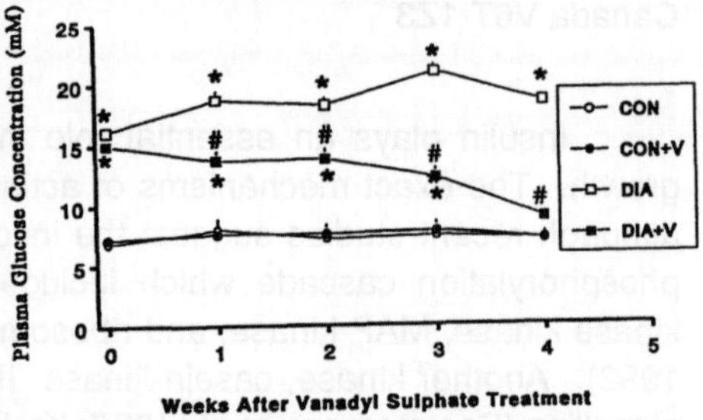

Figure 1. Plasma glucose levels in the one-month control (CON) and diabetic (DIA) rats. V indicates vanadyl sulphate treatment. *$p<0.05$ vs CON, #$p<0.05$ vs DIA, n=7-13.

2. Protein kinase activities in the one-month diabetic rats. (1) As MBP serves as a useful substrate for MAP kinases, we determined the phosphotransferase activities of unfractionated cytosolic skeletal muscle extracts toward MBP. We observed that the basal MBP kinase (MBPK) activities were significantly decreased in the one-month diabetic rats (21%) and the treatment with vanadyl sulphate produced a slight restoration (about 10%) in the basal MBPK activities (Figure 2, panel A). No significant activation of MBPK was observed 20 min after insulin injection, which may have resulted from the fact that MAP kinase activation follows a very short time course (Ahn *et al.*, 1990). (2) CK-II activity was determined as the phosphotransferase activity of the skeletal muscle extracts toward phosvitin. The basal level of CK-II activity was not significantly changed and treatment with vanadyl sulphate did not have any effects on CK-II activity (Figure 2, panel B). Also, insulin injection failed to elicit activation of CK-II. As the stimulation of CK-II by insulin was very modest even in cultured cells (Sommercorn *et al*, 1987), it is likely that CK-II may not play a major role in the signalling mechanism of insulin action and the defects of insulin action in diabetes. (3) Similar to CK-II, the basal level of S6 peptide kinase (S6PK) was not significantly altered in the one-month diabetic rats. Insulin injection, however, slightly activated S6PK in the diabetic rats (Figure 2, panel C). Vanadyl sulphate treatment did not exert significant effects on S6PK in the control rats, but vanadyl slightly reduced the insulin-stimulation of S6PK in the one-month

diabetic rats (Figure 2, panel C). When 40S ribosome was used as a substrate, insulin injection significantly stimulated the S6 kinase activities (>6-fold), with the controls having a higher insulin-stimulated S6 kinase activity than the diabetic rats (Figure 2, panel D). The treatment with vanadyl sulphate normalized the activation of S6 kinase by insulin injection (Figure 2, panel D). These data suggest the existence of a defect in diabetic rats in the insulin-signalling pathways leading to the activation of S6 kinases. The discrepancies between S6 peptide kinase and S6 protein kinase may have been due the fact that the peptide is not a very specific substrate, and the statistical insignificance due to the small number of animals in each group.

3. Protein kinase activities in the six-month diabetic rats. . The basal activities of MBPK, CK-II, S6PK and S6 kinase were also determined in the skeletal muscle extracts from six-month diabetic rats. We observed that MBPK activity was signficantly lower in the diabetic rats (about 30%), while CK-II activity was not significantly changed. Additionally, a marked decrease in the S6 kinase activity (about 40%) and a slight decrease in the S6PK activity were detected in the diabetic rats. The mechanism for the decreased basal activities of MAP kinase and S6 kinases in diabetic rats is not clear from the present study. It is evident, however, that the decrease becomes more severe with longer terms of diabetes. Fractionation of MBPK activities by MonoQ column chromatography indicated that more than 5 peaks of MBPK activities exist in rat skeletal muscle extracts, while only two of

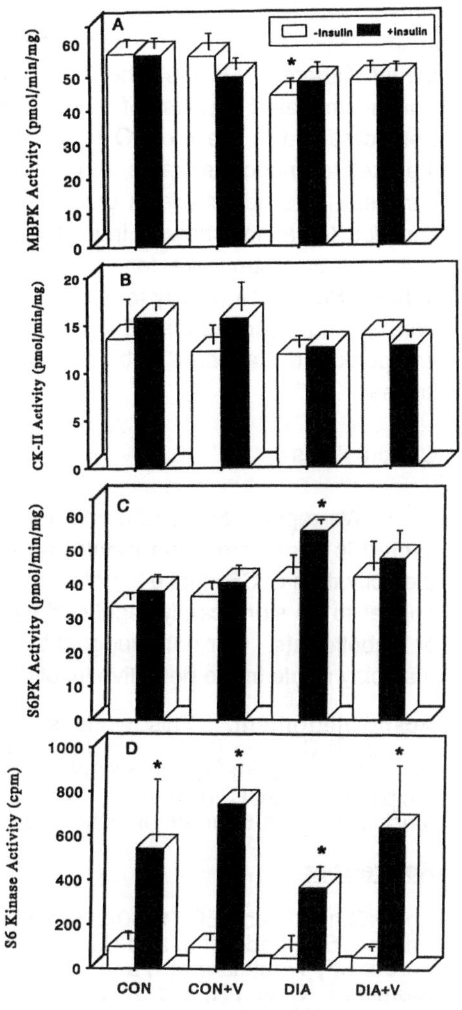

Figure 2. Effects of insulin injection on protein kinase activities in the one-month control (CON) and diabetic (DIA) rats. V indicates the vanadyl sulphate treatment. *$p<0.05$ vs CON or corresponding -Insulin, n=3-6.

them were significantly decreased in the diabetic rats (data not shown), suggesting a differential regulation of these MBPKs during diabetes. On the other hand, two insulin-stimulated peaks of S6 kinase were detected in the MonoQ fractions of the skeletal muscle extracts and only the first peak was detectable in unstimulated samples (data not shown), indicating the first peak is responsible for the changes of basal S6 kinase activity in diabetic rats. We have now purified the S6 kinase in the first peak and it appears to be a novel S6 kinase.

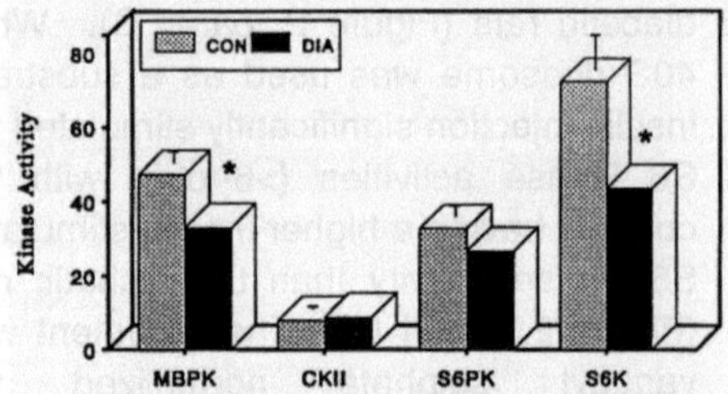

Figure 3. Protein kinase activities in the six-month control (CON) and diabetic (DIA) rats. Kinase activities are expressed as pmol/min/mg, except for S6 kinase which is expressed as fmol/min/mg, *$p<0.05$ vs CON.

Conclusions

We report here for the first time that MAP kinase and S6 kinase activities are altered in short term (one month) and long-term (six month) diabetic rats, and that the activation of S6 kinase is defective in the one-month diabetic rats. CK-II does not appear to be significantly regulated by insulin in rat skeletal muscle in either control or diabetic rats. Our data suggest that alterations in protein serine/threonine kinase may play a role in the defective insulin signalling during diabetes.

Acknowledgment. This work is supported by grants from Medical Research Council of Canada (MRCC) to J.H.M and S.L.P., the Canadian Diabetes Association to J.H.M, and the Heart and StrokeFoundation of B.C. and Yukon to J.D. Y.J.H is a recipient of MRCC fellowship. We thank Ms. Violet Yuen for providing the skeletal muscle from the 6-month diabetic and control rats.

References

Ahn NG and Krebs EG (1990) J Biol Chem 265 : 11495-11501
Brichard SM, Lederer J, and Henquin JC (1991) Diabete & Metabolisme 17: 435-440
Burant CF, Treutelaar MK, and Buse MG (1986) J Clin Invest 77:260-270
Giorgino F, Chen JH, and Smith RJ (1992) Endocrinol 130 : 1433-1444
Heyliger CE Tahiliani AG and McNeill JH (1985) 227:1474-1477
Krebs EG, Eisenman RN, Kuenzel EA, Lichfield DW, Lozeman FJ, Luscher B, and Sommercorn J (1988) Cold Spring Harbor Symposia on Quant Biol 53 : 77-84
Okamoto M, White MF, Maron R, and Kahn CR (1986) Am J Physiol 251 : E542-550
Pelech SL and Krebs EG (1987) J Biol Chem 262 : 11598-11606
Pelech SL and Sanghera JS (1992) Science 257 : 1355-1356
Ramanadham S, Mongold JJ, Brownsey RW, Cros GH, and McNeill JH (1989) Am J Physiol 257 : H904-911
Shechter Y (1990) Diabetes 39 : 1-5
Sommercorn J, Mulligan JA, Lozeman FJ and Krebs EG (1987) Proc Natl Acad Sci USA 84: 8834-8838
Thies RS, Molina JM, Ciaraldi TP, Freidenberg GR, and Olefsky JM (1990) Diabetes 39 : 250-59

PROTEIN PHOSPHORYLATION IN PROKARYOTES — EVIDENCE FOR PKC-LIKE ACTIVITIES IN *ESCHERICHIA COLI*

C. Lippmann[1], S. Güler[1], C. Lindschau[2], C. Alexander[1], H. Urlaub[1] and V.A. Erdmann[1]

[1]Institut für Biochemie, Freie Universität Berlin, Fachbereich Chemie, Otto-Hahn-Bau, Thielallee 63, D-1000 Berlin 33

[2]Abt. für allgemeine Innere Medizin und Nephrologie, Freie Universität Berlin, Klinikum Steglitz, Hindenburgdamm 30, D-1000 Berlin 45

INTRODUCTION

Phosphorylation in eukaryotes for the modulation of enzymatic activities is a common event, often coupled to transmembrane signalling (cell-cell interactions, hormone response) or cell cycle (growth regulation). In microorganisms this chemical modification has long been a matter of controversy due to the lack of conclusive evidence. Different authors detected several phosphoproteins in *E.coli*, depending on growth conditions. In the past decade the occurrence of protein phosphorylation in prokaryotes has been demonstrated in different species but until now very few data are available about the nature of these phosphoproteins and the mechanisms leading to their modification. We clearly identified the elongation factor Tu (EF-Tu) as one of the substrates being post-translationally modified *in vivo* by phosphorylation. The functional aspect of this chemical modulation is still unknown.

In addition, the elongation factor Tu is a substrate for the eukaryotic protein kinase C (Lippmann et al., 1991). The eukaryotic counterpart of EF-Tu, termed EF-1α, seems also to be phosphorylated by PKC *in vitro* and *in vivo* (Venema et al., 1991).

The elongation factor Tu (EF-Tu) is involved in the cyclic elongation of translation, forming a ternary complex with GTP and aminoacyl-tRNAs. The activated complex attaches to the A-site of the ribosome. After peptidyl transfer and concomitant hydrolysis of GTP to GDP, the affinity of EF-Tu to the ribosome and tRNA decreases. EF-Tu·GDP is released and regenerated to EF-Tu·GTP by the elongation factor Ts. EF-Tu belongs to the superfamily of G-proteins, whose functions are regulated by the bound nucleotide.

The site of *in vivo* phosphorylation was determined as threonine 382 by sequencing, and this position could also be phosphorylated *in vitro* by a ribosome associated kinase (see below). The position is conserved in all eukaryotic, archaebacterial and eubacterial organisms (Lippmann et al., 1993, *J. Biol. Chem.* **267**, in press).

METHODS

Three different strategies were applied to search for the eukaryotic type of serine/threonine specific kinases in *E.coli*:

I. Assay for kinase activity in subcellular fractions using EF-Tu as substrate

II. Southern hybridization and PCR of genomic DNA with oligonucleotide probes derived from the catalytic domain of eukaryotic kinases coding for serine/threonine specificity

III. Southern blot analysis with commercial available PKC-specific oligonucleotide probes

NATO ASI Series, Vol. H 76
Tyrosine Phosphorylation/Dephosphorylation and Downstream Signalling
Edited by L. M. G. Heilmeyer Jr.

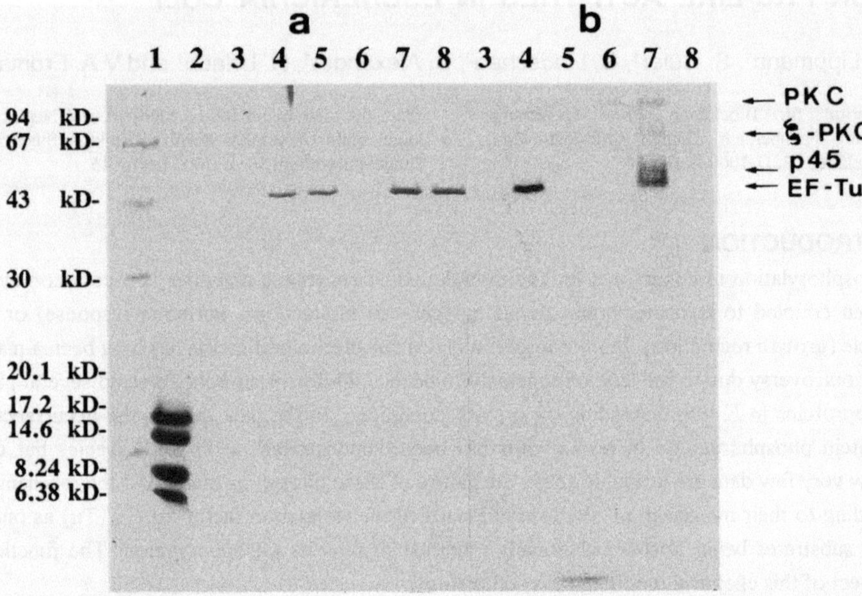

Figure 1: Phosphorylation of *E.coli* EF-Tu by two different preparations of eukaryotic protein kinase C. **Lanes 3 and 4**; PKC from rat brain; **lanes 6 and 7**; PKC from bovine brain. **Lanes 3 and 6** are autophosphorylation controls for the kinases, **lanes 5 and 8** from EF-Tu under the assay conditions without kinase. **Lanes 4 and 7** contain EF-Tu phosphorylated by PKC. **Lanes 1 and 2**; molecular weight marker proteins as indicated. **Panel a**; Coomassie-blue staining; **panel b**; autoradiography.

RESULTS

I. Kinase assay: Kinase activities could only be prepared from fresh grown bacteria cultures. We developed a high speed purification procedure for separating the cleared lysate ("S30") into crude **70S** ribosomes and cytoplasm ("**S100**"). Long term preparation like overnight centrifugation or freezing lead to the complete loss of activities. Kinase activities were assayed within 90 min after cell lysis.

Because EF-Tu like EF-1α (its eukaryotic counterpart) is a substrate for PKC (**figure 1**), which activated form is membrane-associated, we included an additional preparation step in our separation protocol, to prepare crude membranes. All cellular fractions were analyzed under PKC-activating conditions by adding Ca^{2+} and/or phosphatidylserine (PS). The results are displayed in **figure 2**. In the membrane fraction a strong stimulation of the kinase activity can be observed upon addition of phosphatidylserine.

II. PCR: To search for serine/threonine kinase genes two mixed oligonucleotides according to Hanks (1987) were synthesized with length of 17 and 26 nucleotides. Southern blot analysis of restricted genomic DNA of *E. coli* gave positive signals with both probes (data not shown). Some of the positive fragments were cloned and sequence analysis is in progress. In order to get more information about the presence of these sequence(s) not only in *E. coli* but also in other prokaryotes we synthesized a complentary sequence of the 17mer and performed PCR analysis using

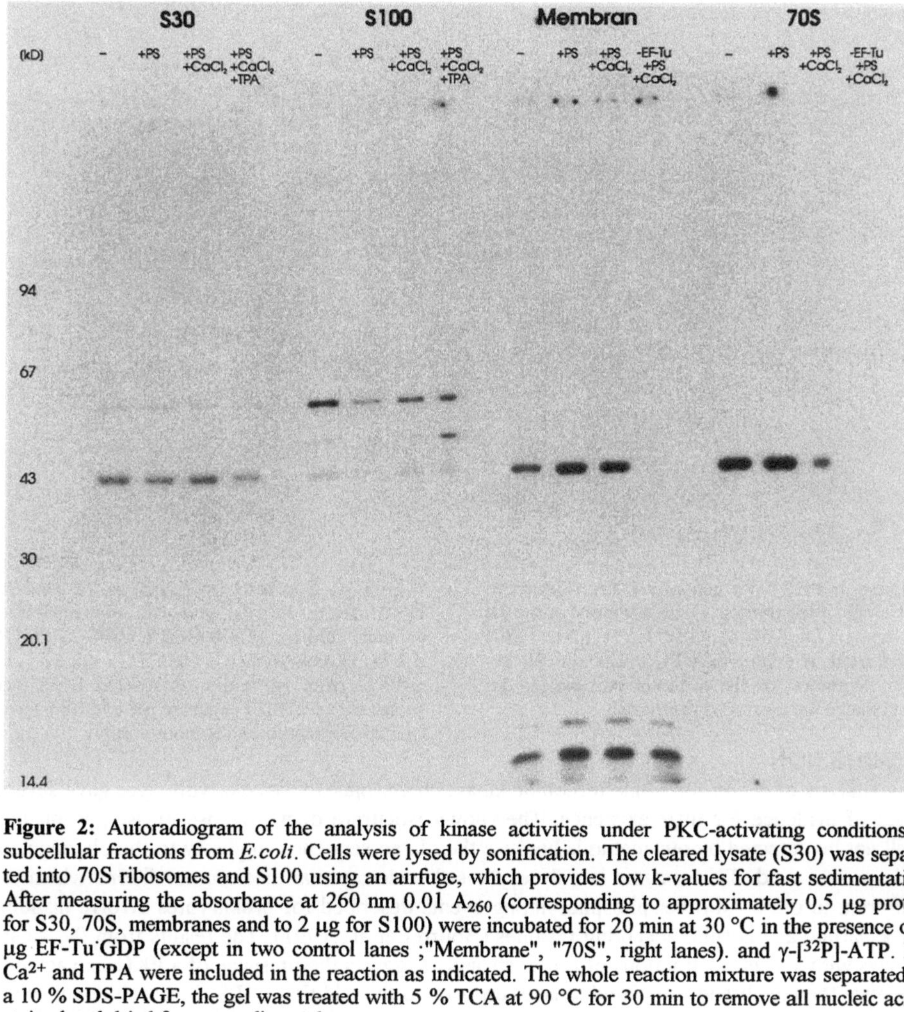

Figure 2: Autoradiogram of the analysis of kinase activities under PKC-activating conditions in subcellular fractions from *E.coli*. Cells were lysed by sonification. The cleared lysate (S30) was separated into 70S ribosomes and S100 using an airfuge, which provides low k-values for fast sedimentation. After measuring the absorbance at 260 nm 0.01 A_{260} (corresponding to approximately 0.5 µg protein for S30, 70S, membranes and to 2 µg for S100) were incubated for 20 min at 30 °C in the presence of 1 µg EF-Tu·GDP (except in two control lanes ;"Membrane", "70S", right lanes). and γ-[^{32}P]-ATP. PS, Ca^{2+} and TPA were included in the reaction as indicated. The whole reaction mixture was separated on a 10 % SDS-PAGE, the gel was treated with 5 % TCA at 90 °C for 30 min to remove all nucleic acids, stained and dried for autoradiography.

the 26mer and the reversed 17mer as primers. A typical result is shown in **figure 3**. In most eukaryotic kinases from which the conserved oligonucleotide sequences are derived, the distance between both sequences is around 25–30 amino acids. This will result in PCR products of 120–150 bp, which are smaller than the observed products from different prokaryotes.

III. Southern blot analysis: In addition, we performed Southern blot analysis of restricted genomic DNA with oligonucleotides specific for different isoenzymes of PKC. For the β- and ζ-isozyme specific probes no signals were detected, but for the α-form the signals persist even under very high stringency washing conditions (final washes 0.2x SSPE, 1 % SDS, 52 °C for 2x 15 min; **figure4**).

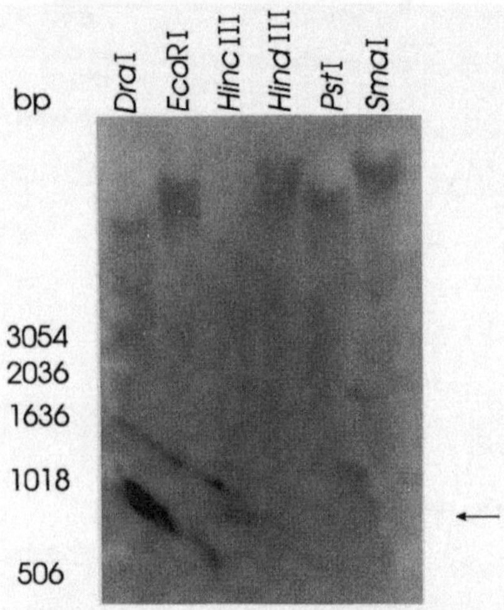

Figure 3: PCR with genomic DNA from three different prokaryotes. The 26mer and the reversed 17mer oligonucleotides (Hanks, 1987) were used as primer and PCR was run 30 cycles. Variation of the different parameters did not change the pattern substancially.

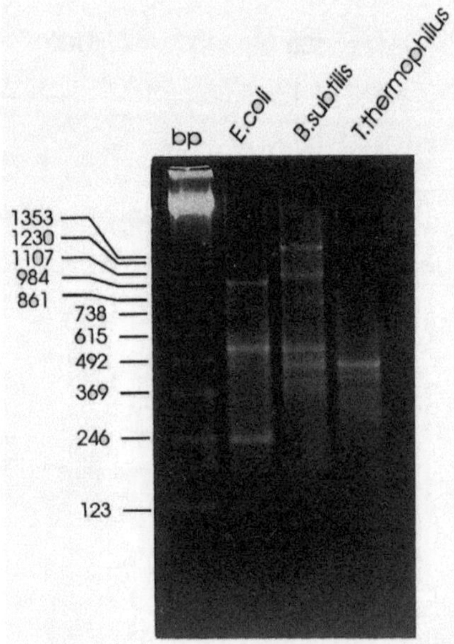

Figure 4: Southern hybridization of genomic DNA from *E.coli* with an oligonucleotide (35mer, Gibco; **CGGGGCCCAGCTTGGC TTTCTCGAACTTCTGCCTG**) specific for α-PKC from rat brain. Restricted DNA was separated on a 0.6 % agarose gel and transferred to a PVDF membrane (Immobilon-N).

DISCUSSION

Our analysis of *E.coli* for serine/threonine kinases homologue to the eukaryotic type gave several lines of evidence for their existence. The known isocitrate dehydrogense kinase/phosphatase will not be detected by our methods because their is no sequence similarity to the oligonucleotide probes used. This enzyme is only expressed under special medium conditions which are different to ours. All other phosphorylating events described for prokaryotes involve histidine or aspartate as phospho-accepting residues (Bourret et al., 1991). The results from the kinase assays with subcelluar fractions of *E.coli* strongly suggests the presence of a PKC-like activity, because the membrane associated kinase responds to the addition of PS. It is well documented that *E.coli* tightly regulates its internal Ca^{2+} concentration (Knight et al., 1991), which is the other second messenger involved in PKC-activation at levels equal to eukaryotic cells. Although we tried to identifiy PKC-like proteins by Western blotting with different mono- and polyclonal antibodies without success, another group reported cross-reactivity of their antibodies with an *E.coli* protein of about 80-85 kd mass (Norris et al., 1991). Currently we are engaged in the isolation of the sequence hybridising with the α-PKC specific oligonucleotid.

REFERENCES

Bourret R.B. et al. (1991) *Ann. Rev. Biochem.*, **60**, 401-441
Hanks S.K. (1987) *Proc. Natl. Acad. Sci.*, **84**, 388-392
Knight M.R. et al. (1991) *FEBS Lett.*, **282**, 405-408
Lippmann C. et al. (1991) *in* Cellular Regulation by Protein Phosphorylation (Heilmeyer, Jr. L.M.G., Ed.), Nato-ASI Serie H56, Springer, Heidelberg, 441-445
Norris V. et al. (1991) *Molec. Microbiol.*, **5**, 2977-2981
Venema R.C.et al. (1991) *J. Biol. Chem.*, **266**, 11993-11998; 12574-12580

PHOSPHORYLATION OF CREATINE KINASE IN MYOGENIC CELLS: EFFECTS OF OKADAIC ACID AND OTHER AGENTS AFFECTING CELLULAR PROTEIN PHOSPHORYLATION

W. Hemmer, M. Skarli[1], H. M. Eppenberger and T. Wallimann
Institute for Cell Biology
ETH-Hönggerberg
CH-8093 Zürich, Switzerland

INTRODUCTION

Creatine Kinase (CK) is a key enzyme of eukaryotic energy metabolism. In chicken, the two cytosolic isoforms, brain- (B-CK) and muscle-type (M-CK), are expressed in a tissue specific manner. During muscle contraction, CK catalyzes the regeneration of ATP from phosphorylcreatine (PCr) and ADP (reviewed in Wallimann et al., 1992). In differentiating chicken embryonic skeletal muscle cells, during myotube formation, a CK isoenzyme transition from B-CK to M-CK takes place, which was also observed in cultured embryonic skeletal muscle cells (Perriard et al., 1978). B-CK is expressed at a high level until shortly after M-CK expression has started. This isoenzyme switch is part of the biochemical differentiation, which accompanies the morphological development of myoblasts. After leaving the cell cycle, postmitotic myoblasts align themselves into arrays and subsequently fuse to form multinucleated myotubes (see references in David et al., 1990). The myotubes show sarcomeric striations and are able to contract in culture, suggesting that their cellular metabolism resembles that of the *in vivo* situation. Essential steps in the process of myoblast fusion depend on the action of various protein kinases, such as the phospholipid/Ca^{2+}-dependent protein kinase (PKC) and the Ca^{2+}/calmodulin-dependent protein (CaM-) kinase (David et al., 1990). Cyclic AMP-dependent protein kinase (PKA) activity is also present during the time of fusion (Rogers et al., 1985).

Several recent reports indicated that rat, chicken and mouse B-CK's are phosphoproteins (Chida et al., 1990a; Mahadevan et al., 1984; Quest et al., 1990). Rabbit M-CK was shown to autophosphorylate *in vitro* (Hemmer et al., 1991). Mouse and human M-CK were phosphorylated *in vitro* by PKA and CaM-kinase (Friedman et al., 1990), and mouse B-CK by PKC (Chida et al., 1990b). The k_m for PCr of the enzyme was altered by phosphorylation (Chida et al., 1990b; Quest et al., 1990). Thus we decided to study the phosphorylation of CK in ^{32}P-labeled cultures of chicken myogenic skeletal muscle cells. The use of okadaic acid (OA), a potent inhibitor of the catalytic subunits of type 1- (PP1) and type 2A phosphatases (PP2A). (Cohen, 1989), and of other substances which cross cell membranes and influence cellular de/phosphorylation processes, enabled us to show that in living skeletal muscle cells, chicken B-CK as well as M-CK are indeed subject to phosphorylation and dephosphorylation.

[1] present adress: Neuromuscular Unit, Department of Pediatrics and Neonatal Medicine, Royal Postgraduate School, Hammersmith Hospital, Du Cane Road, London W12 ONN, UK

NATO ASI Series, Vol. H 76
Tyrosine Phosphorylation/Dephosphorylation
and Downstream Signalling
Edited by L. M. G. Heilmeyer Jr.

MATERIALS and METHODS

Cell culture and ^{32}Pi-labeling: Primary cell cultures were prepared from thigh muscle of 11-day-old white Leghorn chick embryo as described (Entwistle et al., 1986). The cells were plated at 4 x 10^5 cells per 6-cm plate in 3ml standard medium (199 with Earle's salts (Amimed) supplemented with 1mM glutamine, 10% horse serum, 2% chick embryo extract and standard concentrations of penicillin and streptomycin). A complete medium change was carried out 24 hr after plating. At different developmental stages (myoblasts at 42 hr and myotubes at 96 to 120 hr after plating), the cells were starved for 3 hr in phosphate-free medium (MEM with EBS, Amimed), followed by labeling for 150 min with 0.5mCi ^{32}Pi (Amersham, PBS13) per dish and treatment with several agents which influence cellular phosphorylation/dephosphorylation processes. For methionine labeling, 42 hr cultures were starved for 1 hr in serum free 199 medium, lacking methionine but containing antibiotics and glutamin as described above. Then cells were labeled for 2 hr with standard medium supplemented with 100μCi ^{35}S-methionine (Amersham) per dish.

Drug additions: All agents were prepared as ethanolic or aqueous stock solutions and diluted with phosphate-free medium prior to use. The final ethanol concentration was less than 0.1%. Controls were dosed with corresponding amounts of ethanol. The diacylglycerol analog 1-oleoyl-2-acetyl-sn-glycerol (abbreviated as DAG) and 1,2-dioleoyl-sn-glycerol (DAG analog which does not cross membranes) were both used at a final concentration of 15μM. Di-butyryl-3'-5'-cAMP (abbreviated as dbcAMP, Sigma) was used at a final concentration of 10^{-4}M. Okadaic acid (OA, kind gift from Dr. Y. Tsukitani, Fujisawa Chemical Company, Tokyo, Japan) was used at 1 μM, if not indicated otherwise. OA was added only for the last 45 min of labeling, whereas the other agents were present during the complete labeling period.

Immunoprecipitation and gel analysis: After washing in ice cold phosphate buffered saline, cells were harvested in 100 μl of buffer A (1% Triton X-100, 1% sodium deoxycholate, 0.1% SDS, 0.15M NaCl, 20mM Tris [pH 7.4]) and lysed by sonicating (Branson). The lysates were either directly analysed by standard 12% SDS-PAGE or immunoprecipitated as described (Soldati and Perriard, 1991), using affinity purified anti-chicken-B-CK IgG (Quest et al., 1990; Soldati et al., 1990) or anti-chicken-M-CK antisera (Perriard et al., 1978). The immunoprecipitates were analysed by 2D gel electrophoresis, blotting and autoradiography, as previously described (Soldati et al., 1990).

RESULTS and DISCUSSION

To examine the effect of OA on protein phosphorylation, ^{32}Pi-labeled myoblasts were treated with different concentrations of OA. Cell extracts were directly analysed by SDS-PAGE (fig.1). It was not possible to identify the CK isoenzymes as phosphoproteins by this method, since the region were CK migrated (asterisk) was masked by many other phosphorylated proteins. However, the drug treatment specifically increased the phosphorylation states of a variety of proteins, especially in the low molecular weight range (fig.1, arrowheads). Intense phosphorylation of low molecular weight proteins after OA treatment has also been reported when extracts from OA-treated myoblasts were incubated with ^{32}P-γ-ATP (Kim et al., 1991). Likely candidates for some of these proteins are myosin light chains or the ribosomal protein S6 (Cohen, 1989).

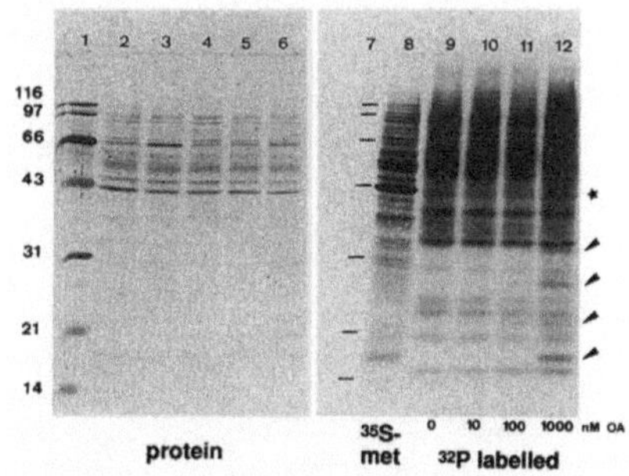

Fig. 1. Effect of OA (at 0, 10, 100 and 1000 nM) on protein phosphorylation in ^{32}Pi-labeled myoblasts, analysed by 12% SDS-PAGE. Lanes 7 to 12 (autoradiogram) correspond to lanes 1 to 7 (protein pattern). Cells analysed in lanes 2,8 were metabolicaly labelled with ^{35}S-methionine, but not with ^{32}Pi. Note the OA-dependent detection of phosphoproteins in lanes 11 and 12, indicated by arrowheads. The position of CK is indicated by an asterisk. Lanes 1,7 correspond to low molecular weight marker; the apparent molecular weights are indicated in kD.

Due to the high complexity of the cellular phosphoprotein pattern, the phosphorylation of CK isoenzymes was investigated by immunoprecipitation of extracts of ^{32}Pi-labeled myotubes using specific affinity purified anti-CK antibodies, followed by 2D gel analysis, blotting and autoradiography (fig.2). Chicken B-CK consists of two major subunit forms (indicated by arrowheads in fig.2) with different pI's but similar molecular weights of about 43 kD (Rosenberg et al., 1981). These subunits, arising by differential splicing, have different N-termini (Wirz et al., 1990). M-CK also consists of two subunits, with a predominant basic form (Schäfer and Perriard, 1988), indicated in figure 2 by long arrows. The molecular basis of the two subunit spots of M-CK, which have identical N-terminal sequences (Hemmer, unpublished), is unknown. CK isoproteins were specifically immunoprecipitated from the cell extracts (fig.2a,c,e,g,i). Phosphoprotein spots derived from phoshorylation of B-CK (fig.2, small arrows) were detected after treatment of cells with dbcAMP, OA, DAG or combinations thereof, but could not be detected in untreated cultures or when preimmune serum was used for immunoprecipitation. Furthermore, when cells were treated with OA, M-CK phosphoprotein spots, showing a slightly more acidic pI than the main M-CK subunit spot, were precipitated. The pI shifts of phosphorylated B- and M-CK relative to the unphosphorylated subunits were very similar, amounting to approximately 0.1 pH units, as calculated from 2D gels. Phosphorylation of CK was found at all developmental stages of myogenic cells investigated. Since the radioactive signals were rather faint, they were harder to detect in early cultures as compared to in differentiated myotubes. The signals assigned as phosphorylated M-CK were only detected in mature cultures, which is consistent with the delayed M-CK expression in these cultures (Perriard et al., 1978). A phosphoprotein of about 40 kD (fig.2d,f,h, and k), which could not yet be identified, was co-precipitated with different batches of anti-B-CK antibodies. This protein was phosphorylated in DAG-stimulated cultures, but no such signal was detectable

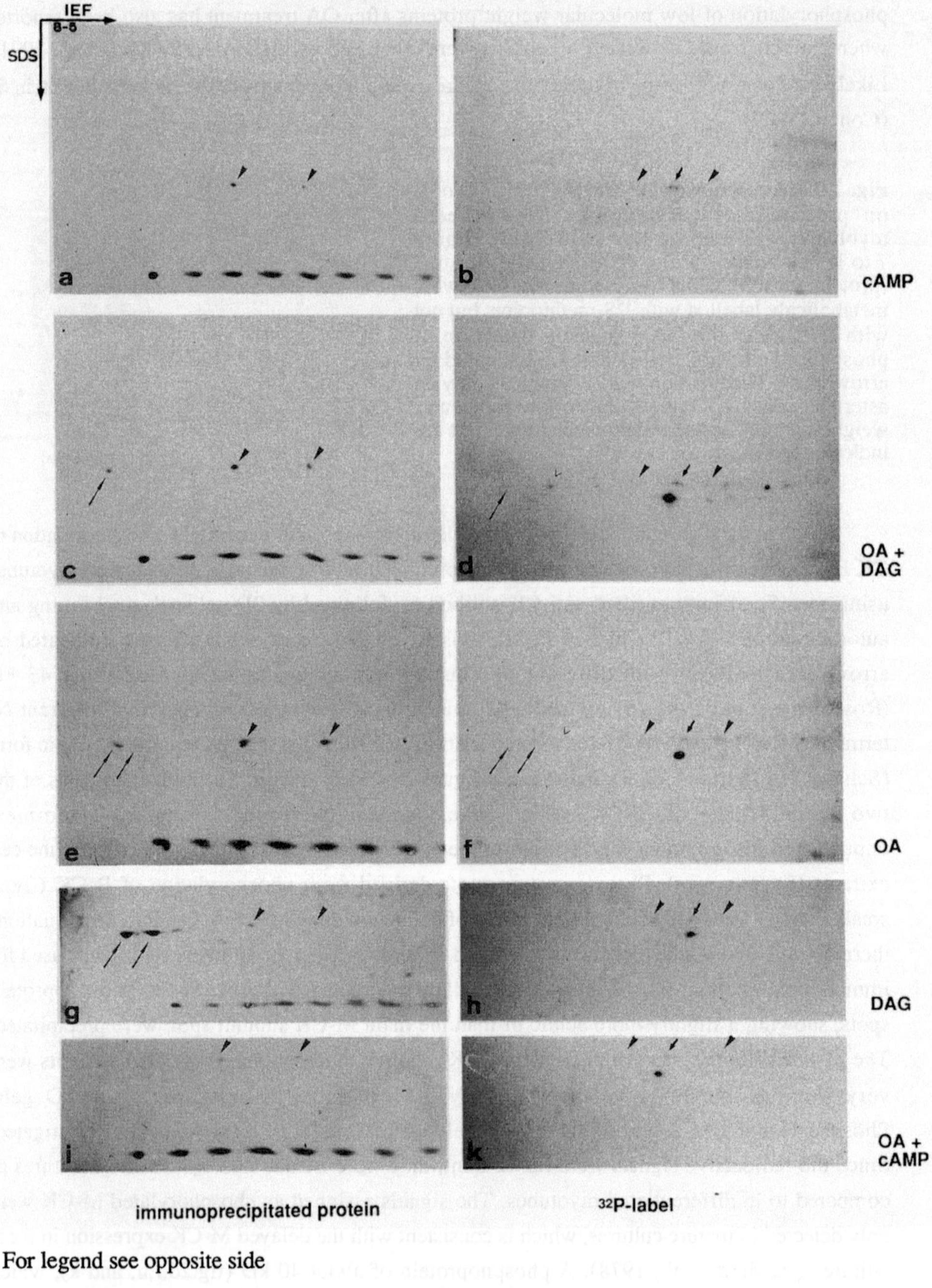

For legend see opposite side

Fig. 2. Lysates of drug-treated, ^{32}Pi-labeled cells, were immunoprecipitated with affinity purified anti-CK antibodies, followed by 2D gel analysis of the immunoprecipitates (pH range 5-8), blotting and autoradiography. Staining for protein with Amidoblack (a,c,e,g,i) revealed the specific precipitation of CK isozymes and the positions of 2D-carbamylate marker spots (carbonic anhydrase, 30 kD, Pharmacia). B-CK protein spots and their respective positions in the autoradiogramms (b,d,f,h,k) were indicated by arrowheads. The position of M-CK is indicated by long arrows, pointing upwards. In (g) 2 µg of purified M-CK were added to confirm its position on the gels. Treatment with dbcAMP (b) revealed a ^{32}Pi-labeled spot (indicated by a short arrow) at the acidic side of the main B-CK subunit spot, likely to arise from phosphorylation of B-CK. The positions of the main subunit spots reveal also a slight radioactive signal. These signals were also seen when the cells were treated with OA (f), DAG (h) or combinations of these substances (d,k), but they could not be detected in 1,2-dioleoyl-sn-glycerol-treated cultures or controls. At a position slightly more acidic than the main M-CK subunit, another phosphoprotein spot, obviously derived from phosphorylation of M-CK, is marked in (b) and (f) by open triangles.

in controls or after treatment with dbcAMP- or 1,2-dioleoyl-sn-glycerol, indicating that this protein was phosphorylated by PKC. OA markedly increased the phosphorylation status of this unknown protein (fig.2d,f), indicating that it is under strong control of phosphatase PP1 or PP2A. Assuming the that dbcAMP-treatment leads to inhibition of PP1 (see below), one might conclude that the P40 protein is dephophorylated by PP2A.

Surprisingly, no obvious quantitative difference in B-CK phosphorylation was found when samples from myogenic cells treated with dbcAMP, DAG or OA were compared. Mouse and rat B-CK were already proposed to be substrates of PKC in mouse keratinocytes (Chida et al., 1990a) or in rat microtubule preparations (Mahadevan et al., 1984), respectively. Thus it may be suggested that DAG stimulated the phosphorylation of B-CK in chicken myogenic cells by activating PKC. Considering that treatment with dbcAMP, similar to that with OA, increased the phosphorylation states of a large number of proteins (not shown), the stimulation of B-CK phosphorylation by dbcAMP may be explained by direct phosphorylation of CK by PKA. But also indirect effects of cAMP-dependent PKA activation, such as inhibition of PP1 through phosphorylation of the heat stable inhibitor-1 (Cohen, 1989), have to be considered for increased phosphorylation of CK. Further pharmacological and biochemical data will be needed to confirm the phosphorylation of CK in chicken myogenic cells by PKC and to decide whether CK is also a substrate of PKA. The primary sequence of chicken B-CK (Wirz et al., 1990) contains putative phosphorylation sites for PKA, PKC and CaM-kinase (Kemp and Pearson, 1990). These protein kinases are all present in chicken myogenic cells (David et al., 1990; Rogers et al., 1985) and thus have to be considered for CK phosphorylation. Phosphorylated forms of chicken B-CK have been shown to display an increased affinity for PCr (Quest et al., 1990). Thus phosphorylation of CK increases the enzyme's ability to transfer the high-energy phosphate from PCr to ADP, thus providing a higher ATP supply for ATP-requiring processes at low PCr levels. Therefore, by phosphorylation of CK, the responsible kinase(s) may help to provide for the increased local energy requirements needed for the intracellular processes that are mediated by these kinases and for their own protein kinase activities.

REFERENCES

Chida, K., K. Kasahara, M. Tsunenaga, Y. Kohno, S. Yamada, S. Ohmi and T. Kuroki. 1990a. Purification and identification of creatine phosphokinase B as a substrate of protein kinase C in mouse skin in vivo. *Biochem. Biophys. Res. Commun.* 173:351-357.

Chida, K., M. Tsunenaga, K. Kasahara, Y. Kohno and T. Kuroki. 1990b. Regulation of creatine phosphokinase B activity by protein kinase C. *Biochem. Biophys. Res. Commun.* 173:346-350.

Cohen, P. 1989. The structure and regulation of protein phosphatases. *Ann. Rev. Biochem.* 58:453-508.

David, J. D., C. R. Faser and G. P. Perrot. 1990. Role of protein kinase C in chick embryo skeletal myoblast fusion. *Dev. Biol.* 139:89-99.

Entwistle, A., D. H. Curtis and R. J. Zalin. 1986. Myoblast fusion is regulated by a prostanoid of the one series independently of a rise in cyclic AMP. *J. Cell Biol.* 103:857-866.

Friedman, D. L., P. R. Puelo and M. B. Perryman. 1990. Phosphorylation of MM creatine kinase in striated muscle. *FASEB J.* 4:A2232.

Hemmer, W., S. J. Glaser, G. R. Hartmann, H. M. Eppenberger and T. Wallimann. 1991. Covalent modification of creatine kianse by ATP: evidence for autophosphorylation. *In* NATO ASI Series: "Cellular Regulation by Protein Phosphorylation." L. M. G. Heilmeyer, eds., Springer-Verlag, Berlin. H 56:143-147.

Kemp, B. E. and R. B. Pearson. 1990. Protein kinase recognition sequence motifs. *Trends Biochem. Sci.* 15:342-346

Kim, H. S., C. H. Chung, M.-S. Kang and D. B. Ha. 1991. Okadaic acid blocks membrane fusion of chick embryonic myoblasts in culture. *Biochem. Biophys. Res. Commun.* 176:1044-1050.

Mahadevan, L. C., S. A. Whatley, T. K. C. Leung and L. Lim. 1984. The brain form of a key ATP-regulating enzyme, creatine kinase, is a phosphoprotein. *Biochem. J.* 222:139-144.

Perriard, J.-C., M. Caravatti, E. R. Perriard and H. M. Eppenberger. 1978. Quantitation of creatine kinase isoenzyme transitions in differentiating chicken embryonic breast muscle and myogenic cell cultures by immunoadsorption. *Arch. Biochem. Biophys.* 191:90-100.

Quest, A. F. G., T. Soldati, W. Hemmer, J.-C. Perriard, H. M. Eppenberger and T. Wallimann. 1990. Phosphorylation of chicken brain-type creatine kinase affects a physiologically important kinetic parameter and gives rise to protein microheterogeneity in vivo. *FEBS Lett.* 269:457-464.

Rogers, J. E., S. Narindrasorasak, G. A. Cates and B. D. Sanwal. 1985. Regulation of protein kinase A and its regulatory subunits during skeletal myogenesis. *J. Biol. Chem.* 260:8002-8007.

Rosenberg, U. B., H. M. Eppenberger and J.-C. Perriard. 1981. Occurence of heterogenous forms of the subunits of creatine kinase in various muscle and nonmuscle tissues and their behaviour during myogenesis. *Eur. J. Biochem.* 116:87-92

Schäfer, B. W. and J.-C. Perriard. 1988. Intracellular targeting of isoproteins in muscle cytoarchitecture. *J. Cell Biol.* 106:1161-1170.

Soldati, T. and J.-C. Perriard. 1991. Intracompartmental sorting of essential myosin light chains: molecular dissection and in vivo monitoring by epitope tagging. *Cell.* 66:277-289.

Soldati, T., B. W. Schäfer and J.-C. Perriard. 1990. Alternative ribosomal initiation gives rise to chicken brain-type creatine kinase isoproteins with heterogeneous amino termini. *J. Biol. Chem.* 265:4498-4506.

Wallimann, T., M. Wyss, D. Brdiczka, K. Nicolay and H. M. Eppenberger. 1992. Intracellular compartmentation, structure and function of creatine kinase isoenzymes in tissues with high and fluctuating energy demands: the 'phosphocreatine circuit' for cellular energy homeostasis. *Biochem. J.* 281:21-40.

Wirz, T., U. Brändle, T. Soldati, J. P. Hossle and J.-C. Perriard. 1990. A unique chicken B-creatine kinase gene gives rise to two B-creatine kinase isoproteins with distinct N termini by alternative splicing. *J. Biol. Chem.* 265:11656-11666.

NON-RADIOACTIVE DETERMINATION OF PHOSPHOAMINO ACIDS BY CAPILLARY ELECTROPHORESIS

M. Heber and H.E. Meyer
Institute of Physiological Chemistry
Ruhr-University
Universitätsstraße 150
4630 Bochum
Federal Republic of Germany

Introduction

Phosphorylation of the specific amino acids serine, threonine or tyrosine is one of the major regulatory mechanisms for control of intracellular function. Up to now, only a few methods are available that allow O-phosphorylated amino acids to be identified after partial hydrolysis without prior ^{32}P-labelling.
A new idea in the non-radioactive determination of phosphoamino acids is the application of capillary electrophoresis, an analytical technique, that achieves a high resolution of soluble components. Due to its high selectivity all phosphoamino acid derivatives can be separated after partial hydrolysis (Hunter and Sefton, 1980) and subsequent coupling with phenylisothiocyanate (Meyer et al., 1991) or dabsyl chloride (Heber et al., 1992).

Materials and Methods

Partial hydrolysis of peptides or proteins as well as derivatization is carried out in the same tube. The sample in 5 x 60 mm test tubes (pyrex glass) is dried under vacuum; 10 μl of 6 N HCl is added to each tube and 400 μl of 6 N HCl and a crystal of phenol are placed beside the tubes on the bottom of the hydrolysis vessel. The vessel is closed, evacuated until HCl starts boiling, then flushed with argon. Both steps are repeated twice. Hydrolysis is performed at 110°C for 2 h. Subsequently the dried samples are derivatized with phenylisothiocyanate or dabsyl chloride.

NATO ASI Series, Vol. H 76
Tyrosine Phosphorylation/Dephosphorylation and Downstream Signalling
Edited by L. M. G. Heilmeyer Jr.

Analysis of PTH-phosphoamino acids by capillary electrophoresis

For derivatization of amino acids with phenylisothiocyanate initially 10 µl conditioning solution 1 (EtOH, TEA, H_2O; 2:2:1 (v/v)) is added to each test tube, containing the dried hydrolysate or 2 nmol of standard, then the samples are dried under high vacuum. This step is repeated and 10 µl of conditioning solution 2 (EtOH, TEA, H_2O; 8:1:1 (v/v)) is added; the sample is flushed with argon and incubated for 5 min at room temperature and 10 µl of coupling solution (ETOH, TEA, H_2O; 7:1:1:1 (v/v)) is added to each test tube. After flushing with argon this mixture is incubated for 5 min at 50°C and dried under high vacuum. 50 µl of 25 % (v/v) TFA is added and the sample is incubated for 30 min at 50°C. After drying, the PTH-phosphoamino acids are dissolved in 5-100 µl of sample solution, containing 20 % (v/v) aqueous acetonitrile, 0.35 % trifluoroacetic acid, 50 mM sodium chloride. Analysis of PTH-phosphoamino acids by capillary electrophoresis is performed under following conditions:

20 mM sodium citrate buffer, pH 2.5; capillary: fused silica; capillary length: 72 cm (50 cm to detector); inner diameter: 50 µm; injection time (vacuum): 10 sec = 30 nl injection volume; voltage: -25 kV; temperature: 30°C; rise time: 1 sec; A.U.F.S. 0.008; polarity: negative (minus mode); detection: 261 nm.

Analysis of dabsyl phosphoamino acids by capillary electrophoresis

For derivatization with dabsyl chloride first 55 µl of 20 mM $NaHCO_3$, pH 9.0 is added to the test tubes containing the dried hydrolysates or 2 nmol of standard phosphoamino acids then followed by 280 µl of 2 mM dabsyl chloride in acetone. The test tubes are sealed with parafilm, stirred and placed in a heat block at 70°C for 15 min. Subsequently, the samples are dried under vacuum. The dabsyl phosphoamino acids are dissolved in 50-200 µl of 70 % (v/v) aqueous ethanol and taken for capillary electrophoresis analysis under these conditions: 25 mM sodium citrate, pH 4.2; capillary: fused silica; capillary length: 72 cm (50 cm to detector); inner diameter: 50 µm; injection time (vacuum): 10 sec = 30 nl injection volume; voltage: -25 kV; temperature: 30°C; rise time: 1 sec; A.U.F.S. 0.008; polarity: negative (minus mode); detection: 463 nm.

Separations of phosphoamino acids were established by using an Applied Biosystems Model 270 A Analytical Capillary Electrophoresis System.

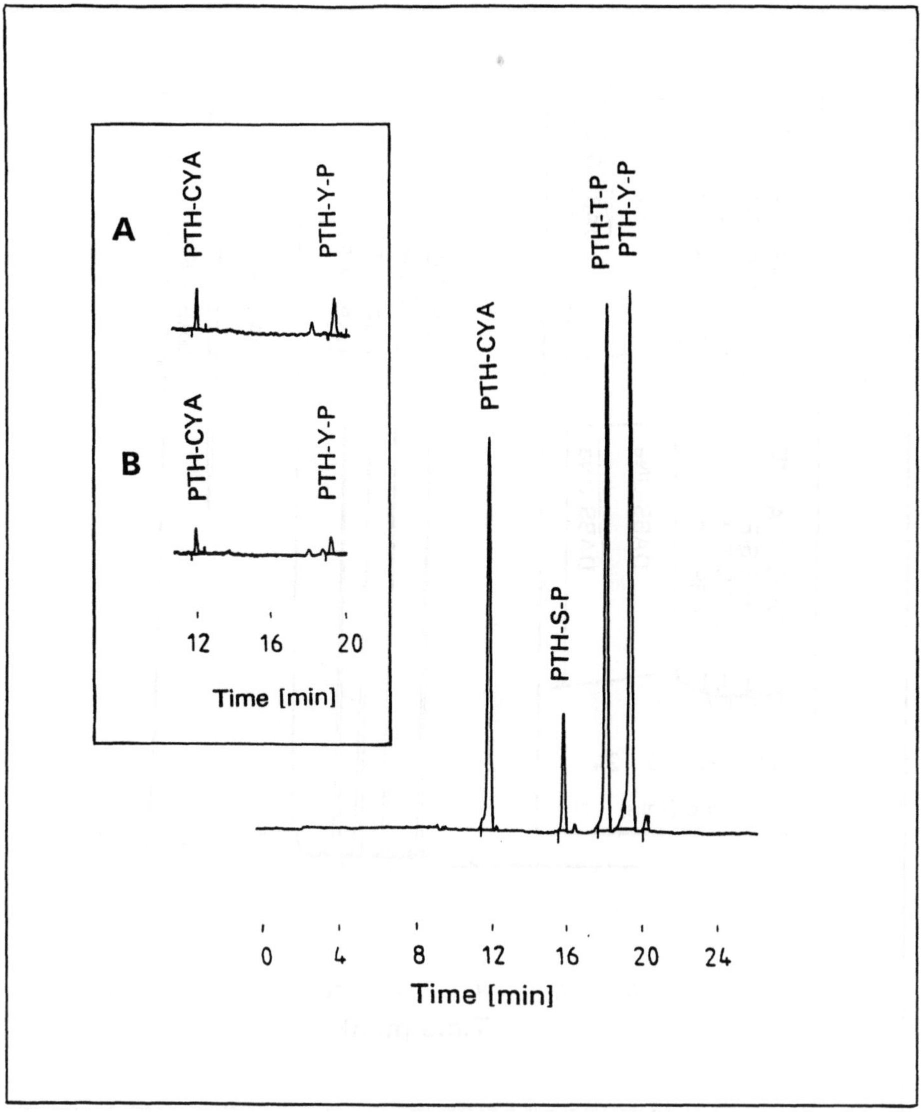

Figure 1
Separation of PTH-amino acids by capillary electrophoresis.
2 nmol of cysteic acid (CYA), phosphoserine (S-P), phosphothreonine (T-P) and phosphotyrosine (Y-P), respectively, are transformed to their PTH-derivatives and dissolved in 100 μl sample solution. 30 nl is analyzed by capillary electrophoresis. Insert, detection limit for PTH-phosphotyrosine: **A**, 1.0 pmol/μl; **B**, 0.25 pmol/μl.

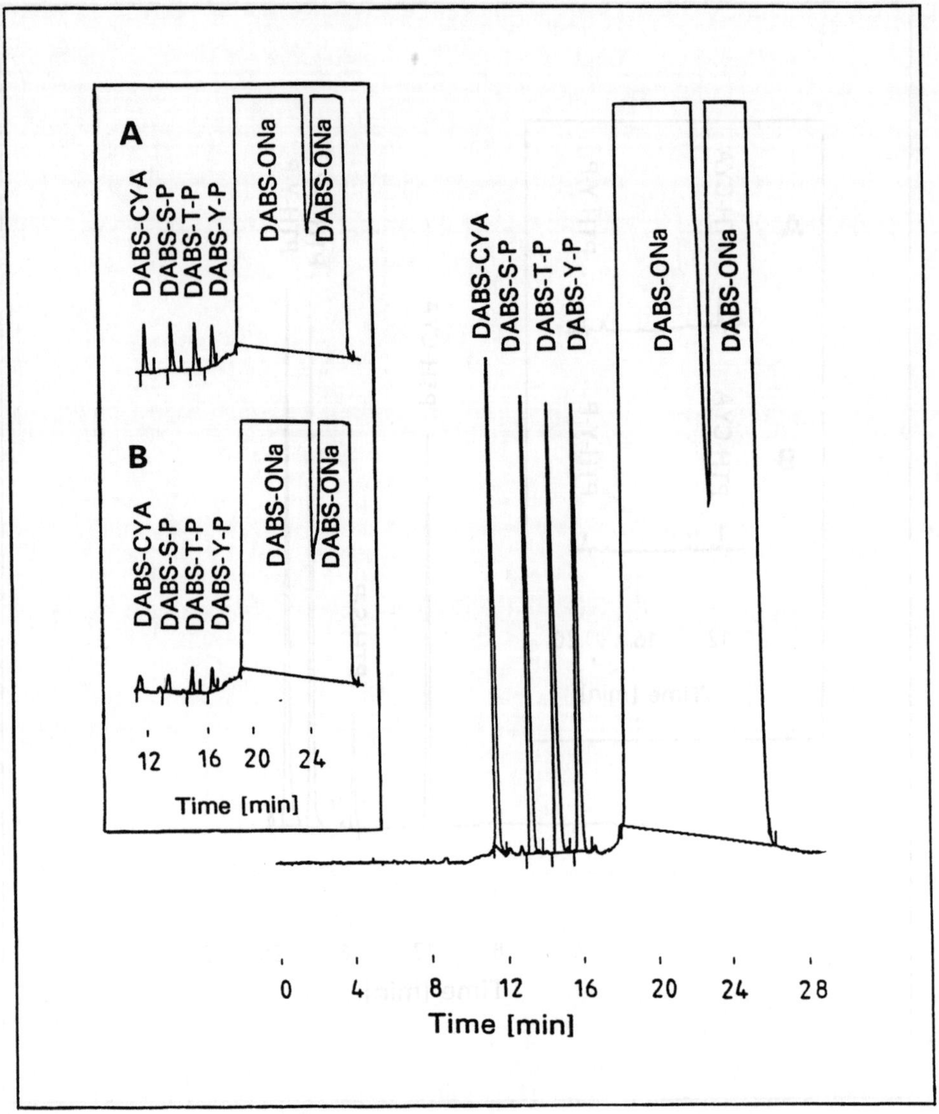

Figure 2
Identification of dabsyl amino acids by capillary electrophoresis
1 nmol of cysteic acid (CYA), phosphoserine (S-P), phosphothreonine (T-P) and phosphotyrosine (Y-P), respectively, are coupled with dabsyl chloride. The dabsyl derivatives are dissolved in 100 µl sample solution. 30 nl is analyzed by capillary electrophoresis. Additional signals are hydrolyzed reagent (DABS-ONa). Insert, detection limit for dabsyl phosphoamino acids: **A**, 1.0 pmol/µl; **B**, 0.25 pmol/µl.

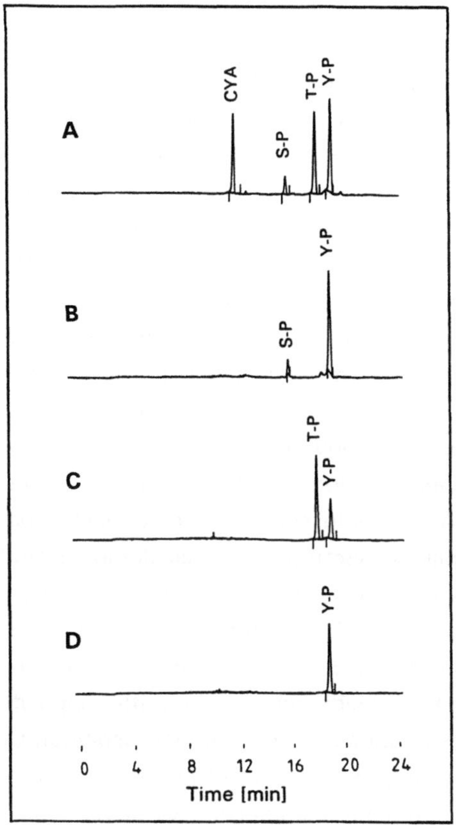

Figure 3

Identification of O-phosphorylated amino acids in phosphopeptides as PTH-derivatives by capillary electrophoresis.

After partial hydrolysis and coupling with phenylisothiocyanate, the samples are dissolved in 50 μl sample solution. 30 nl is analysed by capillary electrophoresis.

A, Analysis of a standard mixture containing 10 pmol /μl of PTH-cysteic acid (CYA), PTH-phosphoserine (S-P), PTH-phosphothreonine (T-P) and PTH-phosphotyrosine (Y-P).

B, 1.8 nmol of the diphosphopeptide A Sp Yp S A is used for hydrolysis. PTH-phosphoserine (S-P) and PTH-phosphotyrosine (Y-P) are recovered with a 2% and a 22% yield, respectively.

C, 1.3 nmol of the diphosphopeptide A Tp Yp S A is used for hydrolysis. PTH-phosphothreonine (T-P) and PTH-phosphotyrosine (Y-P) are recovered with a 27% and 12% yield, respectively.

D, 1.5 nmol of the phosphopeptide L R R A Yp L G is used for hydrolysis. PTH-phosphotyrosine (Y-P) is recovered with 24% yield.

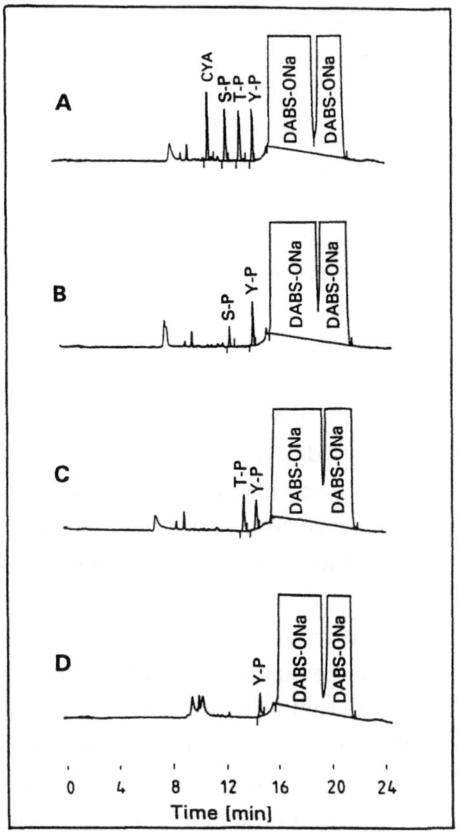

Figure 4

Determination of O-phosphorylated amino acids as dabsyl derivatives by capillary electrophoresis.

After partial hydrolysis and derivatization with dabsyl chloride, the samples are dissolved in 100 μl sample solution. 30 nl is analysed by capillary electrophoresis.

A, Analysis of a standard mixture containing 5 pmol/μl of dabsyl cysteic acid (CYA), dabsyl phosphoserine (S-P), dabsyl phosphopthreonine (T-P) and dabsyl phosphotyrosine (Y-P). Additional signals are hydrolyzed reagent (DABS-ONa).

B, 1 nmol of ths diphosphopeptide A Sp Yp S A is used for hydrolysis. Dabsyl phosphoserine (S-P) and dabsyl phosphotyrosine are recovered with a 10 % and 24 % yield, respectively.

C, 1.5 nmol of the diphosphopeptide A Tp Yp SA is used for hydrolysis Dabsyl phosphothreonine (T-P) and dabsyl phosphotyrosine (Y-P) are recovered with a 16% and 14% yield, respectively.

D, 1.2 nmol of the phosphopeptide A T Yp S A is used for hydrolysıs. Dabsyl phosphotyrosine (Y-P) is recovered with a 17% yield.

Results and Discussion

Coupling of amino acids with phenylisothiocyanate leads to stable derivatives of phosphotyrosine and phosphothreonine (**figure 1**). PTH-phosphoserine is unstable because the phosphate group is splitted off by ß-elimination. Separation is achieved by using an acidic buffer system at pH 2.5 and negative electrical polarity. Under these conditions only the negatively charged PTH-phosphoamino acids and PTH-cysteic acid migrate into the detector direction against the electroendosmotic flow. The positive or neutral standard amino acids migrate into the opposite direction and are not detected. The detection limit is about 0.25 pmol/μl for phosphotyrosine (see **figure 1**, insert B). **Figure 2** demonstrates that phosphotyrosine and phosphothreonine can be determined in phosphopeptides by capillary electrophoresis analysis non-radioactively after partial hydrolysis and modification to their PTH-derivatives.

Labelling of phosphoamino acids with dabsyl chloride achieves stable derivatives of all three phosphoamino acids. A complete separation of dabsyl phosphoamino acids can be obtained by capillary electrophoresis as demonstrated in **figure 3**. Dabsyl phosphoamino acids and dabsyl cysteic acid remain negatively charged and are detected at 463 nm. The detection limit of dabsyl phosphoamino acids is about 0.25 pmol/μl (see **figure 3**, insert B). As shown for the PTH-derivatives, the dabsyl standard amino acids migrate into the opposite direction and are not detected under these conditions. An optimal resolution by capillary electrophoresis can be achieved only within a small pH-range. Dabsyl derivatization has been applied to phosphoamino acid analyses of phosphopeptides. After partial hydrolysis and subsequent coupling with dabsyl chloride phosphoserine, phosphothreonine and phosphotyrosine can be identified in phosphopeptides by capillary electrophoresis (**figure 4**).

As demonstrated, capillary electrophoresis allows the non-radioactive determination of phosphoamino acids and permits a fast and simple screening procedure for phosphoproteins and phosphopeptides in the low picomolar range.

References

Heber M, Liedtke C, Korte H, Hoffmann-Posorske E, Donella-Deana A, Pinna LA, Perich J, Kitas E, Johns RB, Meyer HE (1992) Non-radioactive determination of PTH and DABSYL phosphoamimo acids by capillary electrophoresis. Chromatographia 33: 347-350

Hunter T, Sefton M (1980) Transforming gene product of Rous sarcoma virus phosphorylates tyrosine. Proc. Natl. Acad. Sci. U.S.A. 77: 1311-1315

Meyer HE, Hoffmann-Posorske E, Donella-Deana A, Korte H (1991) Nonradioactive analysis of phosphotyrosine-containing peptides. Methods in Enzymology, Academic Press New York, 201: 206-224

CHARACTERIZATION OF PHOSPHOPROTEINS BY ^{31}P-NMR SPECTROSCOPY

L.M.G. Heilmeyer, Jr, K. Jaquet and K. Korte
Institut für Physiologische Chemie
Abteilung für Biochemie Supramolekularer Systeme
Ruhr-Universität Bochum
Universitätsstr. 150
4630 Bochum 1
Germany

Since the original observation of Fischer and Krebs - nearly 40 years ago - that protein phosphorylation is a means of regulating enzymatic activity it has been well established that hundreds of proteins can be phosphorylated and dephosphorylated in cells and that nearly every aspect of cellular regulation involves this mechanism (for an overview see Heilmeyer (ed.), 1991). Over one hundred protein kinases are known today in their primary structure and it might well be that more than a thousand are encoded in the human genom (Hunter, 1987)). They form a family of proteins comprising several classes which for example can be differentiated on the basis of their substrates as protein serine/threonine kinases and protein tyrosine kinases. Meanwhile, so called dual specificity kinases have been detected which can phosphorylate both serine and tyrosine residues on a particular protein (see G. Thomas, page 31 and 299 in this volume). On the regulation of these protein kinases there is very detailed information available. On a molecular level, however, it is very seldom known how the incorporated phosphates affect the target protein, i.e. catalytic activity, association behavior, localization in the cell, etc. First hand information is obtained from crystallographic studies in which the interactions of incorporated phosphate groups with other residues of the polypeptide backbone can be identified. However, there are only very few examples studied.

The method of choice to study these kinds of interactions is ^{31}P-NMR spectroscopy. It may yield information on the environment and interactions of these phosphate groups and, more interestingly, on the dynamics of these processes. Basic principles are described in several papers (Brauer and Sykes, 1984; Wüthrich, 1976); thus in this article mainly one special case will be described, namely cardiac troponin, which we have studied extensively with this technique.

To demonstrate which information can be obtained by ^{31}P-NMR spectroscopy the structure and the ^{31}P-NMR spectrum of glycogen phosphorylase *a* will be discussed briefly. Each subunit of this dimeric enzyme is known to be phosphorylated on serine 14; the phosphate

NATO ASI Series, Vol. H 76
Tyrosine Phosphorylation/Dephosphorylation
and Downstream Signalling
Edited by L. M. G. Heilmeyer Jr.

group is incorporated by phosphorylase kinase. This phosphorylation occurs in parallel to the conversion of the inactive *b* to the active *a* form. Fig. 1 shows the ribbon diagram of glycogen phosphorylase as it is constructed employing crystallographic data (Acharya et al., 1991). Characteristically, the position of the N-terminus is not well defined, the diagram starts with aminoacid 10. The first α helix begins at residue 18 which together with the helix α_2 forms a contact site of the two subunits. Upon phosphorylation (the approximate position is indicated in Fig. 1 with a "P") a conformational change occurs. In glycogen phosphorylase *b* the N-terminal tail is folded against a hallow in the surface of the subunit (solid trace in Fig. 1). If phosphate is present at serine 14, this N-terminal tail swings around and folds up against the subunit-subunit interface. The newly introduced phosphate makes contact to arginine 69 of the α_2 helix and to arginine 43 present in the cup of the second subunit. Thus, this ion pair formation seems to trigger the conformational change finally leading to the enhanced activity of phosphorylase *a*.

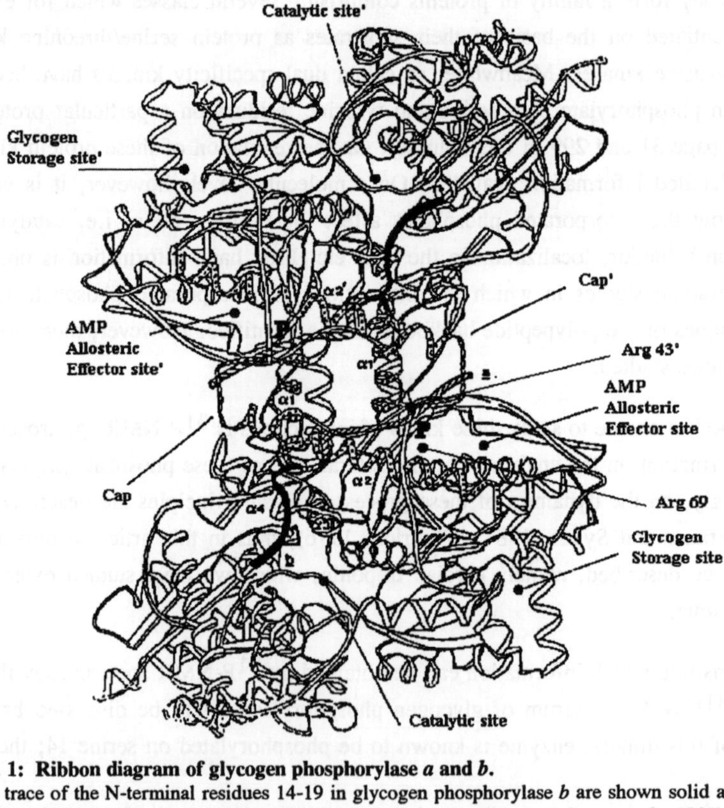

Fig. 1: Ribbon diagram of glycogen phosphorylase *a* and *b*.
The trace of the N-terminal residues 14-19 in glycogen phosphorylase *b* are shown solid and the residues 10-23 in glycogen phosphorylase *a* are shaded (according to Acharya et al., 1991).

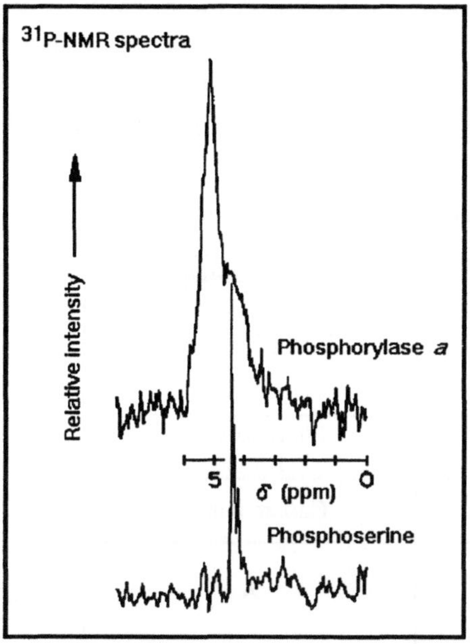

Fig. 2: ^{31}P-NMR spectrum of glycogen phosphorylase *a*
16 mg/ml were dissolved in 0.1 M MOPS pH 7, 0.2 mM EDTA, 10 mM 2-mercaptoethanol, 0.2 M NaCl; 40800 scans were accumulated. The lower spectrum shows free phosphoserine.

The ^{31}P-NMR spectrum of glycogen phosphorylase *a* (reconstituted with pyroxidoxalphosphonate to avoid interference in the interpretation of the spectrum) is shown in Fig. 2. At ca. 5 ppm one major and a minor signal appearing as a shoulder can be seen. The chemical shift identifies both signals as phosphoserine (Sperling et al., 1979). For comparison the spectrum of free phosphoserine is shown. It is solubilized in the same buffer in which phosphorylase *a* is present. Clearly, the shoulder represents a form of phosphorylase *a* in which the phosphoserine is exposed to the solvent. The leftward shift of the major signal is due to deprotonation which causes the shift to lower fields. The deprotonation is readily explained by the ion pair formation with arginine 43' and 69 as can be concluded from the crystallographic data (see above). Furthermore, in comparison to free phosphoserine exhibiting a half line width of ca. 2 Hz a much broader half line width of ca. 20 Hz is observed in both signals of glycogen phosphorylase *a*. This broadening can be interpreted to be due to a dynamic equilibrium between the bound form of this N-terminal region and the buffer exposed form; the bound form represents the major signal and the buffer exposed form the shoulder. This exchange in itself causes line broadening (Hörl et al., 1979). Furthermore, the tight binding of the phosphate onto the protein structure also leads to an enhanced half line width. Thus the chemical shift as well as the half line width of the signal yield information on the environment of these phosphate groups in a protein.

A disadvantage of the method is the high amount of protein needed to record the spectra. Furthermore, the protein must be stable at room temperature over many hours since the spectra have to be accumulated over long times. These limitations allow only a restricted number of proteins to be studied; it is, however, the only method yielding this kind of information.

Table 1: Phosphorylation of Cardiac Troponin (Holocomplex)

Species	Subunit	Protein Kinase
Rabbit Bovine	TnT	Troponin T-Kinase (Casein Kinase)
Rabbit Bovine	TnT	C
Rabbit Bovine	TnT	Phosphorylase Kinase
Rabbit Bovine	TnT	Ca/Cm Kinase
Rabbit Bovine Human	TnI	A
Bovine	TnI	C
Bovine	TnI	G
Rabbit	TnI	Ca/Cm Kinase
Rabbit Bovine	TnC	-

Since approximately 20 years cardiac troponin is known to be a phosphoprotein. Table 1 lists the subunits of cardiac troponin which can be phosphorylated by different protein kinases. The tropomyosin binding subunit, TnT, as well as the inhibitory subunit, TnI, can be phosphorylated by four protein kinases; the Ca^{2+} binding subunit, TnC, does not seem to be phosphorylated. A ^{31}P-NMR spectrum of native freshly isolated cardiac holotroponin reveals four absorption bands between 4.5 and 4.9 ppm; this demonstrates that only serine residues are phosphorylated (Fig. 3). All signals show a half line width of 2 to 4 Hz; i.e. all phosphate groups can freely rotate, they are not tightly bound to the protein backbone.

These signals are affected differentially by saturation of the complex with divalent metal ions (Fig. 4): Ca^{2+} binding to TnC has no effect, Mg^{2+}, however, drastically changes the chemical shift (Swiderek et al., 1990). It is of no importance if first Ca^{2+} is added and then Mg^{2+} or first Mg^{2+} and then Ca^{2+}. These shifts demonstrate that small changes in the environment - due to conformational changes of the protein - can easily be detected by ^{31}P-NMR spectroscopy (Beier et al., 1988).

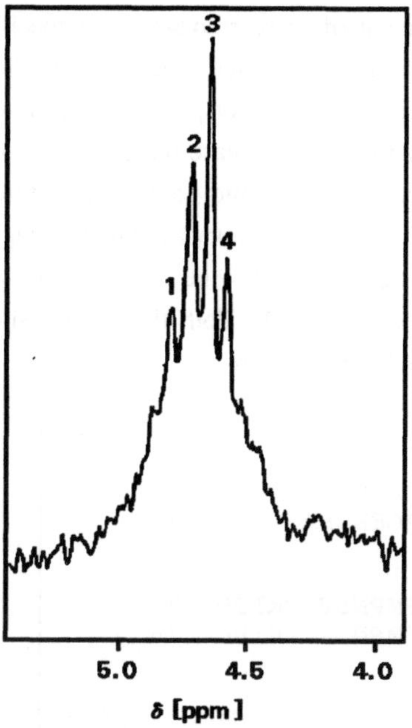

Fig. 3: ^{31}P-NMR spectrum of bovine cardiac troponin (25 mg/ml) containing 2.35 mol of phosphate/mol of troponin.
The signals are detected at 4.81 ppm (1), 4.73 ppm (2), 4.67 ppm (3) and 4.60 ppm (4). 68709 scans were accumulated. The puffer contains 0.5 M KCl, 50 mM MOPS, pH 7.2, 2 mM DTE, 1 mM EGTA, 2 mM $CaCl_2$.

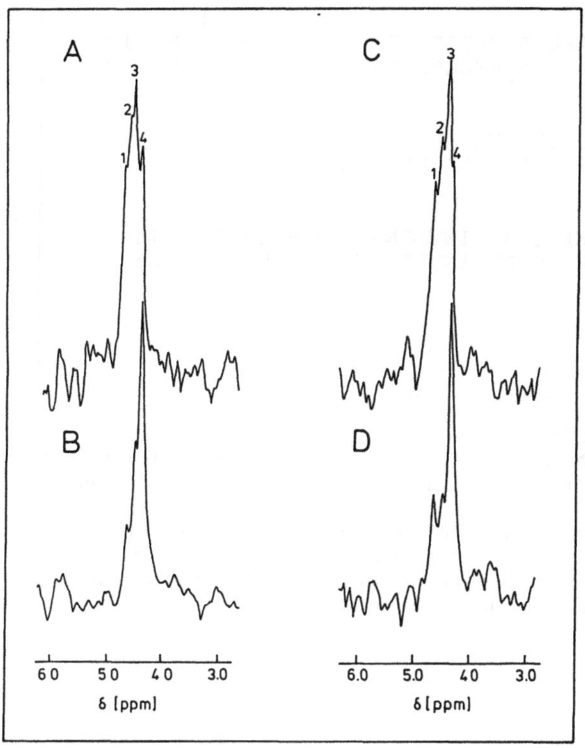

Fig. 4: Dependence of the chemical shifts of ^{31}P-NMR spectra signals of bovine cardiac holotroponin on metal ion binding.
The measured troponin (5.9 mg/ml) contains 2.0 mol of phosphate/mol of troponin (0.7 molP/mol TnT, 1.3 mol P/TnI). The buffer contains 0.5 M KCl, 5.0 mM MOPS, pH 7.2, 2 mM DTE, 1 mM EGTA.
A: 19488 scans were accumulated. Four ^{31}P-NMR signals were detected when the probe was free of Mg^{2+} and Ca^{2+}. B: Saturation of troponin by addition of 10 mM $MgCl_2$ (22133 scans). C: Troponin was saturated with Ca^{2+} by adding 1.1 mM $CaCl_2$ (23975 scans). D: Addition of 1.1 mM $CaCl_2$ and 10 mM $MgCl_2$ (20320 scans).

One of the more difficult tasks consists in the assignment of these resonances to specific phosphorylated sites in the primary structure of the protein. *In vitro* studies have shown that cardiac TnT can be phosphorylated on three sites (Fig. 5). The major phosphorylated site is serine 1. This serine can easily be phosphorylated by troponin T kinase (casein kinase II) (Villar-Palasi and Kumon, 1981) due to the presence of a glutamic acid residue three aminoacids downstream. Serine 176 can be phosphorylated by phosphorylase kinase (Raggi et al., 1989) and serine 194 further downstream only present in the bovine cardiac sequence can be phosphorylated by protein kinase C (Swiderek et al., 1990). Both these latter sites seem to be phosphorylated to a low degree in the isolated holocomplex.

```
               P
Bovine         SDVEEAVEEYEEQEEAAEEEHEEAVEEEAGG---   31
Rabbit         SDLEEVVEEYEEEQE----EQEAGEEEEAGGGAE   30

-EAEAGEPCTAEDGEEEEGREAEDGPVEEFKPKPRPFMPNLVPPKIPD  78
AEAETEETQAEEDGQEEEDKEDEDGPVEESKPKPRPFMPNLVPPKIPD  78

GERVDFDDIHRKRMEKDLNELQTLIEAHFENRKKEEEELVSLKDRIEKRR  128
GERVDFDDIHRKRMEKDLNELQTLIEAHFENRKKEEEELVSLKERIEKRR  128

                                               P
AERAEQQRIRAEREKERQTRLAEERARREEEESRRKAEDEARKKKALSNM  178
AERAEQLRIRAEREKERQNRLAEERARREEEESRRKAEDDARKKKALSNM  178

               P
MHFGGYIQKAQTERKSGKRQTEREKKKKILAERRKVLAIDHLNEDQLRE  227
MHFGGYIQKQA--------QTEREKKKKILAERRKVLAIDHLNEDQLRE  219

KAKELWQNIYDLEAEKFDLQEKFKQQKYEINVLRNRINDNQKVSKTR  274
KAKELWQSIYNLEAEKFDLQEKFKQQKYEINVLRNRINDNQKVSKTR  266

GKAKVTGRWK  284
GKAKVTGRWK  276
```

Fig. 5: Bovine and rabbit cardiac TnT

Both sequences of bovine and rabbit cardiac TnT are shown. The phosphorylation sites of bovine cardiac TnT are marked with "P". The complete sequences of the major isoforms of rabbit cardiac TnT (Pearlstone et al., 1986) and bovine cardiac TnT (TnT-1, Leszyk et al., 1987) are derived solely from protein data. The sequences are about 90 % identical. (A noteworthy difference is a unique eight residue delition in rabbit cardiac TnT, which corresponds to beginning of exon 14. This region of rabbit cardiac TnT is otherwise very similar to all other known cardiac and skeletal TnT sequences.)

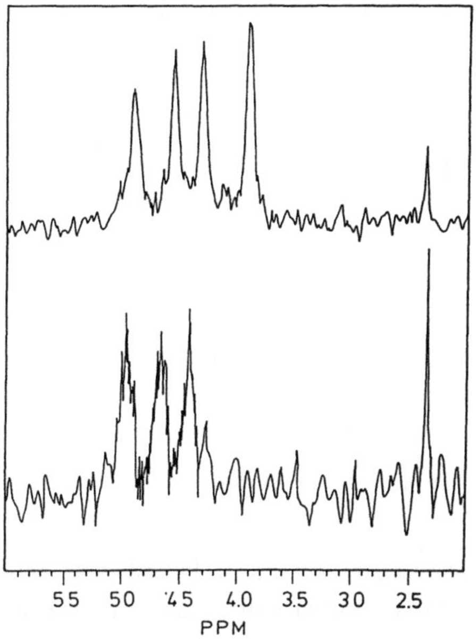

Fig. 6: ^{31}P-NMR spectra of cardiac holotroponin.
The buffer contains 0.5 M KCl, 50 mM MOPS, pH 7.2, 1 mM EGTA and 2 mM DTE. The upper spectrum shows native troponin (20 mg/ml, 86000 scans). In the lower spectrum troponin (30 mg/ml) is dephosphorylated with alkaline phosphatase (37000 scans). Inorganic phosphate as an internal pH standard showed a resonance at a chemical shift of 2.3 ppm.

With alkaline phosphatase TnT can be dephosphorylated specifically (Villar-Palasi and Kumon, 1981). A comparison of the ^{31}P-NMR spectrum of the freshly isolated holocomplex with that containing the completely dephosphorylated T subunit is shown in Fig. 6. Only one signal at the highest field disappears, thus the three remaining signals between 4.4 and 4.9 ppm result from phosphate groups present in troponin I. Based on the sequence previously published by Grand et al. (1976) only one phosphoserine residue (20) should be present in rabbit cardiac troponin I; serine 20 was shown to be a substrate for the cyclic-AMP-dependent protein kinase. Serine 148 can be phosphorylated by phosphorylase kinase. In order to assign these signals to particular phosphoserine residues in the bovine cardiac troponin I sequence analyses had to be performed to localize these phosphoserine residues. Phosphoserine can be determined in peptides with the S-ethylcysteine method (Meyer et al., 1991). The results show clearly that phosphoserine is present only in the N-terminal region and that a bisphosphorylated peptide can be isolated. Sequence analysis reveals that the previously published sequence is incorrect (Mittmann et al., 1990), probably due to difficulties phosphoserine causes during Edman degradation in the spinning cup sequencer. Elimination of phosphoric acid yields dehydroalanine which blocks further Edman degradation. Applying the S-ethylcysteine method on N-termini of several cardiac troponin I isoforms from different species shows that they all contain a characteristic motif of three arginine followed by two serine residues (Table 2). We call this motif a duplicated minimal recognition motif for the cyclic-AMP-dependent protein kinase (Mittmann et al., 1992). This protein kinase recognizes the sequence arginine, arginine, any aminoacid x followed by the serine which is phosphorylated. If this sequence is arranged two times in

Table 2: N-terminal aminoacid sequences of heart specific troponin I from different species

(1) Rabbit		ADESTDA-AG	EARPAPAPV	R RRSS	ANYRAY	ATE
(2) Bovine		ADRSGGSTAG	DTVPAPPPV	R RRSS	ANYRAY	ATE
(3) Man		ADGSSDA-AR	EPRPAPAPI	R RRSS	-NYRAY	ATE
(4) Rat	M	ADESSDA-AG	EPQPAPAPV	R RRSS	ANYRAY	ATE

duplicated minimal recognition motif for cAMP-dependent protein kinase

RRXS
 RRXS

RRRSS

(1) Mittmann et al., 1990
(2) Swiderek et al., 1988; Leszyk et al., 1988
(3) Mittmann et al., 1990; Vallins et al., 1990 (cDNA)
(4) Murphy et al., 1991 (cDNA)

register and added up, this duplicated minimal recognition appears. Indeed, both serine residues can be phosphorylated by the cyclic-AMP-dependent protein kinase as shown in Fig. 7. The cAMP-dependent protein kinase incorporates maximally two mole of phosphate into cardiac troponin I when present in the holocomplex or into isolated troponin I, or into a

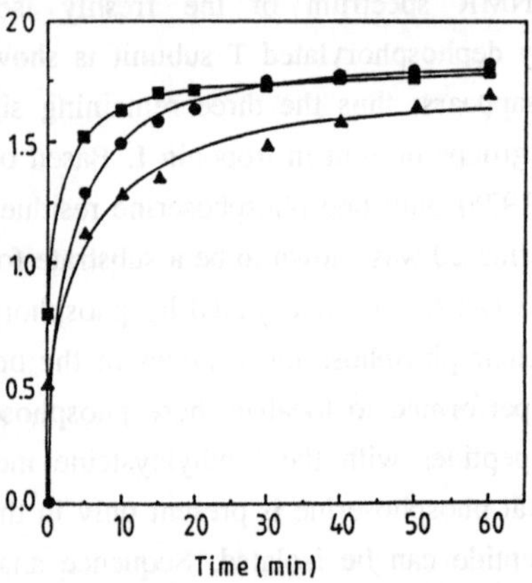

Fig. 7: Phosphorylation of the synthetic peptide PVRRRSSANY, the cardiac holotroponin complex and the cardiac troponin I subunit by cAMP-dependent protein kinase
The endogenous phosphate content of the holocomplex (■) before incubation with protein kinase was 0.8 mol of phosphate/mol of troponin I. The employed troponin I subunit (▲) contained 0.5 mol of phosphate/mol of protein. The synthetic decapeptide (•) was free of endogenous phosphate.

decapeptide, comprising ten aminoacid residues surrounding this duplicated minimal recognition motif. Employing differently phosphorylated species of cardiac troponin present in the holocomplex, an assignment of the observed three ^{31}P-NMR resonances to individual phosphoserine residues is possible. Fig. 8 compares the spectra of the holocomplex carrying phosphate only on troponin I with a complex being phosphorylated to a degree of

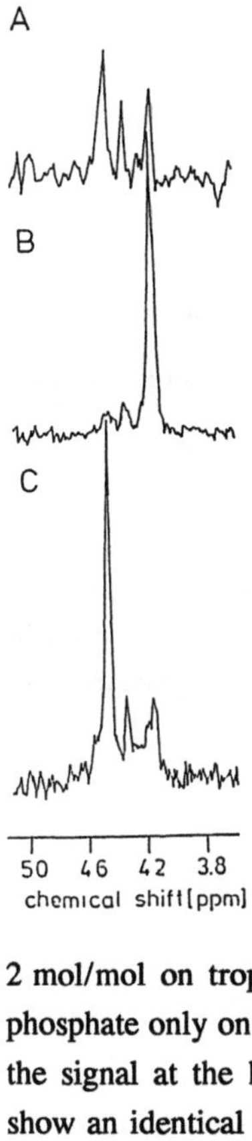

Fig. 8: ^{31}P-NMR spectra of cardiac holotroponin of different phosphorylation degrees
The pH is constant in all three recorded spectra. A: 15 mg/ml cardiac troponin containing 1.4 mol of phosphate/mol of troponin I. B: 20 mg/ml cardiac troponin containing 1.9 mol of phosphate/mol of troponin I after phosphorylation with cAMP-dependent protein kinase. C: 20 mg/ml cardiac troponin containing 0.6 mol of phosphate/mol of troponin I at serine 24.

2 mol/mol on troponin I and a complex which was shown by sequence analysis to carry phosphate only on serine 24 (bovine cardiac troponin I sequence). It can be seen clearly that the signal at the highest field represents the bispospho form, *i.e.* both phosphate groups show an identical chemical shift. The signal at the lowest field is due to the presence of a species phosphorylated only at serine 24, and the signal in the middle, by exclusion, is due to the presence of a species phosphorylated on serine 23. Thus, ^{31}P-NMR spectroscopy clearly differentiates the two monophospho species from each other and from the bisphospho species. The differences in the chemical shifts observed are due to small differences in the respective pKa values of the phosphoserine residues. Upon alkalinisation up to pH 8 all the signals merge, upon acidification they are differentiated again. From these partial titration curves pKa values have been calculated for each phosphoserine residue. In the bisphospho form both phosphates exhibit identical pKa values (Fig. 9) being ca. 0.6 units higher than that of free phosphoserine. Also, both monophospho species show slightly higher pKa values than free phosphoserine. The TnI subunit isolated from the native holotroponin complex exhibits only two signals at chemical shifts identical to those observed for each monophospho species (Fig. 10). Neither reconstitution of isolated TnI with TnC nor with TnT alone reconstitutes the three signal spectrum. However, if the holocomplex is reconstituted from TnT, TnI and TnC the characteristic three signal spectrum reappears; then a separate signal for the bisphospho species can be identified. The

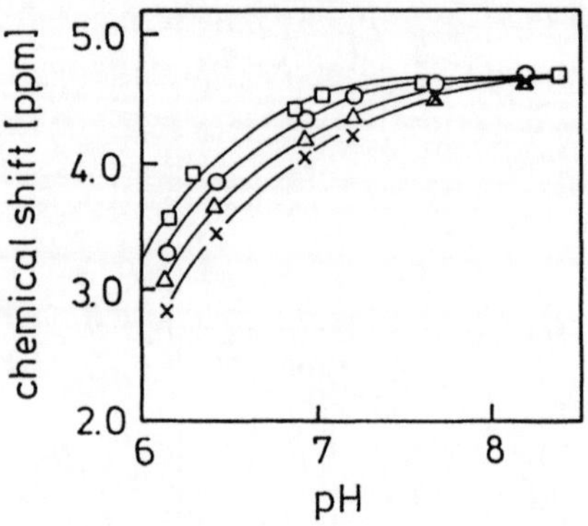

Fig. 9: pH titration of cardiac holotroponin
Cardiac troponin (20 mg/ml) containing 1.4 mol of phosphate/mol of troponin I was titrated. The symbol (o) represents phosphoserine at position 24, (Δ) that at position 23 and (x) the bisphospho form. Free phosphoserine (□) is employed as a reference.

bisphospho form must be protonated to a higher degree than the corresponding monophospho forms or the bisphospho form without subunit-subunit interaction. Probably the TnC subunit carrying acidic groups interacts specifically with the bisphospho form which results in this enhancement of the protonation degree.

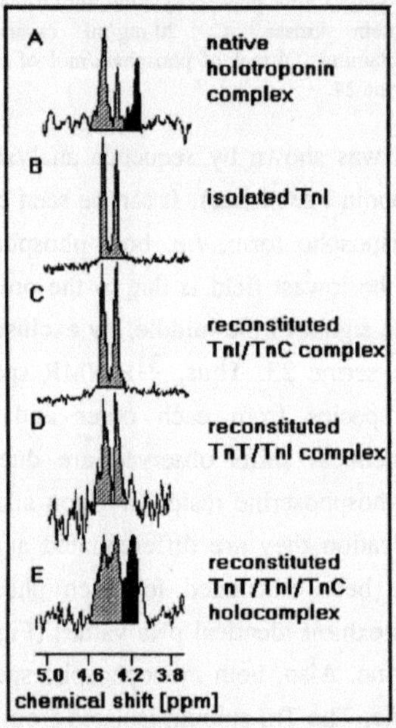

Fig. 10: ^{31}P-NMR spectra of holotroponin, isolated troponin I and reconstituted complexes
All probes contained 1.4 mol of phosphate/mol of troponin I. In each sample the pH is adjusted such that the chemical shift of inorganic phosphate is constant.
A: 15 mg/ml cardiac troponin (25568 scans); B: 10 mg/ml cardiac troponin I isolated from the material used in (A) (7932 scans); C: 20 mg/ml cardiac troponin I reconstituted with cardiac troponin C (18172 scans); D: 14 mg/ml cardiac troponin I reconstituted with cardiac troponin T (34368 scans); E: 15 mg/ml reconstituted cardiac holotroponin complex (T, I, C; 45248 scans).

In summary, ^{31}P-NMR spectroscopy is a powerful method to detect even weak interactions of phosphate groups incorporated into a protein. These interactions then trigger conformational changes bringing about alterations in association, catalytic activity, binding affinities or any other change in function.

References

Acharya, K. R., Stuart, D. I., Varvill, K. M. and Johnson, L. N. (1991) Glycogen phosphorylase b. World Scientific, Singapore, New Jersey, London, Hong Kong

Beier, N., Jaquet, K., Schnackerz, K. D. and Heilmeyer, L. M. G., Jr. (1988) Eur. J. Biochem. 176:327-334

Brauer, M. and Sykes, B. D. (1984) Methods in Enzymol. 107:36-81

Grand, J. A., Wilkinson, J. M. and Mole, L. E. (1976) Biochem. J. 159:633-641

Gusev, N. B., Dobrovolski, A. B. and Severin, S. E. (1980) Biochem. J. 189:219-226

Heilmeyer, L. M. G. (ed.) (1991) Cellular Regulation by Protein Phosphorylation. Springer-Verlag, Berlin, Heidelberg, New York, London, Paris, Tokyo, Hong Kong, Barcelona, Budapest

Hörl, M., Feldmann, K., Schnackerz, K. and Helmreich, E. J. M. (1979) Biochemistry 18:2457-2464

Hunter, T. (1987) Cell 50:823-829

Leszyk, J., Dumaswala, R., Potter, J.D., Gusev, N. B., Verin, A. D., Tobacman, L. S. and Collins, J. H. (1987) Biochemistry 26:7035-7042

Meyer, H. E., Hoffmann-Posorske, E. and Heilmeyer, L. M. G., Jr (1991) in: Methods in Enzymology, Volume 201 *Protein Phosphorylation* (T. Hunter and B. M. Seften, eds.), pp 169-185

Mittmann, K., Jaquet, K. and Heilmeyer, L. M. G., Jr. (1990) FEBS Lett. 273:41-45

Mittmann, K., Jaquet, K. and Heilmeyer, L. M. G., Jr. (1992) FEBS Lett. 302:133-137

Murphy, A. M., Jones II, L., Sims, H. F. and Strauss, A. W. (1991) Biochem. 30:707-712)

Noland, T. A. and Kuo, J. F. (1991) J. Biol. Chem. 266:4974-4978

Pearlstone, J. R., Carpenter, M. R. and Smillie, L. B. (1976) J. Biol. Chem. 261:16796-16810

Raggi, A., Grand, R. J. A., Moir, A. J. G. and Perry, S. V. (1989) Biochimica et Biophysica Acta 997:135-143

Sperling, J. E., Feldmann, K., Meyer, H., Jahnke, U. and Heilmeyer, L. M. G., jr. (1979) Eur. J. Biochem. 101:581-592

Swiderek, K., Jaquet, K., Meyer, H. E., Schächtele, C., Hofmann, F. and Heilmeyer, L. M. G., Jr. (1990) Eur. J. Biochem. 190:575-582

Vallins, W. J., Brand, N. J., Dabbade, N., Butler-Browne, G., Yacoub, M. H. and Barton, P. J. R. (1990) FEBS Lett. 270:57-61

Villar-Palasi, C. andKumon, A. (1981) J. Biol. Chem. 256:7409-7415

Wüthrich, K. (1976) NMR in Biological Research: Peptides and Proteins. North Holland Publishing Company

CK-2, A MULTIFUNCTIONAL PROTEIN KINASE AND ITS ROLE DURING PROLIFERATION

B. Boldyreff, K. Wehr, R. Hecht, and O.-G. Issinger
Institut für Humangenetik
Universität des Saarlandes
W-6650 Homburg
Germany

Casein kinase 2 (CK-2) is a second messenger-independent protein kinase consisting of two or three subunits which form a tetrameric complex, either $\alpha\alpha'\beta_2$ or $\alpha_2\beta_2$. The α subunits (mol. mass 44 - 46 kDa) represent the catalytic subunits of the enzyme, whereas the β subunit is non-catalytic with a molecular mass of 26 kDa. No specific modulator as for the cAMP-dependent protein kinase is known which can dissociate the subunits or regulate the enzyme. CK-2 is unique among the protein kinases in being able to use ATP as well as GTP as a phosphoryl donor. It phosphorylates preferentially serine or threonine residues surrounded by a cluster of acidic amino acid residues (XS/TXXEX).

The level of enzyme activity has been shown to be elevated in transformed cells, during proliferation and during mouse embryogenesis (Table 1), suggesting a functional role in cellular growth, but it also seems to be involved in cellular signalling.

Table 1: Systems with elevated CK-2 activity

elevated CK-2 activity	in comparison to
established cell lines	primary cell cultures
solid tumours	corresponding non-neoplastic tissue
colon carcinoma	colon adenoma
heterotransplanted nude mice tumours	original tumour
cells of the lower part of colon crypts	cells of the upper part of colon crypts
day 12 of mouse embryogenesis (begin of organogenesis)	other stages of mouse embryogenesis

NATO ASI Series, Vol. H 76
Tyrosine Phosphorylation/Dephosphorylation and Downstream Signalling
Edited by L. M. G. Heilmeyer Jr.

In order to study the structure-function relationship of the CK-2 subunits and the role of the individual subunits we have expressed the human CK-2 subunits separately in a prokaryotic expression system using the pT7-7 vector and bacterial strain BL21(DE3). The expressed proteins accounted for approximately 30% of the bacterial protein. We purified them and used them for experiments where we compared their properties to that of the native CK-2 holoenzyme.

One of the first experiments was a sucrose density gradient analysis. We collected 20 fractions and analyzed them by activity measurements and Western blotting. The α subunit sedimented in fraction 15, whereas a mixture of purified α and ß subunit sedimented in fraction 10. This was exactly the position of the tetrameric native CK-2 which we ran in parallel. The main conclusion from this mixing experiment is that as soon as the CK-2 subunits α and ß meet each other they form immediately by a self-assembly mechanism the tetrameric holoenzyme.

A comparison of the activities of the α subunit alone and the holoenzyme showed clearly that the α subunit by itself has a basal activity, but this is increased up to fourfold by adding equimolar amounts of ß subunit.

Next we investigated the influence of denaturing agents on the α subunit and the holoenzyme. As denaturing agents we used high temperature, tryptic digestion, and urea treatment. In the tryptic digestion experiment the enzyme activity was measured in the presence of various amounts of trypsin. In the case of the reconstituted enzyme or the native holoenzyme we found a 50% loss of activity with 1 μg trypsin, but no further loss at higher trypsin concentrations. The α subunit or α subunit + previously trypsinized ß subunit loses 90% of activity with 1 μg trypsin. Similar results were obtained when the enzymes were preincubated at 45°C for 1 to 10 minutes or when the enzyme was treated with urea. Always the holoenzyme was much more stable than the α subunit by itself.

We also looked at the substrate specificity of the α subunit in comparison to the tetrameric holoenzyme. For the synthetic peptide $R_3D_3SD_3$ which we used in most of the activity tests the K_m value was found to be 240 μM for the α subunit alone and 80 μM for the reconstituted enzyme. This is comparable to published results for the native CK-2. We inspected then the substrate specificity for different peptide substrates. The α subunit and the holoenzyme clearly showed differences in the phosphorylation rates toward the different peptides. In particular the activity of the α subunit toward peptides lacking acidic residues in position +1 and +2 was reduced when compared to the activities of either native or recombinant CK-2.

All these results indicate that the ß subunit plays a crucial role in conferring to the α subunit all the characters of the tetrameric holoenzyme. This was also analyzed by CD spectroscopy of the individual CK-2 subunits and of the reconstituted tetrameric complex. Although Chou-Fasman calculations for the α and ß subunit predicted a similar overall conformation, CD spectra showed substantial differences. The α subunit has a higher proportion of α-helical structure than the ß subunit, indicated by strong negative signals at 209 and 222 nm. In contrast the ß subunit is characterized by a high proportion of ß sheet structure. The CD spectrum of the reconstituted CK-2 has again characteristic minimas at 209 and 222 nm. This is quite different from the calculated spectrum obtained as the average of the individual component CD spectra, where neither the minima nor the maxima of the curve were that great than in the experimental spectrum. This indicates that conformational changes occur in the backbone region upon association and that reconstitution results in an increase on the structure of the enzyme. Such changes may explain the increased enzyme activity of the holoenzyme relative to the activity of the α subunit alone. The effect of temperatures between 10 and 60°C on the CD spectra explained, at least in part, the higher stability of the holoenzyme in comparison to the α subunit alone. The spectrum for the reconstituted enzyme remains relatively constant until 50°C. Loss of structure appears then rapidly. In case of the α

subunit loss of structure was observed at lower temperatures. Thus the ß subunit gives a degree of thermal stability to the α subunit and obviously the conformational changes that occur upon association with the α subunit cause a stabilization of the enzyme complex and an increase in α-helical content. The role of the ß subunit in the CK-2 holoenzyme complex can be summarized as following:

1. It is required for maximum activity.
2. It plays a role in substrate recognition since it alters significantly the substrate specificity of the α subunit.
3. It confers stability to the α subunit under denaturing conditions or in the presence of denaturing agents.
4. CD spectra indicate that this protective function results from a stabilization of the conformation of the enzyme complex and an increase in α-helical content.

References

Boldyreff B, Schneider HR, Fritz G, Traffa K, Seitz G, Issinger OG (1989) CKII, a pleiotropic mediator of proliferation and oncogenesis. BioTechForum, Adv Mol Genet 2:91-98

Grankowski N, Boldyreff B, Issinger OG (1991) Isolation and characterization of recombinant human casein kinase II subunits α and ß. Eur J Biochem 198:25-30

Issinger OG, Brockel C, Boldyreff B, Pelton JT (1992) Characterization of the α and ß subunits of casein kinase 2 by far-UV CD spectroscopy. Biochemistry 31:6098-6103

Meggio F, Boldyreff B, Marin O, Pinna LA, Issinger OG (1992) Role of the ß subunit of casein kinase-2 on the stability and specificity of the recombinant reconstituted holoenzyme. Eur J Biochem 204:293-297

Meggio F, Boldyreff B, Marin O, Marchiori F, Perich JW, Issinger OG, Pinna LA (1992) The effect of polylysine on casein kinase-2 activity is influenced by both the structure of the protein/peptide substrates and the subunit composition of the enzyme. Eur J Biochem 205:939-945

Schneider HR, Issinger OG (1989) CKII, a multifunctional protein kinase and its role during proliferation. BiotechEurope 6,1:82-88

IV. Protein Serine/Threonine Phosphate Phosphatases

MOLECULAR COMPLEXITY OF PROTEIN PHOSPHATASE 2A HOLOENZYMES AND DELINEATION OF THEIR FUNCTION

Regina E. Mayer-Jaekel and Brian A. Hemmings
Friedrich Miescher-Institut
P.O. Box 2543
CH-4002 Basel
Switzerland

Introduction

Modification of proteins through the attachment of a phosphate group to serine, threonine or tyrosine residues is a major form of cellular regulation in all eucaryotes (Krebs, 1986). The addition and removal of the phosphate group is mediated by two classes of enzymes, the protein kinases and protein phosphatases, respectively. These regulatory proteins usually respond to second messengers that are generated through hormone, growth factor or neurotransmitter binding to cell surface receptors. Although it is also possible to have intracellular signal transduction pathways involving protein phosphorylation - dephosphorylation networks. However, these are, as yet, poorly understood but will receive considerable attention in the future.

The protein kinases have been extensively studied over the past 20 years, firstly through biochemical techniques, and more recently through the application of molecular biology techniques and reverse genetics. From this work emerges the complex picture of protein kinases modulating protein kinases in a network of signal transduction pathways, that interact both linearly and laterally (Posada & Cooper, 1992; Roberts, 1992; Pelech & Sanghera, 1992). So far over 220 unique protein kinases have been identified, largely due to the power of homology cloning and the PCR technique (Hanks, 1991). Many of the targets of the protein kinase cascades are still to be identified. However, several target molecules have been extensively studied: there are three major groups (a) those involved in acute changes of metabolic pathways, such as, glycogen synthesis and degradation (reviewed in Cohen, 1988), (b) those involved in more long term changes (phenotypic) such as transcriptional activation of target genes (reviewed in Hunter & Karin, 1992), and (c) proteins involved in cell cycle regulation (reviewed in Reed, 1992; Norbury & Nurse, 1992).

Recently it has been appreciated that for reversible phosphorylation to work as a subtle on-off switch mechanism requires the cooperation of the protein phosphatases. For many years this group of enzymes was considered rather dull and even boring. This perception has changed dramatically over the past four years and there has been huge advances in our appreciation of the number and complexity of these enzymes. The spark for this advance was the purification and amino acid sequence analysis of a tyrosine-specific protein phosphatase achieved by Fischer and Tonks, and colleagues at the University of Seattle, Washington (Tonks et al., 1988; Charbonneau et al., 1989). This work led to the identification of the previously cloned leucocyte common antigen CD45 as phosphotyrosine specific phosphatase (Charbonneau et al., 1988). Simultaneously, the discovery that the potent tumor promoter, okadaic acid, acts via inhibition of serine/threonine specific protein phosphatases (Bialojan & Takai, 1988), acted as a catalyst for further studies on the regulation of this type of enzymes. It now emerges that there are around 100 different protein phosphatases when both the tyrosine and serine/threonine subgroups are considered (reviewed by Fischer et al., 1991; Shenolikar & Nairn, 1991; Pot & Dixon, 1992; Chen et al., 1992). Thus their number and complexity are almost equal to the protein kinases. The next major challenge will be the unravelling of the mechanisms of regulation.

In this article we concentrate on the protein phosphatase 2A (PP2A)[1] group of enzymes. This family of phosphoserine/threonine specific enzymes can be distinguished from other protein phosphatases by insensitivity to two heat stable inhibitors, preference for the α subunit of phosphorylase kinase, and the stimulation by polycations (Ingebritsen & Cohen, 1983). More recently different protein phosphatase activities have been classified by their sensitivity to okadaic acid (reviewed in Cohen et al., 1990). However, in the near future this will probably be replaced by the availability of antisera specific for a given phosphatase, thus enabling a detailed characterization of any protein phosphatase holoenzyme.

Protein Phosphatase 2A Holoenzymes

Protein phosphatase 2A holoenzymes have been purified from a variety of species and tissues (reviewed in Ballou & Fischer, 1986; Cohen, 1989; Shenolikar & Nairn, 1991). At least four holoenzymes have been identified that differ in their subunit composition (Figure 1). They all have in common a catalytic subunit of 36 kDa and a 65-kDa regulatory subunit (PR65). These "core" subunits associate with regulatory subunits of either ~54, ~55, ~72 or ~74 kDa to form different heterotrimers. So far it is not clear whether the heterodimeric form exists *in vivo*, or whether it is derived from one of the trimeric holoenzymes by dissociation or proteolysis of the third subunit during purification. The 54- and 55-kDa subunits in the trimeric holoenzymes $PP2A_0$ and $PP2A_1$, respectively, appear to be clearly distinct, since their peptide maps are different and the holoenzymes show different properties (Tung et al., 1985). The trimeric form containing the 72-kDa subunit isolated form rabbit skeletal muscle (Waelkens et al., 1987) may be distinct from another trimeric form isolated from erythrocytes

[1]Abbreviations: PP1, protein phosphatase type 1; PP2A, protein phosphatase type 2A; PR55, PR65 and PR72, 55-, 65- and 72-kDa regulatory subunits of protein phosphatase 2A.

by Usui et al. (1988) which contains a subunit of ~74 kDa. The two holoenzymes have not been directly compared, but they appear to have different chromatographic properties, different sensitivities to polycations and only the 74-kDa subunit is phosphorylated by the cAMP-dependent protein kinase (Usui et al., 1991; Hendrix et al., submitted).

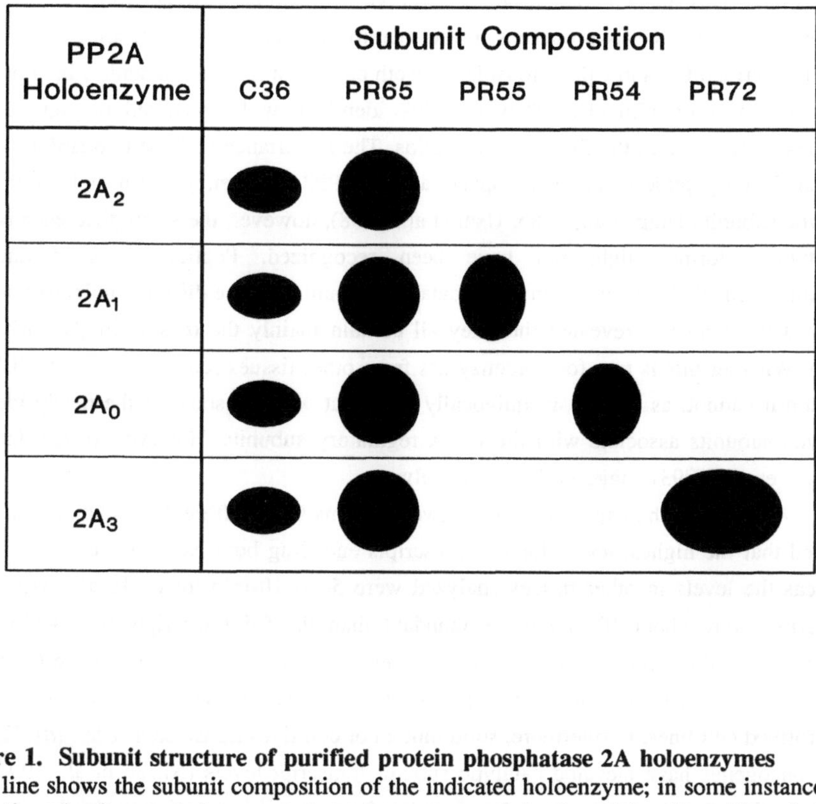

Figure 1. Subunit structure of purified protein phosphatase 2A holoenzymes
Each line shows the subunit composition of the indicated holoenzyme; in some instances two catalytic subunits per holoenzyme have been reported (reviewed by Shenolikar & Nairn, 1991). The nomenclature is taken from Ingebritsen & Cohen (1983). We suggest that the trimeric holoenzyme containing the PR72 subunit be termed $2A_3$. Abbreviations: C36: 36-kDa catalytic subunit; PR54, PR55, PR65 and PR72: 54-, 55-, 65- and 72/74-kDa regulatory subunits.

Molecular Characterization of Protein Phosphatase 2A Subunits

The Catalytic Subunit

The catalytic subunit of PP2A has been cloned from a whole variety of species, including pig (Stone et al., 1987), cow (Green et al., 1987), rabbit (da Cruz e Silva et al., 1987; da Cruz e Silva & Cohen, 1987), human (Arino et al., 1988; Hemmings et al., 1988; Stone et al., 1988), rat (Kitagawa et al., 1988a,b), yeast (Sneddon et al., 1990; Kinoshita et al., 1990), *Drosophila* (Orgad et al., 1990; Mayer-Jaekel et al., 1992), and plants (MacKintosh et al., 1990). This molecule is extremely conserved indicating a strong evolutionary pressure on the maintenance of its properties. From mammals cDNA clones encoding two different isoforms, termed Cα and Cβ, were isolated. Both consist of 309 amino acids with a predicted molecular mass of 35.6 kDa. They are 97% identical, with seven out of eight different residues located within the first 30 amino acids. The importance of these two isoforms is still unclear. From peptide mapping it appears as if all PP2A holoenzyme forms have the same catalytic subunit (Tung et al., 1985; Usui et al., 1988), however, the small difference between the two isoforms might not have been recognized. Peptide sequence data and immunochemical characterization of the catalytic subunit of three different holoenzymes from rabbit skeletal muscle revealed that they all contain mainly the α-isoform (Hendrix et al., 1993). Whether this is true for holoenzymes from other tissues remains to be established. In addition it cannot, as yet, be unequivocally ruled out whether some of the newly identified catalytic subunits associate with the PP2A regulatory subunits. However, recent data from Hendrix et al. (1993) suggests this is unlikely.

Analysis of the expression of the two isoforms from different porcine and rat tissues showed that the highest levels for the transcripts encoding both isoforms occur in the brain whereas the levels in other tissues analyzed were 5- to 10-fold lower. In all cases the Cα transcripts were about 10-fold more abundant than the Cβ transcripts (Khew-Goodall & Hemmings, 1988). Kitagawa et al. (1988a,b) reported increased amounts of the Cβ-isoform transcripts in chemically induced rat liver tumors and of the Cα isoform in several transformed cell lines. Furthermore, solid tumors or cell lines transfected with *ret*II, Ki-*ras* or c-*raf* oncogenes have elevated catalytic subunit transcript levels (Nagao et al., 1990). This suggests that the expression can be modulated by specific signal transduction pathways influenced by these oncogenes.

Characterization of the two genes encoding the human PP2A catalytic subunits (Figure 2) suggests they arose by gene duplication (Khew-Goodall et al., 1991). Both span ~30 kb and are composed of seven exons with the six introns intervening at identical locations. However, the 5' upstream regions, as well as the regions encoding the 5' and 3' untranslated sequences of each mRNA, are different. The promoters of both genes are very G+C rich and lack TATA and CCAAT consensus sequences, characteristic of many housekeeping genes.

Promoter analysis suggests that the steady-state levels of the α and β transcripts in different human cell lines are due, at least in part, to different promoter activities. The Cα promoter contains the consensus sequence of a cAMP-responsive element 26 base pairs upstream of the transcription start site. Although this region is essential for the basal activity of the promoter, transcription could not be induced by increasing intracellular cAMP levels (Y. Khew-Goodall, unpublished data).

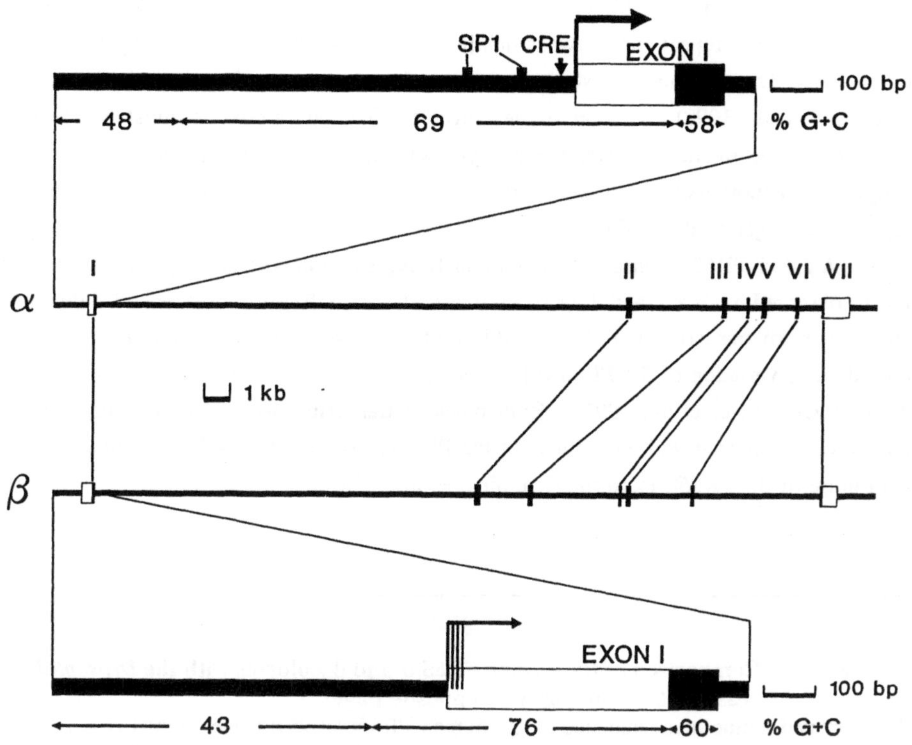

Figure 2. Schematic structure of protein phosphatase 2A catalytic subunit genes
The gene structures are shown in the center with the exons indicated by filled and open boxes for coding and noncoding regions, respectively. Promoter and intron sequences are shown as a solid line. The promoter regions and the first exons are enlarged (top and bottom) to illustrate the transcription start sites (arrows) and the locations of consensus sequences for a cAMP-responsive element (CRE) and Sp1 binding sites (SP1). The G + C nucleotide content of different regions is given in percent.

The 65-kDa Regulatory Subunit (PR65)

Characterization of the PR65 subunit by molecular cloning also revealed the presence of two isoforms, termed PR65α and PR65β, which are 87% homologous (Hemmings et al., 1990). Interestingly, this protein was also isolated by a different approach because of its association with polyoma virus small and middle T antigens and SV40 small t antigen (Walter et al., 1989, 1990; Pallas et al., 1990) (see below). The PR65 protein sequences can be arranged in 15 imperfect repeating units of about 39 amino acids, where certain residues or related residues occur at the same position (Figure 3). The residues conserved between the repeats are also the most highly conserved between species. A molecular model of the protein predicts that each repeat forms two amphiphatic helixes which are paired by hydrophobic interactions. The 15 pairs are stacked on top of each other to form a rod like structure (Ruediger et al., 1992). Deletion studies revealed that the carboxy-terminal 5 repeats are required for the interaction with the catalytic subunit, whereas the polyoma and SV40 T antigens, and most likely also the other "PP2A subunits", interact with the amino-terminal repeats (Ruediger et al., 1992).

A 2.4 kb PR65α transcript is ubiquitously expressed in porcine tissues with the highest levels observed in brain and lowest in liver. The PR65β isoform is encoded by a less abundant 4 kb transcript and shows a different tissue specific expression. This leads to a considerable variation of the PR65 α/β transcript ratio, from approximately 90:1 in heart to 3:1 in liver (Mayer et al., 1991b). Both types of transcripts were also detected in several human cell lines analyzed, with those of the PR65α isoform about 40-times more abundant than those of the PR65β isoform (Hemmings et al., 1990).

Figure 3. Comparison of the human PR65 α and β isoforms with the *Drosophila* and yeast PR65 homologs (see opposite page)

The deduced amino acid sequence of the human PR65α isoform (Hemmings et al., 1990) is arranged to show its repeat structure. The deduced amino acid sequence of the human PR65 β isoform (Hemmings *et al.*, 1990), the *Drosophila* PR65 subunit (Mayer-Jaekel et al., 1992) and the *Saccharomyces cerevisiae TPD3* gene product (Van Zyl et al., 1992) are shown below with identical amino acids indicated by a dash. Gaps, indicated by dots, are introduced to allow optimal alignment with the repeats of the yeast protein. The repeat number is marked on the left of each row and the amino acid position on the right site. The last line gives the consensus residues for that position. The boxed residues correspond to the positions identified earlier as those most highly conserved between the repeats of the mammalian proteins. (Amino acid groups are defined as follows: [1] C; [2] S, T, P, A, G; [3] N, D, E, Q; [4] H, R, K; [5] M, I, L, V; [6] F, Y, W.)

```
   Hα                                                               M A A A D . . G D D S         9
   D                                                                - - - S - K S V - - -        11
   Y           M S G A R S T T A G A V P S A A T T S T T S T T S N S K D S - . . S N E -         35

 1 Hα   L Y P I A V L I D E . . . L R N E D V Q L R L N S I K K L S T I A L A L G V E R T R S E   50
   Hß                     - . . . - - - - - - - - - - - - - - - - - - - - - - - - - - - - - -   50
   D    - - - - - - - - - - . . . - K - - - - - - - - - - - - - - - - - - - - - - E - - - - - -   52
   Y    - - - L - L - M - - . . . - K H D - I A N - V E A M - - - D - - - - - - - P - - - - N -   76

 2 Hα   L L P F L T D T I Y . . . D E D E . V L L A L A E Q L G T F T T L V G G P E Y V H C       88
   Hß   - - - - - - - - - - . . . - V - - . - - - - - - - - - - N - - G - - - - - D F A - -       88
   D    - I - - - - E - - - . . . - - - - . - - - - - - D - - - N - - S - - - - - - F A M Y       90
   Y    - I - - - - E V A QD. .   - - - - . - F A V - - - - - - K - V P Y I - - - Q - A T I      115

 3 Hα   L L P P L E S L A T . . . V E E T V V R D K A V E S L R A I S H E H S P S D L E A H      127
   Hβ   - - - - - - N - - - . . . - - - - - - - - - - - - - - - Q - - Q - - T - V A - - - Y      127
   D    - I - - - - - - - - . . . - - - - - - - - - - - - - - - T V A A - - - A Q - - - I -      129
   Y    - - - V - - I - - S . . . A - - - - - - E - - - D - - N N V A Q - L - Q E Q - F S D      154

 4 Hα   F V P L V K R L A G . . . G D W F T S R T S A C G L F S V C Y P R V S S . A V K A E      165
   Hß   - - - - - - - - - S . . . - - - - - - - - - - - - - - - - - - - - A - N . - - - - -      165
   D    V - - T L Q - - V S . . . - - - - - - - - - - - - - - - - - - - - - T Q . P - - - -      167
   Y    - - - - I E H - - T . . . A - - - S - K V - - - - - - K S V I V - I K D D S L R K N      193

 5 Hα   L R Q Y F R N L C S . . . D D T P M V R R A A A S K L G E F A K V L E L D N V K S E      204
   Hß   T - - Q - - S - - - . . . - - - - - - - - - - - - - - - - - - - - - - - - S - - - -      204
   D    - - A N - - K - - Q . . . - E - - - - - - - - - N - - - - - - - - V - T E Y L - - D      206
   Y    I L A L Y L Q - A Q . . . - - - - - - K - - V G K N - P I L I D L - T Q N L G L - TDEDWDY  238

 6 Hα   I I P M F S N L A S . . . D E Q D S V R L L A V E A C V N I A Q L L P Q E D L E A L      243
   Hß   - V - L - T S - - - . . . - - - - - - - - - - - - - - - S - - - - - S - D - - - - -      243
   D    L - - N - V Q - - Q . . . - D - - - - - - - - - - - - - T - - - - - - - D - V - H -      245
   Y    - S N I - Q K I I N . . . - N - - - - K F - - - D C L I S - L K F F N A K G D - S H      277

 7 Hα   V M P T L R Q A A E . . . D K S W A V R Y M V A D K F T E L Q K A V G P E I T K T D      282
   Hß   - - - - - - - - - - . . . - - - - R - - - - - - - R - S - - - - - M - - K - - L N -      282
   D    - L - - - - - C - S . . . - S - - R - - - - - - E - - V D - - - - - - - - - - R V -      284
   Y    T Q D L - N S - V KLIG    - E A - - - - - - - A - - R - S D - A S Q F S S N Q A Y I D E    320

 8 Hα   L V P A F Q N L M K . . . D C E A E V R A A A S H K V K E F C E N L S A D C R E N V I M S Q  325
   Hß   - I - - - - - L - - . . . - - - - - - - - - - A - - - - - L G - - - P I E D - - T I - - N -  325
   D    - - - - - - Y - L - . . . - A - - - - - - - V A T - - - D - - A - - D K V N Q V Q I - L S S  327
   Y    - - Q P - L - - C E . . . - N - G D L - E - V A K Q - S G - A K F - N D P S . . . T - L N K  360

 9 Hα   I L P C I K E L V S . . . D A N Q H V K S A L A S V I M G L S P I L G K D N T I E H      364
   Hβ   - - - Y - - - - - - . . . - T - - - - - - - - - - - - - - - - T - - - - E - - - - -      364
   D    - - - Y V R D - - - . . . - P - P - - - - - - - - - - - - - - - M - - A Y Q - V - Q      366
   Y    - - - A V E N - S M . . . - E S E T - R - - - - - K - T N I V L L - N - - Q V - N N      399

10 Hα   L L P L F L A Q L K . . . D E C P E V R L N I I S N L D C V N E V I G I R Q L S Q S      403
   Hß   - - - - - - - - - - . . . - - - - D - - - - - - - - - - - - - - - - - - - - - - - -      403
   D    - - - - - - I - - - . . . - - - - - - - - - - - - - - - - - - D - - - - Q - - - - -      405
   Y    F - - I L - N M - R . . . - - F - D - - - - - - A S - K V - - D - - - - E L - - D -      438

11 Hα   L L P A I V E L A E . . . D A K W R V R L A I I E Y M P L L A G Q L G V E F F D E K      442
   Hß   - - - - - - - - - - . . . - - - - - - - - - - - - - - - - - - - - - - - - - - - - -      442
   D    - - - - - - - - - - . . . - S - - - - - - - - - - - - - A - - - - - - Q - - - - Q -      444
   Y    - - - - - T - - - K . . . - V N - - - - M - - - - - I - I - - E - - - M Q - - - Q Q      477

12 Hα   L N S L C M A W L V . . . D H V Y A I R E A A T S N L K K L V E K F G K E W A H A T      481
   Hß   - - - - - - - - - - . . . - - - - - - - - - - - N - - M - - - Q - - - T - - - Q N -      481
   D    - R G - - - G - - N . . . - - - - - - - - - - - L - M - - - - - Q - - A P - - E Q A      483
   Y    - S D - - L S - - W . . . - T - - S - - - - - V N - - - R - T - I - - S D - C R D E      516

13 Hα   I I P K V L A M S G . . . D P N Y L H R M T T L F C I N V L S E V C G Q D I T T K H      520
   Hß   - - - - - - V - A N . . . - - - - - - - - - - - - - - - A - - - A - - - E - - - - Q      520
   D    - - - M I - V - - R . . . N K - - - - - - - C - - - L - - - A - - - - T - - - - - L      522
   Y    - - S R L - K F D LQLL    E . - F V S - F - I - S A L T T - V P - V S L - V V - E Q      557

14 Hα   M L P T V L R M A G . . . D P V A N V R F N V A K S L Q K I G P I L D N S T L Q S E      559
   Hß   - - - I - - K - - - . . . - Q - - - - - - - - - - - - - I - - - - - - T N A - - G -      559
   D    L - - - - - L L - A . . . - Q - - - - - - - - - - T - - - - S - F - E A - V I D A Q      561
   Y    L - - F I S H L - D . . . - G - P - I - - - - - - - Y A V - V K V - I K D E A K Y DALIKNT  602

15 Hα   V K P I L E K L T Q . . . D Q D V D V K Y F A Q E A L T V L S L A                        589
   Hß   - - - V - Q - - G - . . . - E - M - - - - - - - - - I S - - A - -                        589
   D    - - - T - D - - N T . . . - T - M - - - H - - A Q - I A G I A A -                        591
   Y    I L - S - Q T - F - . . . - E - - - - - - - - K K S - A E C Q E L L K N                  635

CONS.   L 5 P - 5 - - L A - . . . D 3 3 - - V R - 2 A - - - 5 - - 5 2 - - L G - 3 - - - - -
```

Protein sequence data and immunological studies revealed that the PR65α subunit is the main isoform in three holoenzymes from rabbit skeletal muscle (Hendrix et al., 1993). Thus, the subunit isoform does not apparently determine which "third" subunit associates with the core dimer. Interestingly, the analysis of different holoenzymes from *Xenopus* oocytes and skeletal muscle indicate changes in the expression of the PR65 isoform during development.

A *Sacchoromyces cerevisiae* gene encoding the PR65 subunit, termed *TPD3*, was isolated from a mutant defective in tRNA synthesis (Van Zyl et al., 1989) and was subsequently identified as PP2A subunit due to its sequence homology with the mammalian protein (Van Zyl et al., 1992). Mutant strains are defective in RNA polymerase III transcription if grown at elevated temperatures. They also have a cold sensitive phenotype revealing defects in cytokinesis (see below).

The 55-kDa Regulatory Subunit (PR55)

Molecular cloning of the 55-kDa regulatory subunit of PP2A revealed three different isoforms, termed PR55α, β and γ (Mayer et al., 1991a; Healy et al., 1991; Pallas et al., 1992). So far only the α and β isoforms have been characterized. The predicted proteins have a molecular mass of approximately 52 kDa and are 86% identical, with the major divergence in the amino-terminal region (see Figure 4).

Analysis of the mRNA expression in human cell lines revealed that the PR55α isoform is encoded by two transcripts of approximately 2.3 and 2.5 kb and a less abundant 4.4 kb mRNA. The 2.3 kb PR55β transcript is expressed at high levels in the neuroblastoma derived cell line LA-N-1, whereas the mRNA level is very low in the other human cell lines analyzed (Mayer et al., 1991a). Neuronal specific expression of the PR55β isoform was also apparent from the analysis of porcine tissue transcripts. Whereas the α isoform appeared ubiquitous, the transcripts encoding the PR55β isoform were only detected in brain (Mayer et al., 1991b). This differential expression of the β-isoform suggested the existence of tissue specific protein phosphatase 2A holoenzymes.

So far only the PP2A holoenzyme containing the α-isoform has been characterized (Hendrix et al., 1993). However, it should be possible to isolate the PR55β containing trimeric PP2A from brain tissue. It would be interesting to analyze whether a previously isolated MAP-2 specific PP2A holoenzyme from bovine brain (Patterson & Flavin, 1986) contains the PR55β isoform.

Interestingly, the genes encoding the PR55 subunit from *Saccharomyces cerevisiae* and from *Drosophila* were both identified through the analysis of mutant strains. The yeast gene, *CDC55*, was isolated from a cold sensitive mutant affected in cytokinesis (Healy et al., 1991) and two alleles of the *Drosophila* gene were found in screens for mitotic mutants (*aar*^{1}) (Mayer-Jaekel et al., 1993) and abnormal imaginal discs (*twins*) (Uemura et al., 1993), respectively (see below).

The 72-kDa Regulatory Subunit (PR72)

Molecular cloning revealed that the PR72 subunit is apparently synthesized as a low and a high molecular weight form (Hendrix et al., submitted). The 72-kDa form corresponds to the purified protein and appears to be mainly expressed in heart and skeletal muscle. A 130-kDa protein (PR130) appears to be encoded by the same gene by the use of alternative promoters and splicing. Although the transcripts encoding the PR130 form are more ubiquitously expressed, the protein has not yet been identified in purified PP2A preparations.

***Drosophila* Protein Phosphatase 2A**

Homology cloning of cDNAs encoding the *Drosophila* PP2A catalytic, PR55 and PR65 subunits revealed extremely high evolutionary conservation (see Figures 3 and 4) (Mayer-Jaekel et al., 1992, 1993). In contrast to mammals, each subunit appears to be encoded by a single gene, located at 28D, 29B and 85F for the catalytic, PR65 and PR55 subunit, respectively. Characterization of the gene encoding the PR55 subunit revealed a complex structure of 7 exons spread over at least 10 kb (Figure 5) (Mayer-Jaekel et al., 1993). Two types of transcripts, termed DPR55-1 and DPR55-4, are synthesized by the use of different promoters and the splicing of two alternative pairs of 5' exons to a set of constant exons.

Analysis of the developmental expression of the different subunit transcripts by Northern blotting revealed multiple transcripts encoding the catalytic and PR65 subunits at all developmental stages, with elevated levels during early embryogenesis (Mayer-Jaekel et al., 1992). High levels of both classes of PR55 transcripts were found predominantly in early embryos, with the DPR55-4-type transcripts being the most abundant. The localization of the different PP2A transcripts was determined by *in situ* hybridization. The catalytic and PR65 subunits, which are found in all PP2A holoenzymes, appear to be coexpressed in all tissues throughout development. However, the transcripts encoding the PR55 subunit show a more restricted distribution and their expression correlates with elevated transcript levels of the other two subunits. The transcripts of all three subunits are maternally expressed and found at very high levels in the early embryos. High zygotic expression of the transcripts is found in the nervous system, starting from the onset of neuroblast formation, and in the gonads. In general, the expression appears highest in highly proliferative tissues. These observations, together with the discovery that a *Drosophila* mutant with abnormalities in mitosis has a defect in the PR55 gene, point towards PP2A being an integral component of cell cycle regulation (Mayer-Jaekel et al., 1993; see below).

```
DPR55-1   MGRWGRQSPVLEPPDPQMQTTPPPPTLPPRTFMRQSSITKIGMMLNTAINTNGAKKPASNCEASWCFSQIKGALDDD.  (77)
DPR55-4                                                   M-G-------------------.  21
HPR55α                                               MAGAG-GNDIQ-----V---V---.  24
HPR55β                                                   MEEDIDTRKINNSFLRDHSY.  20
YCDC55                                                   MAQNNFDFK---CF-DKA-IV  21

                                                                                 (157)
DPR55     VTDADIISCVEFNHDGELLATGDKGGRVVIFQRDPASKAANPRRGEYNVYSTFQSHEPEFDYLKSLEIEEKINKIRWLQQ  101
HPR55α    -AE-----T-----S-------------------QEQEN-IQSHS---------------------------------P-  104
HPR55β    A-E-----T-----T--------------------EQE--NQVH----------------------------------P-  100
YCDC55    --E--L-TA---DYT-NY------------L-E-.....S-S-HC--KFLTE----DA----------------E-K--RP   95

                                                                                 (221)
DPR55     KNPVHFLLSTNDKTVKLWKVSERDKSFGGYNTKEE................NGLIRDPQNVTALRVPSVKQIPLLVEASP  165
HPR55α    --AAQ---------I---I-----RPE---L---................D-RY---TT--T----VFRPMD-M-----  168
HPR55β    Q-AAY-------------------RPE---L-D-................E-RL---ATI-T----VLRPMD-M---T-  164
YCDC55    TQRS----------I-----Y-KNIKLVSQ-NLT-GVTFAKKGKPDNHNSRG-SV-AVLSLQS-KL-QLS-HDKIIA-T-  175

                                                                                 (301)
DPR55     RRTFANAHTYHINSISVNSDQETFLSADDLRINLWHLEVVNQSYNIVDIKPTNMEELTEVITAAEFHPTECNVFVYSSSK  245
HPR55α    --I-------------I---Y--Y---------------ITDR-F-------A-------------------NS--T-------  248
HPR55β    --V-------------------Y--YM-----------NF-IT---F-------A-----------------HH--T-------  244
YCDC55    K-IYS-----------L-----------------------N-DIPD--F---------------------S-----Q---L-M-----  255

                                                                                 (381)
DPR55     GTIRLCDMRSAALCDRHSKQFEEPENPTNRSFFSEIISSISDVKLSNSGRYMISRDYLSIKVWDLHMETKPIETYPVHEY  325
HPR55α    ---------AS--------L-----D-S-----------------F-H-----MT-----V-I---N--NR-V---Q-----  328
HPR55β    ---------AS------T-F-----D-S-----------------F-H----IMT----TV-----N--NR-----Q--D-  324
YCDC55    ---K-----QNS---NKT-T---YLD-I-HN--T--T-----I-F-PN---IA-----TV-I--VN-DN--LK-INI--Q  335

                                                                                 (430)
DPR55     LRAKLCSLYENDCIFDKFECCWNGKDSSIMTGSYNNFFRVF........................DRNSKKDV........  374
HPR55α    --S---------------------S--VV----------M-........................---T-R-I........  377
HPR55β    --S-----------------V---S--V-----------M-........................---T-R--........  373
YCDC55    -KER-SDT----A------VNFS-DS--V-------N-MIYPNVVTSGDNDNGIVKTFDEHNAPNS--N-NIHNSIQNKD  415

                                                                                 (467)
DPR55     ......................TLEASRDIIKP......................KTVLKPRKVCTGGKRK.KDEISVDCLD  411
HPR55α    ......................------ENN--......................R---------AS----.-------S--  414
HPR55β    ......................------ENS--......................RAI-------V---R.-------S--  410
YCDC55    SSSSGNSHKRRSNGRNTGMVGSSNS--SS-AGGEGANSEDSGTEMNEIVLQAD--AFRNKRYGSLAQ-SARNKDWG-DI-  495

                                       (499)
DPR55     FNKKILHTAWHPEENIIAVAATNNLFIFQDKF  443
HPR55α    ------------K-------T----Y-----VN  447
HPR55β    ------------S------------Y-----VN  443
YCDC55    -K-NN--FS---R--S------------SAL  526
```

Figure 4. Comparison of the *Drosophila* PR55 proteins with the human α and β isoforms and the yeast homolog *CDC55*

The deduced amino acid sequence of the two *Drosophila* PR55 splice variants are shown in the single letter code (Mayer-Jaekel et al., 1993). The putative amino-terminal extension of the DPR55-1-type protein is given in italics and the four methionines are shaded. The human PR55α and β sequences (Mayer et al., 1991a) and the *Saccharomyces cerevisiae CDC55* sequence (Healy et al., 1991) are shown below, with identical amino acids indicated by a dash. The amino acid position is given on the right side, with the DPR55-1-type numbering in brackets. Gaps, indicated by dots, were introduced to optimize the alignment with the yeast sequence.

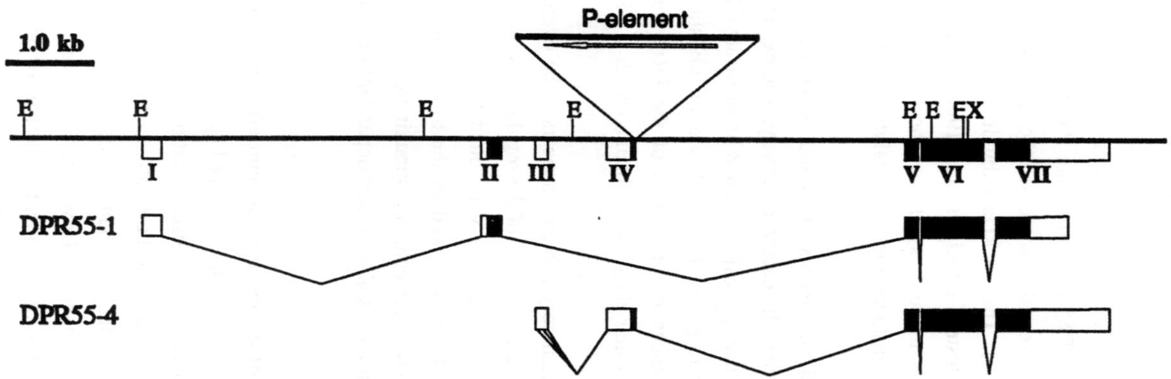

Figure 5. Structure of the *Drosophila* PR55 gene
The gene structure is shown with exons as open and filled boxes for noncoding and coding sequences, respectively. Restriction sites for *Eco*RI [E] and *Xho*I [X] are indicated. Two classes of transcripts are drawn below the gene with the exons shown as boxes and the introns as thin lines. For the DPR55-4-type transcripts varying amounts of the 3' end of exon III were found in the cDNAs which is indicated by alternative lines for intron sequences. The P-element insertion site in the aar^1 mutant at the exon IV / intron 4 splice site is shown.

Modulation of Protein Phosphatase 2A Activity

Purified PP2A holoenzymes differ in their specific activity, substrate specificity and also in their sensitivity to stimulation by polycations. In general, the regulatory subunits appear to inhibit the activity of the catalytic subunit, however, different results can be obtained depending on the substrate used (reviewed by Shenolikar & Nairn, 1991). Reconstitution experiments indicate that the PR65 as well as the PR55 subunits reduce the catalytic activity towards phosphorylase *a*, whereas they increased the activity towards histone H1 (Imaoka et al., 1983; Chen et al., 1989). The 74 kDa subunit appears to reduce the catalytic activity towards all substrates tested (Usui et al., 1988). The holoenzyme composition also modulates the specificity towards synthetic phosphopeptide substrates (Agostinis et al., 1987b, 1990, 1992); in this instance the free catalytic subunit appears to be the least active.

Recent *in vitro* studies using recombinant PR65 subunit and purified catalytic subunit showed that the PR65 subunit generally reduces the catalytic activity. The extent depends on the substrate and appears to be the strongest using phosphorylase *a* as substrate (Kamibayashi et al., 1992) and the least using a synthetic phosphopeptide (B. Favre, unpublished data). The addition of the purified PR55 subunit enhances the inhibition of the PR65 subunit when myosin light chain is used as substrate. Similar data were obtained with a PP2A small t antigen complex (Yang et al., 1991).

The inhibitory effect of the PP2A regulatory subunits appears to be overcome by the presence of polycations (explaining the "activating effect" of the regulatory subunits if histone H1 is used as substrate). The stimulation by polycations was found to be highest with the trimeric holoenzyme containing the PR72 subunit, followed by the trimeric form containing the PR55 subunit and the dimeric form, whereas the catalytic subunit alone was affected the least. The extent of the stimulation by polycations is also largely dependent upon the substrate (Waelkens et al., 1987; Pelech & Cohen, 1985).

Chen et al. (1992) reported that the catalytic subunit itself is subject to reversible phosphorylation. Tyrosine phosphorylation of the catalytic subunit after mitogen stimulation appears to cause a transient inactivation of PP2A which is releaved by autodephosphorylation. The PR55α subunit might also be modulated by phosphorylation (Pallas et al. 1992), however, the physiological relevance of this remains to be established.

Taken together it appears that the regulatory subunits modulate the activity of the catalytic subunit in a substrate dependent manner. However, other functions such as subcellular localization cannot be excluded. The question as to whether, or not, the different holoenzymes might be involved in different signal transduction pathways remains to be established. Studies on mutant yeast or flies defective in specific regulatory subunits will help to solve some of these problems. Whether a second messenger exists to modulate PP2A activity still remains to be elucidated. The fact that an elaborate system of regulatory subunits exists to control and direct its activity strongly argues for the existence of such a molecule (or molecules).

Substrate Specificity

In vitro, PP2A shows a broad substrate specificity, however, not all substrates appear to be physiologically relevant. The preference of PP2A for phosphoresidues in a specific sequence context has been analyzed using synthetic phosphopeptides (reviewed by Pinna et al., 1989). It appears that phosphothreonine residues are the preferred substrates compared to phosphoserine residues. Basic residues on the amino-terminal side of the phosphorylation site have a positive influence on the dephosphorylation efficiency, whereas proline residues on the carboxy-terminal side reduce the dephosphorylation efficiency of the substrate. However, this is not generally applicable to whole proteins since the tertiary structure is also of importance.

The *in vitro* substrate specificity of PP2A towards cellular proteins is quite broad. Comparative analysis of different protein phosphatases *in vitro* indicate that PP2A is involved in the regulation of key enzymes in glycolysis/gluconeogenesis, protein synthesis, fatty acid synthesis and amino acid break down (reviewed in Cohen, 1989). This appears to be supported by studies with okadaic acid on intact cells (Haystead et al., 1989). In addition, several membrane associated proteins such as the β-adrenergic receptor, calcium channels and rhodopsin, and cytoskeleton proteins are potential *in vivo* substrates of PP2A (reviewed by Shenolikar & Nairn, 1991). Interestingly, protein phosphatase 1 (PP1) regulatory subunits also seem to be good substrates for PP2A, which might allow "crosstalk" between these two enzymes. Furthermore, a number of protein kinases are apparently regulated by PP2A *in vitro* and probably also *in vivo*. This includes phosphorylase *b* kinase (Ramachandran et al., 1987), casein kinase 2 (Agostinis et al., 1987a), mitogen-stimulated S6 kinase (Ballou et al., 1988), MAP-2 kinase (Sturgill et al., 1988; Anderson et al., 1990; Haccard et al., 1990), MAP-kinase-kinase (Gomez & Cohen, 1991) and p34^{cdc2} protein kinase (Lee et al., 1991), enabling PP2A to influence indirectly the phosphorylation state of target proteins. More recently it has been appreciated that PP2A participates in the modulation of transcription factor phosphorylation (Hunter & Karin, 1992).

The dimeric form of PP2A was also shown to display considerable phosphotyrosine phosphatase activity under specific assay conditions (Chernoff et al., 1983; Hermann et al., 1988; Goris & Merlevede, 1988; Jessus et al., 1989). The discovery of a polypeptide that activates the protein tyrosine phosphatase activity of PP2A (PTPA) under physiological assay conditions suggest that this may be of physiological relevance (Cayla et al., 1990a,b). Thus PP2A could be a dual specific enzyme, similar to that observed with some protein kinases.

Protein Phosphatase 2A in Cell Cycle Regulation

A regulatory role for PP2A during the cell cycle was first suggested from *in vitro* studies on simian virus 40 (SV40) DNA replication. In this assay system the PP2A catalytic subunit of PP2A stimulates the early stages of DNA replication (Virshup et al., 1989). This effect could be mimicked by crude cell extracts from S phase cells but not from other cell cycle phases, indicating modulation of a substrate directed PP2A activity depending on the cell cycle. The observed stimulation seems to be achieved by the dephosphorylation of the large T antigen, allowing more efficient binding to the DNA template (Virshup & Kelly, 1989; Scheidtmann et al., 1991; Virshup et al., 1992).

Other experimental approaches indicate that PP2A plays a role at the regulation of the entry into mitosis. Injection of okadaic acid in *Xenopus* oocytes triggers the formation of maturation promoting factor, consisting of an active p34^{cdc2} protein kinase / cyclin complex. The okadaic acid concentrations required, and the ineffectiveness of PP1 inhibitors, suggest that PP2A has a negative effect on the entry into mitosis (Goris et al., 1989; Felix et al., 1990). More recently, INH, a factor inhibiting histone H1 protein kinase (p34^{cdc2}) activation, was purified and found to contain the catalytic subunit of PP2A (Lee et al., 1991). The INH effect might either be directly on p34^{cdc2} as shown *in vitro* by Lee et al (1991) or it could be an indirect effect on one of the regulators of p34^{cdc2}. The *cdc25* protein tyrosine phosphatase is required for the activation of MPF and is itself inactivated via serine/threonine dephosphorylation (Kumagai & Dunphy, 1992). Furthermore, the analysis of fission yeast mutants carrying defects in the PP2A catalytic subunit genes revealed a small cell size phenotype if one of the two isoforms is disrupted, whereas a double disruption is lethal (Kinoshita et al., 1990). The premature entry into mitosis in these yeast strains also argues for a function for PP2A as an inhibitor of mitotic induction.

Recent studies on *Drosophila* point towards a function of the PR55 containing trimeric holoenzyme in anaphase (Mayer-Jaekel et al., 1993). In a mutant strain, termed *aar*^{1} for *a*bnormal *a*naphase *r*esolution, the gene encoding the PR55 subunit is disrupted by a P-element insertion (see Figure 5). Mutants die as late pupae / early adults and show abnormal anaphase figures in larval brain cells. The various defects observed include overcondensed chromosomes, lagging chromosomes not segregating to the anaphase poles, chromatin bridges between the poles and also abnormal spindle structures (Gomes et al., 1992). The lethality and the mitotic defects can be restored by introducing the wild type PR55 gene into the mutant genome (Mayer-Jaekel et al., 1993). These results confirm that the PR55 subunit containing holoenzyme plays an essential role in regulating cell cycle progression. The action of PP2A in anaphase is also supported by studies using okadaic acid in mammalian cells (Vandree & Wills, 1992).

Analysis of the *Saccharomyces cerevisiae* mutants *tpd3* and *cdc55* which are defective in the PR65 and PR55 genes, respectively, indicate that the trimeric PP2A is involved in the regulation of morphogenesis (Healy et al., 1991; Van Zyl et al., 1992). If grown at low temperature the mutant strains show abnormal bud formation and are delayed, or blocked, in cell septation and/or separation. However, in these cases the presence of genes encoding proteins with a redundant function which allow growth under normal condition has not been excluded.

Protein Phosphatase 2A and Cell Transformation

As mentioned earlier, the protein phosphatase inhibitor okadaic acid is a potent non-phorbol ester tumor promoter. In a mouse skin bioassay okadaic acid is as effective as phorbol esters (Suganuma et al., 1988). In contrast to the latter class, okadaic acid does not act via the activation of protein kinase C, but exerts its effect by the inhibition of PP1 and PP2A (Takai et al., 1987). At comparable enzyme concentrations, PP2A appears to be the major target since it is 100 times more sensitive than PP1 (Cohen et al., 1990). Experiments with okadaic acid on intact cells showed that the activity of both these enzymes is essential for normal growth, and that their elimination can lead to an abnormal transformed phenotype of the cells. This observation, together with the potent tumor promoting activity of okadaic acid, also suggest that PP2A can function, directly or indirectly, as a tumor suppressor gene. In some instances this is supported by the fact that many proteins phosphorylated by protein kinase C, the target of phorbol ester tumor promoters, are good substrates of PP2A.

In papova virus infected cells PP2A associates with viral proteins (reviewed in Mumby & Walter, 1991). Both the catalytic and the PR65 subunits of PP2A were identified as two of the cellular proteins that associate with polyoma virus middle T antigen and, apparently, the only cellular proteins associating with either the polyoma virus or SV40 small t antigens (Walter et al., 1990; Pallas et al., 1990). These analyses showed that the PR65 subunit is essential for this association, whereas the catalytic subunit associates indirectly by binding to the PR65 subunit (Joshi & Rundell, 1990; Yang et al., 1991). The ability of polyoma virus middle T antigen to associate with the PP2A core enzyme was found to be necessary, but not sufficient, for the transforming activity exerted by middle T antigen. Small t antigen, which only associates with these two cellular proteins, might exert its helper function via PP2A.

The polyoma virus tumor antigens appear to associate only with the dimeric PP2A and not with the trimeric forms. Analysis with SV40 small t antigen indicates that the association of this protein with the dimeric PP2A leads to similar enzymatic characteristics as the trimeric PP2A holoenzyme containing the PR55 subunit (Yang et al., 1991). This might suggest that the PR55 or PR72 subunits can be displaced by the viral proteins. It will be interesting to see how changes in the enzymatic properties of PP2A contribute to the helper activity exerted by small t antigen or, alternatively, if it could also be a function of the free PP2A regulatory subunits. Another interesting aspect of this association could be the tyrosine dephosphorylation and thereby activation of the middle T associated pp60$^{c\text{-}src}$ by PP2A, which was shown to display tyrosine phosphatase activity under specific conditions (Cayla et al., 1990a,b).

Conclusions

From *in vitro* studies PP2A has been implicated in the regulation of numerous cellular processes. In most of these *in vitro* studies the catalytic subunit of PP2A is usually employed (reviewed in Shenolikar & Nairn, 1991). However, this approach to investigating the role of PP2A in cellular processes is far too simplistic and re-analysis will be necessary. This re-evaluation is brought about by the fact that the regulatory subunits of PP2A play a critical role in directing function. Analysis of further mutants from yeast or *Drosophila* should provide additional insights.

PP2A covers a group of holoenzymes that appear to contain the same core structure of the catalytic and PR65 subunits, which then associate with a "variable" subunit. This picture was built up by systematically purifying PP2A holoenzymes from rabbit skeletal muscle, separating the individual subunits by SDS-polyacrylamide gel electrophoresis followed by Western blotting and *in situ* digestion to obtain tryptic peptides for amino acid sequence analysis (Hemmings et al., 1990; Mayer et al., 1991a; Hendrix et al., 1993 and submitted). The sequence data served to design oligonucleotide probes for the molecular cloning of the cDNAs encoding these subunits. The characterization of other holoenzymes was achieved by either peptide-specific antisera or antisera raised against bacterially expressed recombinant proteins.

Molecular cloning of the individual subunits revealed a far more complex picture. cDNA analysis indicates that most of the subunits are encoded by two distinct genes, except in the case of the PR72 subunit. In this latter instance two proteins of predicted molecular weight of 62 and 130 kDa are generated from the same gene by alternative splicing (Hendrix et al., submitted). From the data so far available there appears to be about 10 genes (Hemmings et al., unpublished data) that encode PP2A regulatory subunits but the isolation and characterization of the regulatory subunits is not yet complete. This vast array of regulatory subunits allows many different holoenzymes to be assembled *in vivo.* This probably adds an additional dimension for flexibility in PP2A mediated regulation.

From our recent studies on the role of PP2A holoenzymes from *Drosophila* it appears that the third, or "variable", subunit plays a "director" or "specifier" function. In the mutant *aar*[1] flies the expression of the core subunits appears to be normal as judged by Northern blot analysis, but the expression of the PR55 subunit is knocked out by the P-element insertion which apparently prevents the proper processing of the transcript. Thus, although the *aar*[1] flies have the core subunits, the loss of the PR55 subunit results in a lethal phenotype, albeit at a late stage of development. The results suggest that when the PR55 subunit associates with the "core" dimer (catalytic and PR65) it fulfills two roles: (a) it contributes towards the substrate specificity of the holoenzyme, and (b) it also targets the enzyme to the appropriate cellular compartment or structure.

Thus the analysis of a mutant fly has provided considerable insight into the role and regulation of PP2A from *Drosophila*. Further analysis of *Drosophila* mutants could provide insights into the roles of the other "variable" subunits. These "specifier / director" subunits represent the key to understanding the *in vivo* regulation of PP2A.

REFERENCES

Agostinis P, Goris J, Pinna LA, Merlevede W (1987a) Regulation of casein kinase 2 by phosphorylation/dephosphorylation. Biochem J 248:785-789

Agostinis P, Goris J, Waelkens E, Pinna LA, Marchiori F, Merlevede W (1987b) Dephosphorylation of phosphoproteins and synthetic phosphopeptides. J Biol Chem 262:1060-1064

Agostinis P, Goris J, Pinna LA, Marchiori F, Perich JW, Meyer HE, Merlevede W (1990) Synthetic peptides as model substrates for the study of the specificity of the polycation-stimulated protein phosphatases. Eur J Biochem 189:235-241

Agostinis P, Derua R, Sarno S, Goris J, Merlevede W (1992) Specificity of the polycation-stimulated (type-2A) and ATP,Mg-dependent (type-1) protein phosphatases toward substrates phosphorylated by p34cdc2 kinase. Eur J Biochem 205:241-248

Anderson NG, Maller JL, Tonks NK, Sturgill TW (1990) Requirement for integration of signals from two distinct phosphorylation pathways for activation of MAP kinase. Nature 343:651-653

Arino J, Woon CW, Brautigan DL, Miller TBJr, Johnson GL (1988) Human liver phosphatase 2A: cDNA and amino acid sequence of two catalytic subunit isotypes. Proc Natl Acad Sci USA 85:4252-4256

Ballou LM, Siegman M, Thomas G (1988) S6 kinase in quiescent Swiss 3T3 cells is activated by phosphorylation in response to serum treatment. Proc Natl Acad Sci USA 85:7154-7158

Ballou LM, Fischer EH (1986) Phosphoprotein phosphatases. In: Boyer PD, Krebs EG (eds)The Enzymes, Vol.17. Academic Press, Orlando, pp 311-361

Bialojan C, Takai A (1988) Inhibitory effect of a marine-sponge toxin, okadaic acid, on protein phosphatases. Biochem J 256:283-290

Cayla X, Goris J, Hermann J, Hendrix P, Ozon R, Merlevede W (1990a) Isolation and characterization of a tyrosyl phosphatase activator from rabbit skeletal muscle and Xenopus laevis oocytes. Biochemistry 29:658-667

Cayla X, Goris J, Hermann J, et al (1990b) Phosphotyrosyl phosphatase activity of the polycation-stimulated protein phosphatases and involvement of dephosphorylation in cell cycle regulation. In: Weber G (ed)Adv. Enzyme Regulation 30. Pergamon Press, Oxford, pp 265-285

Charbonneau H, Tonks NK, Kumar S, Diltz CD, Harrylock M, Cool DE, Krebs EG, Fischer EH, Walsh KH (1989) Human placenta protein-tyrosine-phosphatase: amino acid sequence and relationship to a family of receptor-like proteins. Proc Natl Acad Sci USA 86:5252-5256

Charbonneau, H, Tonks NK, Walsh KA, Fischer EH (1988) The leucocyte common antigen (CD45): A putative receptor-linked protein tyrosine phosphatase. Proc Natl Acad Sci USA 85:7182-7186

Chen J, Martin BL, Brautigan DL (1992) Regulation of protein serine-threonine phosphatase-2A by tyrosine phosphorylation. Science 257:1261-1263

Chen MX, Chen YH, Cohen PTW (1992) Polymerase chain reactions using Saccharomyces, Drosophila and human DNA predict a large family of protein serine/threonine phosphatases. FEBS Lett 306:54-58

Chen S-C, Kramer G, Hardesty B (1989) Isolation and partial characterization of an Mr 60,000 subunit of a type 2A phosphatase from rabbit reticulocytes. J Biol Chem 264:7267-7275

Chernoff J, Li H-C, Chen Y-S-E, Chen LB (1983) Characterization of a phosphotyrosyl protein phosphatase activity associated with a phosphoseryl protein phosphatase of Mr = 95,000 from bovine heart. J Biol Chem 258:7852-7857

Cohen P (1988) Protein phosphorylation and hormone action. Proc R Soc Lond B 234:115-144

Cohen P (1989) The regulation of protein phosphatases. Annu Rev Biochem 58:453-508

Cohen P, Holmes CFB, Tsukitani Y (1990) Okadaic acid : a new probe for the study of cellular regulation. Trends Biochem Sci 15:98-102

da Cruz e Silva OB, Alemany S, Campbell DG, Cohen PTW (1987) Isolation and sequence analysis of a cDNA clone encoding the entire catalytic subunit of type-2A protein phosphatase. FEBS Lett 221:415-422

da Cruz e Silva OB, Cohen PTW (1987) A second catalytic subunit of type-2A protein phosphatase from rabbit skeletal muscle. FEBS Lett 226:176-178

Felix M-A, Cohen P, Karsenti E (1990) Cdc2 H1 kinase is negatively regulated by a type 2A phosphatase in the Xenopus early embryonic cell cycle: evidence from effects of okadaic acid. EMBO J 9:675-683

Fischer EH, Charbonneau H, Tonks NK (1991) Protein tyrosine phosphatases: a diverse family of intracellular and transmembrane enzymes. Science 253:401-406

Gomes R, Karess RE, Ohkura H, Glover DM, Sunkel CE (1992) Abnormal anaphase resolution (aar): a locus required for progression through mitosis in Drosophila. J Cell Sci, in press

Gomez N, Cohen P (1991) Dissection of the protein kinase cascade by which nerve growth factor activates MAP kinases. Nature 353:170-173

Goris J, Hermann J, Hendrix P, Ozon R, Merlevede W (1989) Okadaic acid, a specific protein phosphatase inhibitor, induces maturation of MPF formation in Xenopus laevis oocytes. FEBS Lett 245:91-94

Goris J, Merlevede W (1988) Stimulation of the ATP,Mg-dependent protein phosphatase by p-nitrophenyl phosphate. Analytical Biochem 171:423-428

Green DD, Yang S, Mumby MC (1987) Molecular cloning and sequence analysis of the catalytic subunit of bovine type 2A protein phosphatase. Proc Natl Acad Sci USA 84:4880-4884

Haccard O, Jessus C, Cayla X, Goris J, Merlevede W, Ozon R (1990) In vivo activation of a microtubule-associated protein kinase during meiotic maturation of the Xenopus oocyte. Eur J Biochem 192:633-642

Hanks SK (1991) Eukaryotic protein kinases. Current Biology 1:369-383

Haystead TAJ, Sim ATR, Carling D, Honnor RC, Tsukitani Y, Cohen P, Hardie DG (1989) Effects of the tumour promoter okadaic acid on intracellular protein phosphorylation and metabolism. Nature 337:78-81

Healy AM, Zolnierowicz S, Stapleton AE, Goebl M, DePaoli Roach AA, Pringle JR (1991) CDC55, a Saccharomyces cerevisiae gene involved in cellular morphogenesis: identification, characterization, and homology to the B subunit of mammalian type 2A protein phosphatase. Mol Cell Biol 11:5767-5780

Hemmings BA, Wernet W, Mayer RE, Maurer F, Hofsteenge J, Stone SR (1988) The nucleotide sequence of the cDNA encoding the human lung protein phosphatase 2Aβ catalytic subunit. Nucl Acids Res 16:11366

Hemmings BA, Adams-Pearson C, Maurer F, Müller P, Goris J, Merlevede W, Hofsteenge J, Stone SR (1990) α- and β-forms of the 65-kDa subunit of protein phosphatase 2A have a similar 39 amino acid repeating structure. Biochemistry 29:3166-3173

Hendrix P, Mayer-Jaekel RE, Goris J, Cron P, Hofsteenge J, Merlevede W, Hemmings BA Structure and expression of a 72-kDa regulatory subunit of protein phosphatase 2A. Evidence for different size isoforms produced by alternative splicing. J Biol Chem, submitted

Hendrix P, Turowsky P, Mayer-Jaekel RE, Goris J, Hofsteenge J, Merlevede W, Hemmings BA (1993) The isoform of the 65-kDa regulatory and the 36-kDa catalytic subunit does not determine the subunit composition of protein phosphatase 2A. J Biol Chem, in press

Hermann J, Cayla X, Dumortier K, Goris J, Ozon R, Merlevede W (1988) Modulation of the substrate specificity of the polycation-stimulated protein phosphatase from Xenopus laevis oocytes. Eur J Biochem 137:17-25

Hunter T, Karin M (1992) The regulation of transcription by phosphorylation. Cell 70:375-387

Imaoka T, Imazu M, Usui H, Kinohara N, Takeda M (1983) Resolution and reassociation of three distinct components from pig heart phosphoprotein phosphatase. J Biol Chem 258:1526-1535

Ingebritsen TS, Cohen P (1983) The protein phosphatases involved in cellular regulation. 1. Classification and substrate specificities. Eur J Biochem 132:255-261

Jessus C, Goris J, Cayla X, Hermann J, Hendrix P, Ozon R, Merlevede W (1989) Tubulin and MAP2 regulate the PCSL phosphatase activity. Eur J Biochem 180:15-22

Joshi B, Rundell K (1990) Association of simian virus 40 small-t antigen with the 61-kilodalton component of a cellular protein complex. J Virology 64:5649-5651

Kamibayashi C, Lickteig RL, Esters R, Walter G, Mumby MC (1992) Expression of the A subunit of protein phosphatase 2A and characterization of its interactions with the catalytic and regulatory subunits. J Biol Chem 267:21864-21872

Khew-Goodall Y, Mayer RE, Maurer F, Stone SR, Hemmings BA (1991) Structure and transcriptional regulation of protein phosphatase 2A catalytic subunit genes. Biochemistry 30:89-97

Khew-Goodall Y, Hemmings BA (1988) Tissue-specific expression of mRNAs encoding α- and β-catalytic subunits of protein phosphatase 2A. FEBS Lett 238:265-268

Kinoshita N, Ohkura H, Yanagida M (1990) Distinct, essential roles of type 1 and 2A protein phosphatases in the control of the fission yeast cell division cycle. Cell 63:405-415

Kitagawa Y, Sakai R, Tahira T, Tsuda H, Ito N, Sugimura T, Nagao M (1988a) Molecular cloning of rat phosphoprotein phosphatase 2Aβ cDNA and increased expression of phosphatase 2Aα and 2Aβ in rat liver tumors. Biochem Biophys Res Commun 157:821-827

Kitagawa Y, Tahira T, Ikeda I, Kikuchi K, Tsuiki S, Sugimura T, Nagao M (1988b) Molecular cloning of cDNA for the catalytic subunit of rat liver type 2A protein phosphatase, and detection of high levels of expression of the gene in normal and cancer cells. Biochem Biophys Acta 951:123-129

Krebs EG (1986) The enzymology of control by phosphorylation. In: Boyer PD, Krebs EG (eds)The Enzymes, Vol. 17. Academic Press, Orlando, pp 3-20

Kumagai A, Dunphy WG (1992) Regulation of the cdc25 protein during the cell cycle in Xenopus extracts. Cell 70:139-151

Lee TH, Solomon MJ, Mumby MC, Kirschner MW (1991) INH, a negative regulator of MPF, is a from of protein phosphatase 2A. Cell 64:415-423

MacKintosh C, Haycox G, Hardie DG, Cohen PTW (1990) Identification by molecular cloning of two cDNA sequences from the plant Brassica napus which are very similar to mammalian protein phosphatases-1 and -2A. FEBS Lett 276:156-160

Mayer RE, Hendrix P, Cron P, Matthies R, Stone SR, Goris J, Merlevede W, Hofsteenge J, Hemmings BA (1991a) Structure of the 55 kDa regulatory subunit of protein phosphatase 2A: evidence for a neuronal specific isoform. Biochemistry 30:3598-3597

Mayer RE, Khew-Goodall Y, Stone SR, et al (1991b) Structure and expression of protein phosphatase 2A. In: Merlevede W, Di Salvo J (eds)Adv.Prot.Phosphatases Vol. 6. Leuven University Press, Leuven, pp 265-286

Mayer-Jaekel RE, Baumgartner S, Bilbe G, Ohkura H, Glover DM, Hemmings BA (1992) Molecular cloning and developmental expression of the catalytic and 65-kDa regulatory subunits of protein phosphatase 2A in Drosophila. Mol Biol Cell 3:287-298

Mayer-Jaekel RE, Ohkura H, Gomes R, Sunkel CE, Baumgartner S, Hemmings BA, Glover DM (1993) The 55-kDa regulatory subunit of Drosophila protein phosphatase 2A is required for anaphase. Cell, in press

Mumby MC, Walter G (1991) Protein phosphatases and DNA tumor viruses: transformation through the back door? Cell Regul 2:589-598

Nagao M, Sakai R, Kitagawa Y, et al (1990) Role of protein phosphatases in malignant transformation. In: Knudson AG et al (ed) Genetic Basis for Carcinogenesis: Tumor Suppressor Genes and Oncogenes. Taylor & Francis Ltd., London, pp 177-184

Norbury C, Nurse P (1992) Animal cell cycles and their control. Annu Rev Biochem 61:441-470

Orgad S, Brevis ND, Alphey L, Axton JM, Dudai Y, Cohen PTW (1990) The structure of protein phosphatase 2A is as highly conserved as that of protein phosphatase 1. FEBS Lett 275:44-48

Pallas DC, Shahrlik LK, Martin BL, Jaspers S, Miller TB, Brautigan DL, Roberts TM (1990) Polyoma small and middle T antigens and SV40 small t antigen form stable complexes with protein phosphatase 2A. Cell 60:167-176

Pallas DC, Weller W, Jaspers S, Miller TB, Lane WS, Roberts TM (1992) The third subunit of protein phosphatase 2A (PP2A), a 55-kilodalton protein which is apparently substituted for by t antigens in complexes with the 36- and 63-kilodalton PP2A subunits, bears little resemblance to T antigens. J Virology 66:886-893

Patterson CLJr, Flavin M (1986) A brain phosphatase with specificity for microtubule-associated protein-2. J Biol Chem 261:7791-7796

Pelech S, Cohen P (1985) The protein phosphatases involved in cellular regulation. 1. Modulation of protein phosphatase-1 and 2A by histone H1, protamine, polylysine and heparin. Eur J Biochem 148:245-251

Pelech S, Sanghera JS (1992) MAP Kinases: Charting the Regulatory Pathways. Science 257:1355-1356

Pinna LA, Agostinis P, Donella-Deana A, et al (1989) Molecular bases for protein dephosphorylation. A study with phosphorylated peptide substrates. In: Merlevede W, Di Salvo J (eds)Advances in Protein Phosphatases 5. Leuven University Press, Leuven, pp 51-74

Posada J, Cooper JA (1992) Molecular signal integration. Interplay between serine, threonine, and tyrosine phosphorylation. Mol Biol Cell 3:583-592

Pot. DA, Dixon JE (1992) A thousand and two protein tyrosine phosphatases. Biochem Biophys Acta 1136:35-43

Ramachandran C, Goris J, Waelkens E, Merlevede W, Walsh DA (1987) The interrelationship between cAMP-dependent α and β subunit phosphorylation in the regulation of phosphorylase kinase activity. J Biol Chem 262:3210-3218

Reed SI (1992) The role of p34 kinases in the G1 to S-phase transition. Ann Rev Cell Biol 8:529-561

Roberts TM (1992) A signal chain of events. Nature 360:534-535

Ruediger R, Roeckel D, Fait J, Berqvist A, Magnusson G, Walter G (1992) Identification of binding sites on the regulatory A subunit of protein phosphatase 2A for the catalytic C subunit and for tumor antigens of simian virus 40 and polyomavirus. Mol Cell Biol 12:4872-4882

Scheidtmann KH, Virshup DM, Kelly TJ (1991) Protein phosphatase 2A dephosphorylates simian virus 40 large T antigen specifically at residues involved in regulation of DNA-binding activity. J Virology 65:2098-2101

Shenolikar S, Nairn AC (1991) Protein phosphatases: recent progress. In: Greengard P, Robinson GA (eds)Advances in Second Messengers and Phosphoprotein Research Vol. 25. Raven Press, New York, pp 1-121

Sneddon AA, Cohen PTW, Stark MJ (1990) Saccharomyces cerevisiae protein phosphatase 2A performs an essential cellular function and is encoded by two genes. EMBO J 9:4339-4346

Stone SR, Hofsteenge J, Hemmings BA (1987) Molecular cloning of cDNAs encoding two isoforms of the catalytic subunit of protein phosphatase 2A. Biochemistry 26:7215-7220

Stone SR, Mayer RE, Wernet W, Maurer F, Hofsteenge J, Hemmings BA (1988) The nucleotide sequence of the cDNA encoding the human lung protein phosphatase 2Aα catalytic subunit. Nucl Acids Res 16:11365

Sturgill TW, Ray LB, Erikson E, Maller JL (1988) Insulin-stimulated MAP-2 kinase phosphorylates and activates ribosomal protein S6 kinaseII. Nature 334:715-718

Suganuma M, Fujiki H, Suguri H, Yoshizawa S, Hirota M, Nakayasu M, Ojika M, Wakamatsu K, Yamada K, Sugimura T (1988) Okadaic acid: An additional non-phorbol-12-tetradecanoate-13-acetate-type tumor promoter. Proc Natl Acad Sci USA 85:1768-1771

Takai A, Bialojan C, Troschka M, Ruegg JC (1987) Smouth muscle myosin phosphatase inhibition and force enhancement by black sponge toxin. FEBS Lett 217:81-84

Tonks NK, Diltz CD, Fischer EH (1988) Characterization of the major protein-tyrosine-phosphorylation. J Biol Chem 263:6734-6737

Tung HYL, Alemany S, Cohen P (1985) The protein phosphatases involved in cellular regulation. 2. Purification, subunit structure and properties of protein phosphatases-$2A_0$, $2A_1$ and $2A_2$ from rabbit skeletal muscle. Eur J Biochem 148:253-263

Uemura T, Shiomi K, Togashi S, Takeichi M (1993) Mutation of twins encoding a regulator of protein phosphatase 2A leads to pattern duplication in the Drosophila imaginal disc. Genes Develop, in press

Usui H, Imazu M, Maeta K, Tsukamoto H, Azuma K, Takeda M (1988) Three distinct forms of type 2A protein phosphatase in human erythrocyte cytosol. J Biol Chem 263:3752-3761

Usui H, Ariki M, Miyauchi T, et al (1991) Functions and properties of subunits in type 2A protein phosphatases from human erythrocyte. In: Merlevede W (ed)Advances in Protein Phosphatases Vol. 6. Leuven University Press, Leuven, pp 287-306

van Zyl WH, Wills N, Broach JR (1989) A general screen for mutants of Saccharomyces cerevisiae deficient in tRNA biosynthesis. Genetics 123:55-68

van Zyl WH, Huang W, Sneddon AA, Stark MJ, Camier S, Werner M, Marck C, Sentenac A, Broach JR (1992) Inactivation of the protein phosphatase 2A regulatory subunit A results in morphological and transcriptional defects in Saccharomyces cerevisiae. Mol Cell Biol 12:4946-4959

Vandre DD, Wills VL (1992) Inhibition of mitosis by okadaic acid: possible involvement of a protein phosphatase 2A in the transition from metaphase to anaphase. J Cell Sci 101:79-91

Virshup DM, Kauffman MG, Kelly TJ (1989) Activation of SV40 DNA replication in vitro by cellular protein phosphatase 2A. EMBO J 8:3891-3898

Virshup DM, Russo AAR, Kelly TJ (1992) Mechanism of activation of simian virus 40 DNA replication by protein phosphatase 2A. Mol Cell Biol 12:4883-4895

Virshup DM, Kelly TJ (1989) Purification of replication protein C, a cellular protein involved in the initial stages of simian SV40 DNA replication in vitro. Proc Natl Acad Sci USA 86:3584-3588

Waelkens E, Goris J, Merlevede W (1987) Purification and properties of polycation-stimulated phosphorylase phosphatases from rabbit skeletal muscle. J Biol Chem 262:1049-1059

Walter G, Ferre F, Espiritu O, Carbone-Wiley A (1989) Molecular cloning and sequence of cDNA encoding polyoma medium tumor antigen-associated 61-kDa protein. Proc Natl Acad Sci USA 86:8669-8672

Walter G, Ruediger R, Slaughter C, Mumby M (1990) Association of protein phosphatase 2A with polyoma virus medium tumor antigen. Proc Natl Acad Sci USA 87:2521-2525

Yang S-I, Lickteig RL, Estes R, Rundell K, Walter G, Mumby MC (1991) Control of protein phosphatase 2A by simian virus 40 small-t antigen. Mol Cell Biol 11:1988-1995

DEPHOSPHORYLATION IN VITRO INHIBITS LIGAND AND DNA BINDING ACTIVITIES OF HUMAN ESTROGEN RECEPTOR

Petr Folk* and James L. Wittliff
Hormone Receptor Laboratory
James G. Brown Cancer Center
Department of Biochemistry
University of Louisville
Louisville, KY 40292
USA

INTRODUCTION

Phosphorylation and/or dephosphorylation of estrogen receptor (ER) triggered by parallel signalling pathways is likely to be an important mechanism regulating the transcription of estrogen responsive genes (Herschman, 1991). Phosphorylation is implicated in regulating the hormone binding of estrogen, glucocorticoid, as well as other steroid receptors (Moudgil, 1990; Rao et al., 1987). In our laboratory we have studied the molecular polymorphism and structure function relationships of ER in human breast cancer (Hyder et al., 1987). Estrogen receptor bears phosphorylation sites in its hormone binding domain (Migliaccio et al., 1989), which also harbors one of its two transcription activation functions (Gronemeyer, 1992). We have demonstrated that human ER expressed in E. coli, despite being depraved of the phosphorylation reactions of its cell of origin, retains its ligand binding properties (Wittliff et al., 1990). Here we show that both the ligand- and DNA-binding activities of ER can be abolished by in vitro dephosphorylation of ER with alkaline phosphatase (AP), suggesting that dephosphorylation plays a role during and after receptor transformation to a DNA-binding form.

MATERIAL AND METHODS

Preparation of ER in cytosol from human breast cancer biopsies was done as described (Hyder et al., 1987). The content

* Department of Physiology, Charles University, Vinicna 7, Praha 2, 128 00, Czechoslovakia

NATO ASI Series, Vol. H 76
Tyrosine Phosphorylation/Dephosphorylation and Downstream Signalling
Edited by L. M. G. Heilmeyer Jr.

of ER was assayed either by multipoint titration or by EIA as specified in Wittliff et al. (1987). Gel retardation assays were performed according to Wittliff et al. (1990).

RESULTS AND DISCUSSION

Alkaline phosphatase (AP) inactivates ligand binding of ER in vitro and this effect can be reversed by addition of 10 mM molybdate + 1 mM DTT (see Tab. I). The effect of AP can be

TABLE I
The Influence of Alkaline Phosphatase (AP) on the Ligand Binding Capacity of ER Partially Purified from Human Breast Cancer Cytosol

PROT.	INCUBATION CONDITIONS	ER CONCENTRATION (fmol/ml) control	ER CONCENTRATION (fmol/ml) AP added	RELATIVE ACTIVITY (% of control)
A.	- Crude cytosol	235 ± 8	---	---
	- 1 h at 30°C	218 ±12	214 ± 8	98
	- 1 h at 37°C	176 ± 4	182 ± 9	103
B.	- Purified cytosol	255 ±14	---	---
	- 1 h at 30°C	232 ± 9	186 ±11	80
	- 10min at 37°C	205 ±19	163 ± 7	68
	- 1 h at 37°C	194 ± 8	62 ± 9	32
C.	- Purified cytosol	174 ± 6	---	---
	- 1 h at 37°C	127 ±11	48 ± 3	38
	- +molybdate, then 1h at 37°C	149 ±12	136 ± 8	91
	- 1h at 37°C, then +molybdate	154 ±11	163 ±11	106

The human breast cancer cytosol was prepared in T_{10}EGD buffer (10 mM Tris, 1.5 mM EDTA, 10% glycerol, 1 mM DTT, pH 7.4) and used immediately (sect. A) or partially purified on heparin agarose (sect. B, C). Elution was performed with T_{10}EG buffer containing 6 mg/ml heparin. AP was dissolved in T_{10}E buffer containing 0.5 mM $MgCl_2.6H_2O$ and used at 10 U/ml final concentration. After the incubation (10 min to 1 h at 37°C) the reaction mixture was cooled on ice and pyrophosphate was added. After the incubations ER binding capacity was estimated using ^{125}IE. The effect of 10 mM molybdate was measured with molybdate added before (+ molybdate; 1h at 37°C) or after (1h at 37°C; + molybdate) the incubation. Alternative experiments were done with ER purified with DES-affinity chromatography (results not shown) that gave similar results as those in Sect. B and C. The results are expressed as means ± S.D. (n=3 in A,C; n=4 in B).

demonstrated on purified ER only. ER in crude cytosol or ER complexed with antiER monoclonal antibody D547 (data not shown) retains its ligand binding properties even after 1 h incubation with AP. Incubation of ER with AP or nuclear extract <u>in vitro</u> prevents also estrogen response element (ERE) binding to ER in a

F I G U R E 1
Binding of ER to its cognate regulatory element DNA (ERE) can be abolished by alkaline phosphatase (AP) treatment.

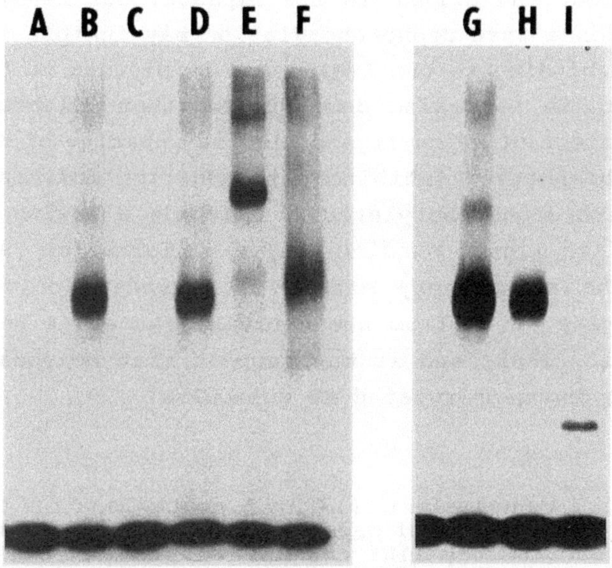

Gel retardation assay showing ER binding to labeled ERE under various conditions. Cytosols were prepared in DNA-binding buffer (H_{10}EGD) and incubated for 3 h with 100 nM unlabeled estradiol to stabilize the receptor protein. The estrogen response element (ERE) from Xenopus vitellogenin A2 gene was used. Assays (30 µl total volume) contained 10 µl of extract, 6.25 µg of poly-dIdC and 50-100 000 cpm of labelled ERE in H_{10}EGD+ 65 mM KCl. Mixtures were incubated for 15 min on ice prior to the addition of ERE and then again for 25 min at 20°C. Samples were loaded onto 4% PAGE gels. A - free probe, no cytosol added; B - control, ER + ^{32}P-ERE; C - ER + ^{32}P-ERE + excess of unlabelled ERE; D - ER + ^{32}P-ERE + excess of unlabeled GRE (glucocorticoid response element); E - ER + ^{32}P-ERE + antiER monoclonal antibody D547; F - ER + ^{32}P-ERE + monoclonal antibody specific to progestin receptor; G to I - the samples were preincubated with or without AP for 30 min at 37°C prior to the addition of the probe; G - ER + ^{32}P-ERE; H - ER + ^{32}P-ERE + AP + pyrophosphate; I - ER + ^{32}P-ERE + AP only.

gel retardation assay. The incubation in the presence of pyrophosphate has no effect on ERE binding (Fig. 1), showing that the effect of AP is due to dephosphorylation of ER. The lane "I" in Fig. 1 shows that ERE binding was lost (the faster migrating band in lane "I" is not ERE specific) after 30 min preincubation of ER with AP in crude cytosol. This sensitivity is in contrast to the effect of AP on ligand binding of ER, where the binding activity is retained even after 1 h incubation of ER with AP in crude cytosol. This suggests that in vitro the dephosphorylation may occur to various extent and differ in its impacts. Our results provide further evidence that dephosphorylation may inhibit the action of ER both in binding to the ligand and in binding to the DNA response element. We have also demonstrated that molybdate may counteract the effect of AP on ligand binding capacity of ER. Molybdate acts as phosphatase inhibitor; in addition molybdate may also interact with dephosphorylated ER molecule allowing it to resume or retain its ligand binding-active conformation (see Tab. I, part C). The stimulatory effect of molybdate on the ligand binding capacity of ER from human breast cancer is well known (Gaubert et al., 1982) and it may suggest that molybdate acts by reactivating dephosphorylated ER molecules.

REFERENCES

Gaubert CM, Bianucci S, Shyamala G (1982) A comparison of the cytoplasmic estrogen receptors of mammary gland from virgin and lactating mice. Endocrinology 110: 683-685

Gronemeyer H (1992) Control of transcription activation by steroid hormone receptors. FASEB J 6: 2524-2529

Herschman HR (1991) Primary response genes induced by growth factors and tumor promoters. Annu Rev Biochem 60: 281-319

Hyder SM, Sato N, Wittliff JL (1987) Characterization of estrogen receptors and associated protein kinase activity by high-performance hydrophobic-interaction chromatography. J Chromatogr 397: 251-267

Migliaccio A, Rotondi A, Auricchio F (1989) Phosphorylation of steroid hormone receptors. Proc Nat Acad Sci USA 81: 5921-5925

Moudgil VK (1990) Phosphorylation of steroid hormone receptors. Biochim Biophys Acta 1055: 243-258

Rao KVS, Peralta WD, Greene GL, Fox CF (1987) Progesterone receptor phosphorylation in response to ligands activating protein kinases. Biochem Biophys Research Commun 146: 1357-1365

Wittliff JL, Wenz LL, Dong J, Nawaz Z, Butt TR (1990) Expression and characterization of an active human estrogen receptor as a ubiquitin fusion protein from Escherechia coli. J Biol Chem 265: 22016-22022

CHARACTERIZATION OF NOVEL NUCLEAR INHIBITORS OF PROTEIN PHOSPHATASE - 1

A. Van Eynde, M. Beullens, M. Bollen and W. Stalmans
Afdeling Biochemie, Faculteit Geneeskunde
Katholieke Universiteit Leuven
B-3000 Leuven
Belgium

Introduction

Based on their substrate specificity and on the effect of inhibitors and activators, four major families of Ser/Thr-specific protein phosphatases have been distinguished (Cohen, 1989; Bollen and Stalmans, 1992). One of these families, termed type-1 protein phosphatases (PP-1), comprises a group of enzymes that all contain an isoform of the same catalytic subunit. They differ, however, in their noncatalytic subunit(s), which determine the activity and subcellular location of the enzyme.

Of all cellular compartments, the nucleus contains by far the highest concentration of PP-1. As discussed by Bollen et al. (1993), recent studies point to an essential role of nuclear protein phosphatase-1 (PP-1N) in the entry and completion of mitosis. In addition, nuclear protein phosphatase-1 has been implicated in interphase chromosome condensation and in the inactivation of a transcription factor, termed CREB, that mediates transcription of cAMP-inducible genes.

In spite of its evident importance, very little is known about the structure and regulation of PP-1N. Preliminary evidence indicates that the catalytic subunit ($PP\text{-}1_C$) in the nucleus is associated with other polypeptides, yielding high-molecular-weight holoenzymes (Bollen et al., 1993). The observation that $PP\text{-}1_C$ is associated with chromosomes at mitosis may suggest the existence of a "chromosome-binding subunit". Genetic studies have resulted in the identification of a nuclear 30-kDa polypeptide in fission yeast that serves as a "positive" regulator of PP-1N.

Results and Discussion

During a comparative study of the properties of PP-1 holoenzymes, we noted that the phosphorylase phosphatase activity of PP-1 in bovine thymus

NATO ASI Series, Vol. H 76
Tyrosine Phosphorylation/Dephosphorylation
and Downstream Signalling
Edited by L. M G. Heilmeyer Jr.

nuclei was increased 2-6 times by a preincubation with trypsin (Beullens et al., 1992). Further investigations indicated that this effect of trypsin could be largely explained by proteolysis of inhibitory polypeptides of PP-1_C. These nuclear inhibitors of PP-1 (termed NIPP-1) were mainly associated with the insoluble nuclear fraction, and could be solubilized by a preincubation with 0.3 M NaCl. During chromatography on Mono Q, NIPP-1 was resolved into three species, designated NIPP-1a, NIPP-1b and NIPP-1c according to their order of elution. The first two of these heat-stable and acid-resistant inhibitors have been further purified. Starting from the salt extract of the nuclear particulate fraction, this represented a purification of 20,000-30,000 fold. Fig. 1 illustrates the last purification step of NIPP-1b, which consisted consecutively of SDS/PAGE, blotting onto Immobilon membranes and elution of the individual polypeptides with 70% formic acid. The inhibitors remained active under these circumstances and could directly be identified as polypeptides of 18 kDa (NIPP-1a) and 16 kDa (NIPP-1b).

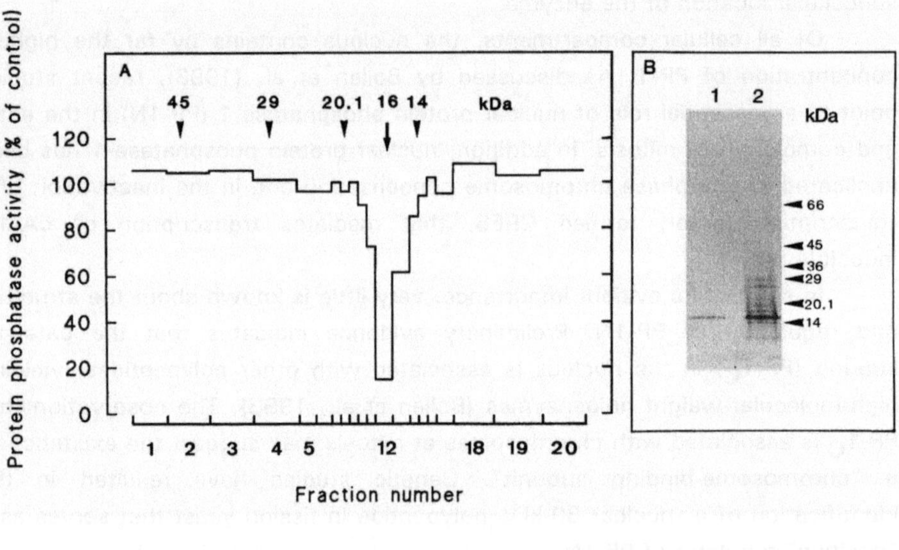

Fig. 1. Purification of NIPP-1 by denaturing electrophoresis. After SDS/PAGE of a partially purified sample of NIPP-1b, the separated polypeptides were blotted onto Immobilon membranes and subsequently visualized by staining with Amido Black. The membrane was then cut into small strips as shown (panel A), and the polypeptides were eluted from each strip with 70% formic acid. After lyophilization, the dissolved fractions were assayed for NIPP-1. Panel B shows a Coomassie-stained gel (SDS-PAGE) of NIPP-1 before (lane 2) and after (lane 1) this purification step. Reprinted by permission from *J. Biol. Chem.* (Beullens et al., 1992).

The maximal extent of inhibition of PP-1_C by NIPP-1 was very much dependent on the nature of the substrate (Beullens et al., 1992). Complete inhibition was obtained with phosphorylase and casein as substrate, but the dephosphorylation of basic substrates was either only partially inhibited by NIPP-1 (myelin basic protein) or not at all (histone IIA). Further, the casein phosphatase activity of PP-1_C was less efficiently inhibited by NIPP-1 in the presence of nonphosphorylated histones. These data may indicate that NIPP-1, which is an acidic protein (pI = 4.4), is inactivated by complexation with basic proteins.

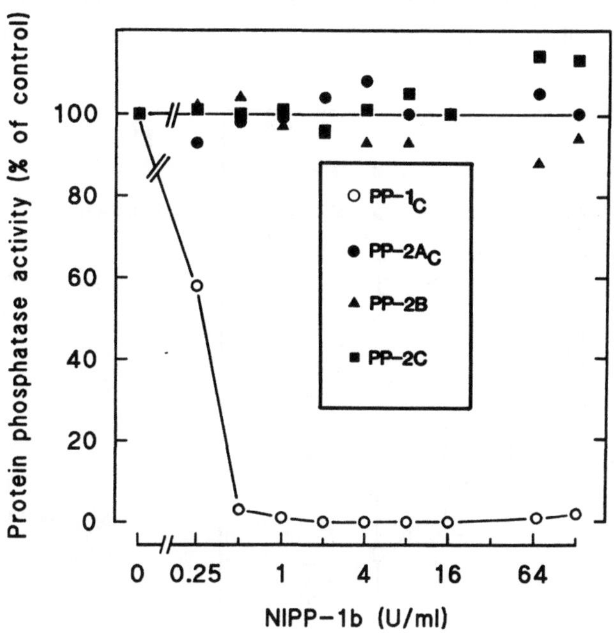

Fig. 2. Effect of NIPP-1b on Ser/Thr-protein phosphatases. PP-1_C and PP-$2A_C$ were measured with phosphorylase as substrate, while the activities of PP-2B and PP-2C were determined from the rate of dephosphorylation of casein. The concentration of NIPP-1b is shown on a log scale. The results represent the means of four experiments. Reprinted by permission from *J. Biol. Chem.* (Beullens et al. 1992).

As shown in Fig. 2 for NIPP-1b, the nuclear inhibitors completely blocked the phosphorylase phosphatase activity of PP-1_C, but had no effect on the activities of PP-2A, PP-2B and PP-2C, even at concentrations that were 250-fold higher than those required for complete inhibition of PP-1_C.

In the range of phosphatase concentrations between 60 pM and 15,000 pM, inhibition of PP-1_C was obtained at stoichiometric concentrations of NIPP-1 (Beullens et al., 1992). Extrapolation of the NIPP-1 concentration that caused halfmaximal inhibition to infinite dilutions of the phosphatase yielded an apparent K_i value of about 1 pM. NIPP-1 emerges therefore as by far the best inhibitor of PP-1 ever described.

The cytosol contains heat-stable polypeptides (inhibitor-1 and/or DARPP-32) that become inhibitors of PP-1 upon phosphorylation by cAMP-dependent protein kinase (see Bollen and Stalmans, 1992). We have found that cAMP-dependent protein kinase phosphorylates NIPP-1 with very high affinity, as compared to histone 2A or the model substrate kemptide. Maximal

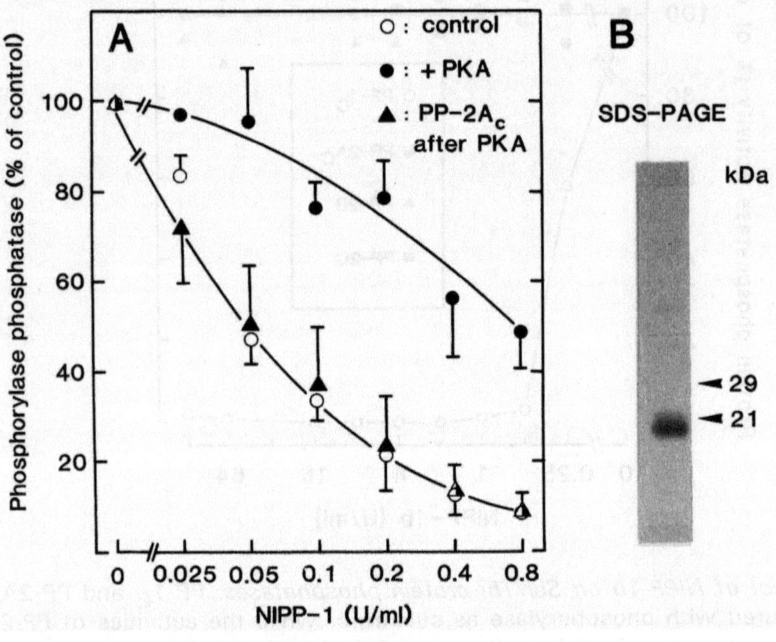

Fig. 3. Phosphorylation and inactivation of NIPP-1a by cAMP-dependent protein kinase (PKA). Purified NIPP-1a (5 U/ml) was incubated for 1h at 30°C in the absence or presence of PKA (0.15 U/ml) plus 0.2 mM ATP and 1mM Mg acetate. After a heat treatment, the inhibitory potency of NIPP-1 was assayed (panel A). NIPP-1 was also assayed after an incubation of the PKA-phosphorylated sample with PP-$2A_C$ (100 U/ml) for 30 min, followed by a heat inactivation of the phosphatase. The results represent the means ± S.E.M. for 3-4 experiments. Panel B shows an autoradiogram of NIPP-1a after phosphorylation with PKA as in panel A, except for the presence of γ-^{32}P-labeled ATP and subsequent denaturing electrophoresis.

phosphorylation of NIPP-1 was associated with a tenfold decrease in inhibitory potency (Fig. 3). It is also shown that the effect of phosphorylation could be reversed by incubation with the catalytic subunit of PP-2A; the type-1 catalytic subunit was at least 300 times less efficient.

Our findings suggest that cAMP-dependent protein kinase can enhance the activity of PP-1 in the nucleus. This may be a mechanism to curtail cAMP-induced signalling in the nucleus (e.g. by a reversal of phosphorylation of the transcription factor CREB). It may also serve to achieve a selective cAMP-dependent dephosphorylation of intranuclear proteins that are phosphorylated by cAMP-independent protein kinases.

References

Beullens M, Van Eynde A, Stalmans W, Bollen M (1992) The isolation of novel inhibitory polypeptides of protein phosphatase-1 from bovine thymus nuclei. J. Biol. Chem. 267:16538-16544

Bollen M, Stalmans W (1992) The structure, role, and regulation of type-1 protein phosphatases. Crit. Rev. Biochem. Mol. Biol. 27:227-281

Bollen M, Beullens M, Van Eynde A, Stalmans W (1993) Identification and regulation of nuclear inhibitors of protein phosphatase-1. Adv. Protein Phosphatases 7, in press

Cohen P (1989) The structure and regulation of protein phosphatases. Annu. Rev. Biochem. 58:453-508

V. S6-Phosphorylation

THE MITOGEN-ACTIVATED $p70^{s6k}$ / $p85^{s6k}$

G. Thomas
Friedrich Miescher Institute
P.O. Box 2543
4002 Basel, Switzerland

Introduction

The addition of growth factors to quiescent cells in culture induces their reentry into the cell cycle, followed by DNA synthesis and cell division (Pardee AB, 1989). The approach we have taken in unraveling the mechanisms by which growth factors regulate this response has been to begin with an early obligatory step in cell growth, the activation of protein synthesis (Brooks RF, 1977; Rossow PW et al., 1979), which is not only required for the G_0/G_1 transition, but also for cells to progress through G_1 and enter S (Brooks RF, 1977; Rossow PW et al., 1979). The activation of protein synthesis is controlled at the level of initiation and appears to be triggered by the phosphorylation of a number of key components of the translation machinery including 40S ribosomal protein S6 (Hershey, JWB, 1989; Morley S and Thomas G, 1991). In the case of S6, five moles of phosphate are incorporated into the protein in response to mitogens (Krieg J et al., 1988). The five sites of phosphorylation reside at the carboxy end of the protein and these appear to be phosphorylated in the order $Ser_{236} > Ser_{235}/Ser_{240} > Ser_{244} > Ser_{247}$ (Krieg et al. 1988).

In this lecture I will review recent progress in the field with regard to (i) the identification and purification of the S6 kinase, (ii) the cDNA cloning of this enzyme, (iii) its mechanism of regulation, and (iv) the multiple signalling pathways which control its activity. These findings are discussed with regard to future studies which will be required to examine the essentiality of both the kinase and S6 phosphorylation in the mitogenic response.

NATO ASI Series, Vol. H 76
Tyrosine Phosphorylation/Dephosphorylation and Downstream Signalling
Edited by L. M. G. Heilmeyer Jr.

The p70^{s6k}

The kinase responsible for mediating the S6 phosphorylation response was first identified in extracts of quiescent Swiss 3T3 cells stimulated to proliferate with either epidermal growth factor or serum (Novak-Hofer I and Thomas G, 1984). Purification of this enzyme following classical biochemical approaches revealed a polypeptide of M_r 70,000 (Jenõ P et al., 1988), subsequently denoted p70^{s6k} (Reinhard C et al., 1992). The enzyme is highly specific for ribosomal protein S6 (Jenõ P et al., 1989), is biphasically activated (Šuša M et al., 1989; Šuša M and Thomas G, 1990), and selectively inactivated by a type 2A phosphatase (Ballou LM et al, 1988a). In addition, the enzyme will phosphorylate four and possibly all of the five S6 phosphorylation sites observed *in vivo* (Ferrari S et al., 1991), with recognition of the substrate dependent on Arg_{231} and Arg_{233} (Flotow H and Thomas G, 1992).

Although the studies above provided sufficient amounts of the p70^{s6k} for its identification and characterization, they were inadequate for generating protein sequence data (Jenö P et al, 1989: Jenö P et al, 1989). Thus we turned to the whole animal and examined a number of approaches which were known to induce S6 phosphorylation in the rat (Kozma SC et al., 1990; Lane HA and Thomas G, 1991). The most potent induction of p70^{s6k} was obtained by administration of cycloheximide. The drug is known to induce S6 phosphorylation in a number of tissues including liver (Nielsen PJ et al., 1982) via a stress response which is directly elicited through the inhibitory effects of cycloheximide on protein synthesis (Blenis J et al., 1991). Employing this approach, sufficient amounts of the p70^{s6k} were purified from the livers of rats treated with the drug (Kozma SC et al., 1990; Lane HA and Thomas G, 1991) to provide protein sequence data for both the generation of specific antibodies (Lane HA et al., 1992) and the cloning of the p70^{s6k} (Kozma SC et al., 1990).

The $p70^{s6k}$ / $p85^{s6k}$ cDNA clones

Using the protein sequence data described above, advantage was taken of the conserved catalytic domains of serine/threonine kinases to generate a 0.4 kb cDNA fragment by a reverse transcriptase polymerase chain reaction. This product was used to screen a rat liver cDNA library and five clones were isolated, two of which were full-length and identical in sequence (Kozma SC et al., 1990). This $p70^{s6k}$ cDNA, referred to as clone 1, was 2.8 kb long and contained a 133 nucleotide GC-rich, 5'-untranslated region followed by a 502 amino acid residue coding sequence. The nucleotide sequence surrounding the initiation codon conformed with a strong consensus translational start site, and the 3'end, though AT-rich, was devoid of a poly A tail (Kozma SC et al., 1990). Shortly after these observations were made, a second $p70^{s6k}$ cDNA was identified. This clone was 2.3 kb long and contained a short, 21 nucleotide, 5'-untranslated region followed by a weak consensus translational start site and referred to as clone 2 (Banerjee P et al., 1990). In contrast to clone 1, clone 2 contained an open reading frame capable of encoding a protein of 525 amino acids (Banerjee P et al., 1990). A second AUG start codon in clone 2 was identified at nucleotide 91 which is identical to the strong translational start site in clone 1. In fact, the common coding sequences are identical with the exception of a single amino acid. In addition, the 3'-untranslated tail of clone 2 was identical to that of clone 1 but ended in a poly A tail at nucleotide 2283. A cDNA probe covering the common coding region revealed four distinct transcripts of 6.0, 4.0, 3.2 and 2.5 kb by Northern blot analysis (Kozma SC et al., 1990). By employing specific probes it was possible to demonstrate that both clones 1 and 2 were derived from the larger transcript, whereas only clone 1 was derived from the 4.0 and 3.2 kb transcripts and clone 2 from the 2.5 kb transcript (Reinhard C et al., 1992). Finally, genomic PCR and Southern blot analyses demonstrated that a single gene was responsible for encoding both transcripts (Reinhard C et al., 1992).

To determine which transcript was encoding the p70^{s6k}, both clones were expressed *in vitro* and the products were analyzed with specific antibodies (Reinhard C et al., 1992). The results demonstrated that both clones encoded the p70^{s6k} but that clone 2 encoded a novel isoform of the kinase containing a 23 amino acid NH_2 terminal extension, referred to as the p85^{s6k} (Reinhard C et al., 1992). Employing a specific antibody it was shown that this isoform of kinase is also activated when quiescent cells in culture are stimulated to proliferate (Reinhard C et al., 1992). Preliminary data also show that this form of the kinase is targeted to the nucleus (C. Reinhard, A. Fernandez, N. Lamb and G.T., unpublished data), a compartment in the cell where S6 has also been shown to be phosphorylated in response to mitogens (Franco R and Rosenfeld MG, 1990).

p70^{s6k} phosphorylation sites

Earlier studies had shown that the p70^{s6k} is activated by phosphorylation and removal of these phosphates leads to inactivation of the kinase (Ballou LM et al., 1988b). During the protein sequencing of the p70^{s6k} three phospho-peptides were identified that resided close to one another in the primary sequence (Ferrari S et al., 1992). A frequent hallmark of regulatory phosphorylation is the clustering of phosphorylation sites (Roach PJ, 1991). To determine whether the sites identified by direct sequencing were associated with p70^{s6k} activation, the kinase was labeled to high specific activity with $^{32}P_i$ following serum stimulation of quiescent Swiss 3T3 cells (Ferrari S et al., 1992). Sequential cleavage with cyanogen bromide and endoproteinase Lys-C followed by tryptic peptide analysis demonstrated that all the sites of phosphorylation resided in a single endoproteinase Lys-C peptide of M_r 2400 (Ferrari S et al., 1992). By taking advantage of a computer program which predicts the mobility of phospho-tryptic peptides following two-dimensional thin-layer electrophoresis / chromatography, a single endoproteinase Lys-C peptide emerged as the most likely candidate for containing all the sites of phosphorylation. This peptide spanned amino acids 409 to 427. To ensure that this was the correct phosphopeptide, synthetic peptides were made, phosphorylated either enzymatically or chemically, and tested for their ability to comigrate on thin-layer plates with tryptic peptides derived from the *in vivo*-labeled $p70^{s6k}$. In this way it was shown that three of the four phosphorylation sites identified were equivalent to those sequenced in the rat liver $p70^{s6k}$ (Ferrari S et al, 1992). The sites of phosphorylation are Ser_{411}, Ser_{418}, Thr_{421} and Ser_{424}. These findings are summarized in the figure below.

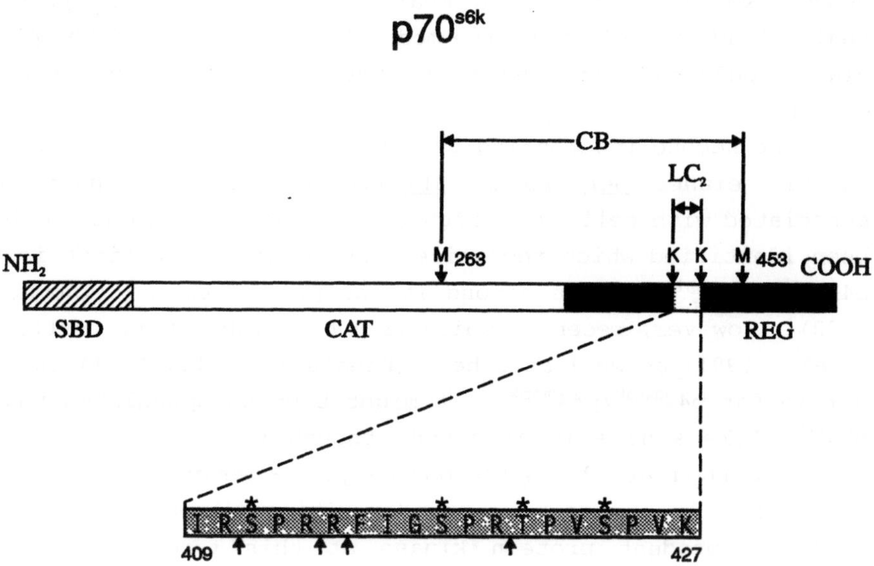

Domain of the $p70^{s6k}$ containing sites of phosphorylation. Hatched bars, putative substrate binding domain (SBD); white bar, conserved catalytic domain (CAT); black bar, regulatory domain (REG); stippled bar, autoregulatory domain containing sites of phosphorylation. Cyanogen bromide peptide (CB); endoproteinase Lys-C peptide (LC_2). Arrows ( ) indicate cleavage sites for cyanogen bromide, endoproteinase Lys-C and trypsin. Single-letter code is used for amino acids. Asterisks (*) indicate phosphorylated amino acids. Numbering of amino acid residues according to Kozma et al. (1990).

The sequence containing these phosphorylation sites exhibits two interesting features (fig.). First, all the sites of phosphorylation are clustered within a domain, Lys_{400} to Gly_{431}, which has been speculated to operate as a pseudosubstrate or autoinhibitory domain (Banerjee P et al., 1990). We initially questioned this hypothesis because peptides spanning this sequence do not serve as a substrate for the kinase, even though Ser_{418} aligns with the preferred site of phosphorylation in ribsomal protein S6, Ser_{236} (Flotow H and Thomas G, 1992). Moreover, a peptide spanning this region was 60-fold less effective in inhibiting kinase activity than a peptide based on the S6 phosphorylation sites (Flotow H and Thomas G, 1992). Location of these mitogen-stimulated phosphorylation sites on or near this putative autoregulatory region leads to an attractive model for activation of the $p70^{s6k}$ via a conformational change in the potential autoinhibitory domain.

The second interesting feature of these sites is that they exhibit either <u>Ser</u>-Pro or <u>Thr</u>-Pro motifs which have been associated with cell cycle regulatory events. Two enzymes have been identified which recognize such motifs. The first is the $p42^{mapk}/p44^{mapk}$ and the second is the $p34^{cdc2}$ kinase (Thomas G, 1992). However, recent studies from this laboratory (Ballou LM et al., 1991) as well as others (Blenis J et al., 1991) exclude a role for $p42^{mapk}/p44^{mapk}$. It might then be speculated that a $p34^{cdc2}$-like kinase is involved, though no such activity has been reported at the G_o/G_1 boundary. However, it should be noted that in a recent report describing the activation of a proline-dependent protein kinase at this time in the cell cycle, it was also found that this enzyme appears to be a $p34^{cdc2}$-cyclin A type kinase (Hall FL et al., 1991). With the expanding number of cyclins and CD kinases (CDKs) being identified, it is possible that an as yet unidentified CDK is fully or partially responsible for phosphorylation and activation of the $p70^{s6k}$.

Signalling pathways

We have demonstrated that treatment of quiescent Swiss 3T3 cells with epidermal growth factor results in time-dependent biphasic activation of the $p70^{s6k}$ (Šuša M et al., 1989). The second phase of this response was inhibited by greater than 85% in cells in which protein kinase (PKC) had been down-regulated by long-term treatment with 12-0-tetra-deconoyl-phorbol-13-acetate (TPA) (Šuša M et al 1989). This loss of the second phase was also associated with an inhibition of S6 phosphorylation, initiation of protein synthesis, and cell growth (Šuša M et al 1989). This finding was consistent with the ability of the ligand-activated EGF tyrosine kinase receptor to bind, phosphorylate, and stimulate phospholipase C_γ (Meisenhelder J et al., 1989), leading to the production of diacylglycerol, the stimulation of PKC (Nishizuka Y, 1988) and subsequent activation of late phase $p70^{s6k}$ activity. However, the use of TPA in distinguishing PKC signalling pathways recently has been placed in doubt (Sha'afi RI, 1989; Hockberger P. et al., 1989). In addition, others in the field have failed to establish a link between EGF-induced $p70^{s6k}$ activation and PKC by using long-term TPA pretreatment (Pelech SL and Krebs EG, 1987; Chen RH et al., 1991; Blackshear PJ et al., 1991). These observations prompted us to further test our earlier findings by the use of PKC inhibitors and/or two different mitogens, one of which is strongly coupled to the PKC signalling pathway in these cells, platelet-derived growth factor (PDGF) (Meisenhelder J et al., 1989) and one of which is not, insulin (Blackshear PJ et al., 1991). The results showed that the late phase of $p70^{s6k}$ activation is largely abolished by preincubation with CGP 41 251 (Šuša M et al., 1992), a derivative of staurosporine (Meyer T et al., 1989). Under these conditions there was no effect on the first phase of activation (Šuša M et al., 1992). In contrast, staurosporine inhibited both phases of S6 kinase activation, whereas the inactive analogue CGP 42 700 (Meyer T et al., 1989) shows no effect on either phase (Šuša M et al., 1992). PDGF and insulin also induced biphasic activation of the $p70^{s6k}$, but in neither case was there an effect on either phase of activation by CGP 41 251. This observation was very surprising in the case of PDGF, which is a potent activator of PKC and is known to modulate the phosphorylation state of many proteins through this kinase (Kazlauskas A and Cooper JA, 1988; Rossomando AJ et al., 1989). However, this result was also supported by almost identical findings when PKC action was blocked by prolonged TPA pretreatment (Šuša M et al., 1992). Thus, EGF appears to use PKC activation to signal the $p70^{s6k}$. In contrast, even though PDGF strongly activates the PKC pathway, it apparently does not use this pathway for $p70^{s6k}$ activation.

Perspectives

A great deal of attention has been directed at the signalling pathways which control S6 phosphorylation during the early mitogenic response. It now will be essential to determine the role of each of these components in the overall growth response. In addition, it will be important to determine whether the novel $p85^{s6k}$ isoform resides on the same signalling pathway as the $p70^{s6k}$ and the basis for the generation of two $p70^{s6k}$ from the same gene.

Acknowledgements:

I would like to thank Drs. R. Pearson and M. Stewart for their critical reading of the manuscript and C. Wiedmer for her typing and editing expertise.

References

Banerjee P, Ahamad MF, Grove JR, Kozlosky C, Price DJ, Avruch J (1990) Molecular structure of a major insulin/mitogen-activated 70 kDa S6 protein kinase. Proc Natl Acad Sci USA 87:8550-8554

Ballou LM, Jenö P, Thomas G (1988a) Protein phosphatase 2A inactivates the mitogen stimulated S6 kinase from Swiss mouse 3T3 cells. J Biol Chem 263:1188-1194

Ballou LM, Siegmann M, Thomas G (1988b) S6 kinase in quiescent Swiss mouse 3T3 cells is activated by phophorylation in response to serum treatment. Proc Natl Acad Sci USA 85:7154-7158

Ballou LM, Luther H, Thomas G (1991) MAP2 kinase and 70K S6 kinase lie on distinct signalling pathways. Nature 349:348-350

Blackshear PJ, Haupt DM, Stumpo DJ (1991) Insulin activation of protein kinase C: a reassessment. J Biol Chem 266:10946-10952

Blenis J, Chung J, Erikson E, Alcorta DA, Erikson RL (1991) Distinct mechanisms for the activation of the RSK kinases/MAP2 kinase/$pp90^{rsk}$ and pp70-S6 kinase signalling systems are indicated by inhibition of protein synthesis. Cell Growth Differ 2:279-285

Brooks RF (1977) Continuous protein synthesis is required to maintain the probability of entry into S phase. Cell 12:311-317

Chen RH, Chung J, Blenis J (1991) Regulation of $pp90^{rsk}$ phosphorylation and S6 phosphotransferase activity in Swiss 3T3 cells by growth factor-, phorbol ester-, and cyclic AMP-mediated signal transduction (1991) Mol Cell Biol 11:1861-1867

Ferrari S, Bandi HR, Hofsteenge J, Bussian BM, Thomas G (1991) Mitogen-activated 70K S6 kinase. Identification of *in vitro* 40 S ribosomal S6 phosphorylation sites. J Biol Chem 266:22770-22775

Ferrari S, Bannwarth W, Morley SJ, Totty NF, Thomas G (1992) Activation of $p70^{s6k}$ is associated with phosphorylation of four clustered sites displaying Ser/Thr motifs. Proc Natl Acad Sci USA in press

Flotow H, Thomas G (1992) Substrate recognition determinants of the mitogen activated 70K S6 kinase from rat liver. J Biol Chem 267:3047-3078

Franco R, Rosenfeld MG (1990) Hormonally inducible phosphorylation of a nuclear pool of ribosomal protein S6. J Biol Chem 265:4321-4325

Hall FL, Braun RK, Mihara K, Fung YKT, Berndt N, Carbonaro-Hall DA, Vulliet R (1991) Characterization of the cytoplasmic proline-directed protein kinase in proliferative cells and tissues as a heterodimer comprised of $p34^{cdc2}$ and $p58^{cyclin\ A}$. J Biol Chem 266:1743-1740

Hershey JWB (1989) Protein phosphorylation controls translation rates. J Biol Chem 264:20823-20826

Hockberger P, Toselli M, Swandulla D, Lux HD (1989) A diacylglycerol analogue reduces neuronal calcium currents independently of protein kinase C activation. Nature 338:340-342

Jenö P, Ballou LM, Novak-Hofer I, Thomas G (1988) Identification and characterization of a mitogen-activated S6 kinase. Proc Natl Acad Sci USA 85:406-410

Jenö P, Jäggi N, Luther H, Siegmann M, Thomas G (1989) Purification and characterization of a 40S ribosomal protein S6 kinase from vanadate-stimulated Swiss 3T3 cells. J Biol Chem 264:1293-1297

Kazlauskas A, Cooper JA (1988) Protein kinase C mediates platelet-derived growth factor-induced tyrosine phosphorylation of p42. J Cell Biol 106:1395-1402

Kozma SC, Lane HA, Ferrari S, Luther H, Siegmann M, Thomas G (1989) A stimulated S6 kinase from rat liver: identity with the mitogen activated S6 kinase of Swiss 3T3 cells. EMBO J 8:4125-4132

Kozma SC, Ferrari S, Bassand P, Siegmann M, Totty N, Thomas G (1990) cloning of the mitogen activated S6 kinase from rat liver reveals an enzyme of the second messenger subfamily. Proc Natl Acad Sci USA 87:7365-7369

Krieg J, Hofsteenge J, Thomas G (1988) Identification of the 40S ribosomal protein S6 phosphorylation sites induced by cycloheximide. J Biol chem 263:11473-11477

Lane HA, Thomas G (1991) Purification and properties of mitogen-activated S6 kinase from rat liver and 3T3 cells. In Meth Enzymol 200:268-291

Lane HA, Morley SJ, Dorée M, Kozma SC, Thomas G (1992) Identification and early activation of a *Xenopus laevis* $p70^{s6k}$ following progesterone-induced meiotic maturation. EMBO J 11:1743-1749

Meyer T, Regenass U, Fabbro D, Alteri E, Rösel J, Müller M, Caravatti G, Matter A (1989) A derivative of staurosporine (CGP 41 251) shows selectivity for protein kinase C inhibition and *in vitro* anti-proliferative as well as *in vivo* anti-tumor activity. Int J Cancer 43:851-856

Morley S, Thomas G (1991) Intracellular messengers and the control of protein synthesis. Pharmac Ther 50:291-319

Meisenhelder J, Suh PG, Rhee SG, Hunter T (1989) Phospholipase C-γ is a substrate for the PDGF and EGF receptor protein-tyrosine kinases *in vivo* and *in vitro*. Cell 57:1109-1122

Nielsen PJ, Manchester KL, Towbin H, Gordon J, Thomas, G (1982) The phosphorylation of ribosomal protein S6 in rat tissues following cycloheximide injection, in diabetes, and after denervation of diaphragm. A simple immunological determination of the extent of S6 phosphorylation on protein blots. J Biol Chem 257:12316-12321

Nishizuka Y (1988) The molecular heterogeneity of protein kinase C and its implications for cellular regulation. Nature 334:661-665

Novak-Hofer I, Thomas G (1984) An activated S6 kinase in extracts from serum- and epidermal growth factor-stimulated Swiss 3T3 cells. J Biol Chem 259:5995-6000

Pardee AB (1989) G_1 events and regulation of cell proliferation. Science 246:603-608

Pelech SL, Krebs EG (1987) Mitogen-activated S6 kinase is stimulated via protein kinase C-dependent and -independent pathways in Swiss 3T3 cells. J Biol Chem 262:11598-11606

Reinhard C, Thomas G, Kozma SC (1992) A single gene encodes two isoforms of the $p70^{s6k}$: activation upon mitogenic stimulation. Proc Natl Acad Sci USA 89:4052-4056
Roach PJ (1991) Multisite and hierarchal protein phosphorylation. J Biol Chem 266:14139-14142
Rossow P, Riddle VGH, Pardee AB (1979) Synthesis of labile, serum-dependent protein in early G_1 controls animal cell growth. Proc Natl Acad Sci USA 76:4446-4450
Rossomando AJ, Payne DM, Weber MJ, Sturgill TW (1989) Evidence that pp42, a major tyrosine kinase target protein, is a mitogen-activated serine/threonine protein kinase. Proc Natl Acad Sci USA 86:6940-6943
Sha'afi RI (1989) Some effects of phorbol esters are not mediated by protein kinase C. Biochem J 261:688
Šuša M, Olivier AR, Fabbro D, Thomas G (1989) EGF induces biphasic S6 kinase activation: late phase is protein kinase C-dependent and contributes to mitogenicity. Cell 57:817-824
Šuša M, Thomas G (1990) Identical M_r 70,000 S6 kinase is activated biphasically by epidermal growth factor: a phosphopeptide which characterizes the late phase. Proc Natl Acad Sci USA 87:7040-7044
Šuša M, Vulević, Lane, HA, Thomas G (1992) Inhibition or down-regulation of protein kinase C attenuates late phase $p70^{s6k}$ activation induced by epidermal growth factor but not by platelet-derived growth factor or insulin. J Biol Chem 267:6905-6909
Thomas G (1992) Map kinase by any other name smells just as sweet. Cell 68:3-6

EXPRESSION OF THE $p70^{s6k}$ / $p85^{s6k}$ ISOFORMS FROM A SINGLE GENE AND ACTIVATION FOLLOWING MITOGENIC STIMULATION

C. Reinhard, G. Thomas and S.C. Kozma
Friedrich Miescher Institute
P.O. Box 2543, 4002 Basel, Switzerland

An early obligatory response elicited by growth factors on quiescent cells is a 2- to 3-fold increase in protein synthesis, concomitant with the multiple phosphorylation of 40S ribosomal protein S6 (Thomas et al., 1979). The responsible kinase phosphorylates 5 of 7 serines at the C-terminus of the protein, has a M_r 70,000, and is referred to as $p70^{s6k}$. Mitogenic activation of $p70^{s6k}$ itself is associated with Ser/Thr phosphorylation of the kinase, and is reversed through dephosphorylation by a type 2A phosphatase (Ballou et al., 1988).

The $p70^{s6k}$ has been purified on a large scale from the livers of rats injected with cycloheximide (Kozma et al., 1989, Price et al., 1989). Sequencing of CNBr and tryptic peptides was used to derive oligonucleotides which were employed in polymerase chain reactions using rat cDNAs as template. The product obtained was a 0.4 kb DNA fragment specific for $p70^{s6k}$, and was subsequently used to screen cDNA libraries. This approach led to the isolation of two distinct cDNA clones, one from a rat liver cDNA library and one from a rat hepatoma cDNA library (Kozma et al., 1990; Banerjee et al., 1990), referred to here as clones 1 and 2, respectively (Fig. 1). Clone 1 is 2.8 kb in length and contains a 133 nucleotide (nt) 5′ untranslated region (UR), an open reading frame (ORF) encoding 502 amino acids (aa) and a 3′ UR devoid of a polyA tail. Clone 2 is 2.3 kb in length and its sequence is identical to clone 1 except for the 5′ end. Clone 2 contains a 21 nt 5′ UR and the ORF encodes a 525 aa protein, corresponding to the protein encoded by clone 1 with a 23 aa extension of the N-terminus (Fig. 1). The 3′ UR of clone 2 is shorter than in clone 1 but identical in the shared common sequence. A cDNA corresponding to clone 2 has been isolated from a rabbit cDNA library

NATO ASI Series, Vol. H 76
Tyrosine Phosphorylation/Dephosphorylation and Downstream Signalling
Edited by L. M. G. Heilmeyer Jr.

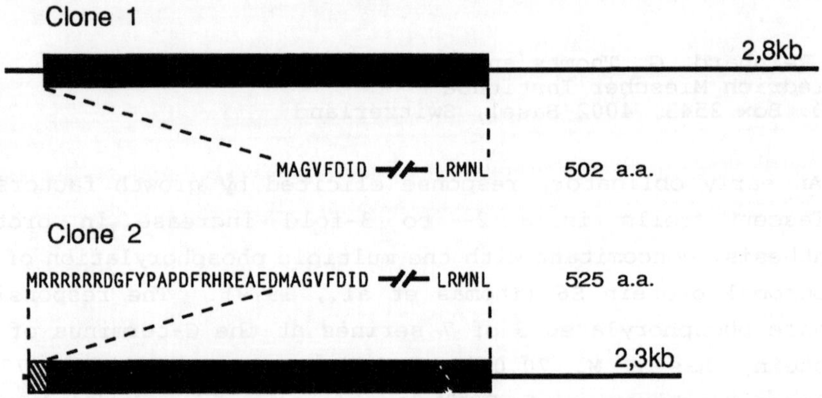

Fig. 1. Rat s6k cDNA clones 1 and 2. Solid thick bar indicates common coding region; shaded thick bar indicates 5′ coding region specific for clone 2; thin bar indicates untranslated regions. Amino acids (aa) are denoted by a single-letter code.

(Harmann and Kilimann, 1990), and clone 1 and 2 type cDNAs have been obtained from a human cDNA library (Grove et al., 1991).

Northern blots from rat liver mRNA revealed four transcripts of 6.0, 4.0, 3.2 and 2.5 kb, with probes common to both clones. Using probes unique to either clone 1 or clone 2 it was demonstrated that the 6.0 kb transcript contains sequences specific for both clones and that clone 1 corresponds to the 4.0 or the 3.2 kb transcript and clone 2 to the 2.5 kb transcript (Reinhard et al., 1992). Furthermore, Southern blot analysis of genomic rat DNA showed that the two clones are transcribed from a single gene.

In vitro translation of clones 1 and 2 indicated that both cDNA types encode the $p70^{s6k}$ polypeptide but clone 2 yields an additional polypeptide of M_r 85,000, representing a $p70^{s6k}$ isoform referred to as $p85^{s6k}$. Expression of two polypeptides from clone 2 can be explained by an unfavorable consensus sequence for translational initiation at the first Met, allowing transition to the second initiation translation start site (Kozak, 1991).

In the unique 23 aa N terminal extension of the $p85^{s6k}$, the first Met is followed by a cluster of six consecutive Arg, reminiscent of a nuclear targeting motif (Garcia-Bustos et al., 1991). The 23 aa extension also contains residues which correspond to a bipartite nuclear targeting motif (Dingwall and Laskey, 1991), characterized by two basic aa, followed by ten intervening residues and a cluster of five aa in which three are basic (see Fig. 1). Thus nuclear compartmentalization is plausible for the $p85^{s6k}$ isoform, substantiating the results of Franco and Rosenfeld (1990) who have found phosphorylated S6 in the nucleus. The catalytic domain of $p70^{s6k}/p85^{s6k}$ resembles that of Ser/Thr kinases regulated by second messengers and typifed by cAMP-dependent protein kinase (Hanks et al., 1988). Within this subfamily, it has highest homology with the first catalytic domain of S6 kinase II (56%) (Jones et al., 1988), yeast protein kinase I (49%) (Maurer, 1988) and protein kinase C_ε (44%) (Ono et al., 1988). The homology to S6 kinase II, protein kinase C and cAMP-dependent protein kinase extends 50-70 aa carboxyl to the catalytic domain but drops drastically in the last 120 aa. In this latter segment reside the four sites of phosphorylation associated with kinase activation (Ferrari et al., 1992). These sites surround a sequence homologous to

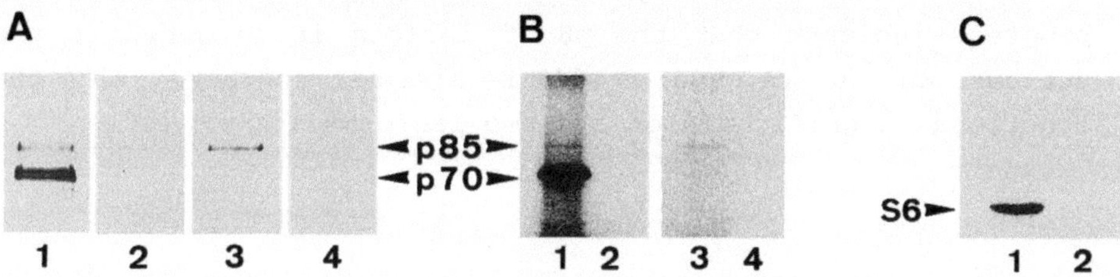

Fig. 2. Rat liver $p70^{s6k}/p85^{s6k}$. (A) Western blot of purified rat S6 kinase using antisera against either both isoforms or $p85^{s6k}$ alone, in the absence (lanes 1 and 3, respectively) or presence (lanes 2 or 4, respectively) of competing peptide. (B) Auto-phosphorylated S6 kinase immunoprecipitated by antisera as in A. (C) Activity of rat S6 kinase immunoprecipitated by anti-$p85^{s6k}$ antiserum in the absence (lane 1) or presence (lane 2) of competing peptide.

the sequence recognized by the kinase on the S6 protein, which may have an autoinhibitory function (Banerjee et al., 1990; Flotow and Thomas, 1992).

In vivo expression of the two isoforms was analyzed by immunoblotting; highly purified S6 kinase from cycloheximide-treated rat livers with antibodies recognizing either both isoforms or the $p85^{s6k}$ only (Fig. 2A). The antibody recognizing both isoforms decorated a major band at M_r 70,000 and a minor one at M_r 85,000 whereas the $p85^{s6k}$ specific antibody recognized only the latter minor band. Incubation of the purified S6 kinase fraction with $[^{32}P]\gamma$-ATP (Fig. 2B) and immunoprecipitation followed by S6 kinase assay (Fig. 2C) demonstrated that in the livers of cycloheximide-treated rat, both isoforms are present, autophosphorylate and have S6 kinase activity.

The detection of active $p85^{s6k}$ raised the question of its contribution to S6 kinase activation following mitogenic stimulation of quiescent cells. Cell extracts, prepared from either quiescent or 60-min serum-stimulated Swiss 3T3 cells, were assayed for S6 kinase activity after immunoprecipitation with antibodies recognizing either both isoforms or the $p85^{s6k}$ isoform. The results showed a 5- to 10-fold increase in S6 kinase activity after serum stimulation with both antibodies, arguing that $p85^{s6k}$ also contributes to increased S6 kinase activity following mitogenic stimulation. Experiments in progress indicate that the $p85^{s6k}$ isoform is targeted to the nucleus, and future studies will be aimed at characterizing the functional significance of the two isoforms.

Acknowledgements: We are grateful to Dr. Heidi A. Lane for her critical reading of the manuscript and to Carol Wiedmer for her editing and typing expertise.

References

Ballou LM, Jenö P, Thomas G (1988) S6 kinase in quiescent 3T3 cells is activated by phosphorylation in response to serum treatment. Proc Natl Acad Sci USA 85:7154-7158

Banerjee P, Ahmad MF, Grove JR, Kozlosky C, Price DJ, Avruch J (1990) Molecular structure of a major insulin/mitogen-activated 70-kDa S6 protein kinase. Proc Natl Acad Sci USA 87:8550-8554

Dingwall C and Laskey RA (1991) Nuclear targeting sequences - a consensus? Trends Biochem Sci 16:478-481

Ferrari S, Bannwarth W, Morley S, Totty NF and Thomas, G (1992) Activation of $p70^{s6k}$ is associated with phosphorylation of four clustered sites displaying Ser/Thr-Pro motifs. Proc Natl Acad Sci USA 89:7282-7286

Flotow H, Thomas G (1992) Substrate recognition determinants of the mitogen-activated 70K S6 kinase from rat liver. J Biol Chem 267:3074-3078

Franco R, Rosenfeld MG (1990) Hormonally inducible phosphorylation of a nuclear pool of ribosomal protein S6. J. Biol. Chem. 265:43321-4325

Garcia-Bustos J, Heitman J and Hall MN (1991) Nuclear protein localization. Biochim Biophys Acta 1071:83-101

Grove JR, Banerjee P, Balasubramanyam A, Coffer PJ, Price DJ, Avruch J, Woodget JR (1991) Cloning and expression of two human 70 S6 kinase polypeptides differing only at their amino termini. Mol Cell Biol 11:5541-5550

Hanks SK, Quinn AM, Hunter T (1988) The protein kinase family: conserved features and deduced phylogeny of the catalytic domain. Science 241:42-52

Harmann B, Kilimann MW (1990) cDNA encoding a 59 kDa homolog of ribosomal protein S6 kinase from rabbit liver. FEBS Lett. 273:248-252

Jones SW, Erikson E, Blenis J, Maller JL, Erikson RL (1988) A *Xenopus* ribosomal protein S6 kinase has two apparent kinase domains that are each similar to distinct protein kinases. Proc Natl Acad Sci USA 85:3377-3381

Kozak M (1991) Structural features in eukaryotic mRNA that modulate the initiation of translation. J Biol Chem 266:19867-19870

Kozma SC, Lane HA, Ferrari S, Luther H, Siegmann M and Thomas G (1989) A stimulated S6 kinase from rat liver: identity with the mitogen-activated S6 kinase of 3T3 cells. EMBO J 8:4125-4132

Kozma SC, Ferrari S, Bassand P, Siegmann M, Totty N and Thomas G (1990) Cloning of the mitogen-activated S6 kinase from rat liver reveals an enzyme of the second messenger subfamily. Proc Natl Acad Sci USA 87:7365-7369

Maurer RA (1988) Isolation of a yeast protein kinase gene by screening with a mammalian protein kinase cDNA. DNA 7:469-474

Ono Y, Fujii T, Ogita K, Kibbawa U, Igarashi K, Nishizuka Y (1988) The structure, expression and properties of additional members of the protein kinase C family. J Biol Chem 263:6927-6932

Price DJ, Nemenoff RA and Avruch J (1989) Purification of a hepatic S6 kinase from cycloheximide-treated rats. J Biol Chem 264:13825-13833

Reinhard C, Thomas G, Kozma SC (1992) A single gene encodes two isoforms of the p70 S6 kinase: activation upon mitogenic stimulation. Proc Natl Acad Sci USA 89:4052-4056

Thomas G, Siegmann M and Gordon J (1979) Multiple phosphorylation of ribosomal protein S6 during transition of quiescent 3T3 cells into early G_1 and cellular compartmentalization of the phosphate donor. Proc Natl Acad Sci USA 76:3952-3956

Control of S6 phosphorylation in normal and v-src transformed cells

Borja Belandia, Mª Victoria Carretero, Juan Angel Fresno and Jorge Martín-Pérez.

Instituto de Investigaciones Biomédicas, C.S.I.C., Madrid, Spain.

Introduction

Reversible protein phosphorylation is an essential mechanism for the control of cellular proliferation, differentiation as well as in tumorogenesis (Aaranson et al., 1991; Cantley et al., 1991; Kozma and Thomas, 1992). This post-translational modification is regulated by the levels of activity of both protein kinases and protein phosphatases. Since the multiple phosphorylation of the 40S ribosomal protein S6 is an ubiquitous events upon cellular stimulation (Kozma et al., 1989), we use S6 as a target to study if, as do serum growth factors, v-src transformation alters the mechanisms regulating its phosphorylation.

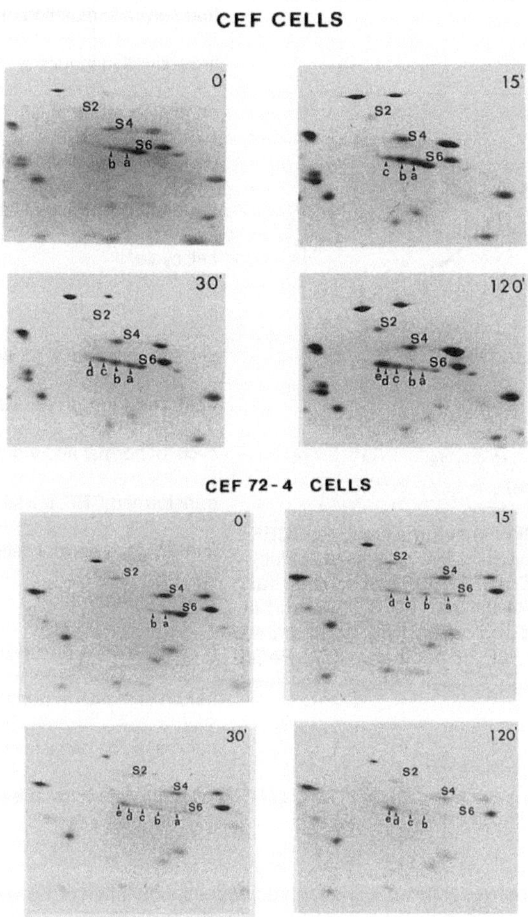

Fig. 1. Two-dimensional gel electrophoresis analysis of S6 phosphorylation in normal (CEF) and ts-NY72-4 infected (CEF 72-4) cells.

Results and Discussion

Serum stimulation of CEF or activation of the v-src oncogene by shifting cells infected with the temperature sensitive mutant ts-NY72-4, to permissive temperature, induce the multiple protein

NATO ASI Series, Vol. H 76
Tyrosine Phosphorylation/Dephosphorylation and Downstream Signalling
Edited by L. M. G. Heilmeyer Jr.

phosphorylation of the 40S ribosomal protein S6 (Blenis and Erikson, 1985; Martín-Pérez et al., 1991). The analysis of the S6 phosphorylation by two-dimensional gel electrophoresis (Siegmann and Thomas, 1987) shows that in quiescent CEF and CEF 72-4, at non-permissive temperature, S6 protein was mainly dephosphorylated (Fig. 1). As soon as 15 min after stimulation of both CEF and CEF 72-4, an increment in S6 phosphorylation was already detected by the expression of the S6 derivatives a, b, c and d containing 1, 2, 3 and 4 mole of phosphate respectively. After two hours, the most abundant derivative was S6e, containing 5 mole of phosphate (Fig. 1). These results indicate that serum stimulation or v-src activation induce a parallel increase of S6 phosphorylation during the early G1 phase of the cell cycle.

When the phosphorylation of S6 was analyzed through the cell cycle of normal and v-src transformed CEF, it was found that in normal cells S6 phosphorylation reached a peak after 8 h of serum stimulation and then decreased to basal levels at mitosis. Transformed cells, in contrast, maintained the high levels of S6 phosphorylation for at least 16 h after the ts-v-src induction (Fig. 2).

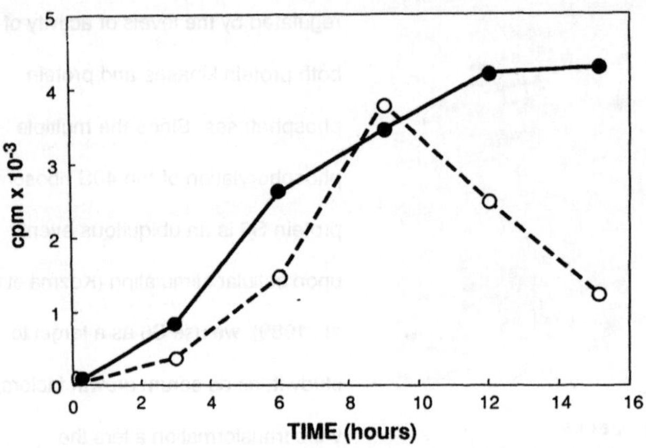

Fig. 2. Time course of S6 phosphorylation in serum stimulated (CEF, ---, open circles) and v-src transformed (CEF 72-4, ___, filled circles) cells. Quiescent cultures of CEF and CEF 72-4 were stimulated (as described in the text) in the presence of 0.4 mCi/ml of 32Pi, and at indicated times, ribosomes were isolated and radioactivity incorporated into S6 was determined by counting the 32P-S6 band after SDS-PAGE and autoradiography.

This result suggests that the equilibrium between phosphorylation and dephosphorylation of S6 was altered by v-src.

It has been shown previously that S6 phosphatase is a type-1 phosphatase (PP1; Olivier et al., 1988; Andres and Maller, 1989, Martín-Pérez et al., 1992). Since after stimulation of CEF proliferation

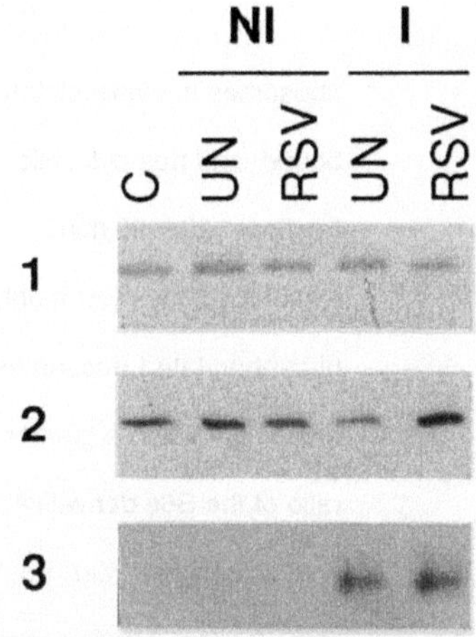

Fig. 3. Activity and amounts of PP1 in normal and v-src transformed CEF. Panel 1: Coomassie blue stain of the 32P-phosphorylase a; Panel 2 autoradiography of the 32P-phosphorylase a , line C shows the amount of 32P-phosphorylase a used in the assay; Panel 3, Immunoblot of the PP-1c released from the immunoprecipitates. UN, uninfected cells; RSV, transformed cells; I immune serum; NI, non immune serum.

by serum or v-src, the S6 kinase activity was similar (Blenis and Erikson, 1986), we predicted that the S6 phosphatase activity was inhibited by v-src transformation. To test this hypothesis, PP-1 was immunoprecipitated from exponentially-growing cultures of uninfected and wt-RSV-transformed CEF with an antipeptide antibody specific for PP-1 (Olivier and Thomas, 1990). The phosphatase was released from the immune-complexes with an excess of the antigenic peptide, incubated with okadaic acid (1 nM) to inhibit any phosphatase 2A that might have been present, and then PP-1 activity was assayed for the dephosphorylation of ^{32}P-*phosphorylase a* (Cohen et al., 1988). The PP1 recovered from normal CEF was 2.5 times more active than the one obtained from transformed cells (Fig. 3, panels 1 and 2). Since the amount of PP-1 immunoprecipitated and then released was the same in both cell types, as determined by immunoblotting (Fig. 3, panel 3) , we can conclude that v-src inhibits PP1 activity. This inhibition was associated with an increase in the Km of PP1 for *phosphorylase a.* The catalytic subunit of PP1 was found phosphorylated in serine and threonine residues in both normal and v-src transformed CEF. However, the comparison of tryptic phosphopeptide maps of PP1 between them, showed specific differences (data not shown). This result could help to explain the inhibition of PP1 in v-src transformed CEF.

Although the phosphorylation of S6 is a highly conserved response to mitogenic signals, its role remains unclear. In serum stimulated Swiss 3T3 cells, the most highly phosphorylated ribosomes were detected in polyribosomes (Thomas et al., 1982). Since the components of the protein synthesis machinery are associated with the cytoskeleton , and oncogenic transformation alters its structure (Ambros et al., 1975), we analyzed if v-src modified the distribution of S6 phosphorylated ribosomes in cytoskeleton bound and free cytosolic fractions. After 30 min stimulation, the most highly phosphorylated ribosomes, those containing a greater ratio of the S6e derivative, were associated with cytoskeleton in both normal and v-src transformed CEF (Fig. 4).

This result corroborates the observations with Swiss 3T3 cells (Thomas et al., 1982) and support the idea that phosphorylation of S6 is involved in the initiation of protein synthesis.

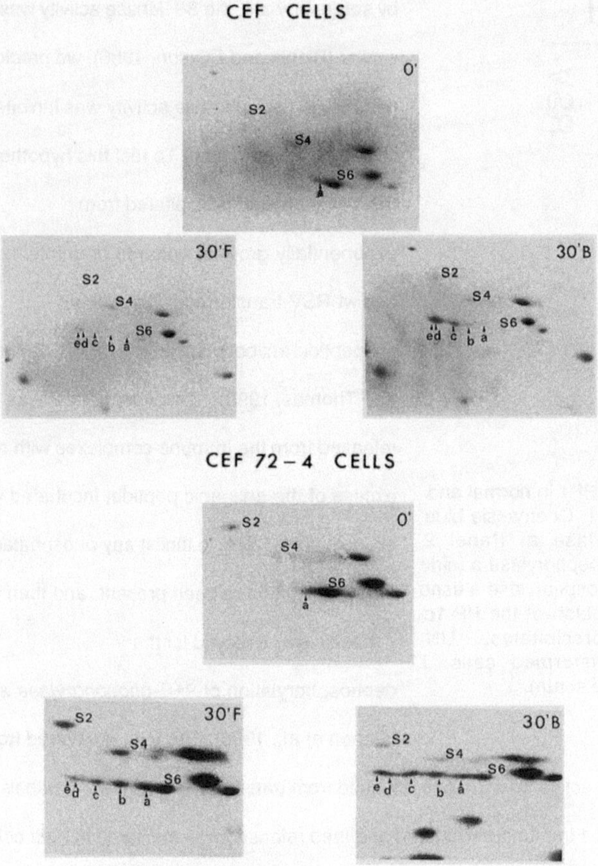

Fig. 4 Levels of S6 phosphorylation in cytosolic free (F) and cytoskeleton bound (B) ribosomes of normal (CEF) and v-src transformed (CEF 72-4) cells. 0", quiescent cells.

Acknowledgments
This work was supported by grants from DGICYT and CAM to J. M-P. B. B. and J.A. Fresno are supported by fellowships of Gobierno Vasco, Mª.V. Carretero is supported by a fellowship of CAM.

Literature references

Aaranson,S.A. 1991. Science, **254**, 1146.
Ambros,V.R.et al. 1975. Proc.Natl.Acad.Sci.USA, **72**, 3144.
Andres,J.L. and Maller,J.L. 1989. J.Biol.Chem., **264**, 151.
Blenis,J. and Erikson,R.L. 1985. Proc.Natl.Acad.Sci.USA., **82**, 7621.
Blenis,J. and Erikson,R.L. 1986. Proc.Natl.Acad.Sci.USA., **83**, 1733.
Cantley,L.C. et al. 1991. Cell, **64**, 281.
Cohen,P. et al. 1988. Meth.Enzymol., **159**, 390.
Martín-Pérez,J. et al. 1991. In: Bases Moleculares del Cancer, aplicaciones clinicas y desarrollo tecnologico. Eds. M. Barbacid y J.C. Lacal. Farmaindustria. Madrid. pp265.
Martín-Pérez,J. et al. 1992. Submitted.
Olivier,A.R. et al. 1988. Proc.Natl.Acad.Sci.USA., **85**, 4720.
Olivier,A.R. and Thomas,G. 1990. J. Biol.Chem., **265**, 22460.
Siegmann,M. and Thomas,G. 1987. Meth.Enzymol., **146**, 362.
Thomas et al. 1982. Cell, **30**, 235.
Kozma et al. 1989. Cell.Signall., **1**, 219.
Kozma,S. and Thomas,G. 1992. Rev.Phys.Biochem.Pharm., **119**, 123.

THE PHOSPHORYLATED RIBOSOMAL PROTEIN IN *TETRAHYMENA* IS HOMOLOGOUS WITH MAMMALIAN RIBOSOMAL PROTEIN S4

Lisbeth Palm, Peter Højrup and Karsten Kristiansen
Department of Molecular Biology
University of Odense
Campusvej 55
DK-5230 Odense M
Denmark

INTRODUCTION

Phosphorylation of a restricted number of ribosomal proteins has been described for all eukaryotic cells analyzed to date. The actual number of phosphorylated ribosomal proteins varies, but a single basic ribosomal protein with a M_r around 30,000 is the most prominent phosphoprotein in eukaryotic ribosomes (Wool, 1979; Cuny, 1985; Bratholm, 1986). In mammals up to five phosphate groups can be incorporated into this protein, designated S6 (Thomas *et al.*, 1980), and increased S6 phosphorylation is an apparent prerequisite for activation of protein synthesis in response to growth factors (Thomas *et al.*, 1982).

In lower eukaryotes the number of phosphate groups incorporated into the major basic ribosomal phosphoprotein varies from two to four (for reviews see, Cuny, 1985, Bratholm, 1986). In most cases the conditions leading to increased phosphorylation in lower eukaryotes resemble those leading to increased S6 phosphorylation in mammals, and generally, it has been assumed that the major basic ribosomal phosphoprotein in lower eukaryotes is homologous with mammalian S6. In yeast molecular cloning has unambiguously demonstrated that the basic ribosomal phosphoprotein, S10, is indeed a true S6 homolog (Leer *et al.*, 1982). However, the functional significance of S10 phosphorylation in yeast is unclear (Johnson and Warner, 1987).

In the ciliated protozoan *Tetrahymena* phosphorylation of ribosomal proteins is confined to a single basic 40S ribosomal protein originally named S6, now S7 (Kristiansen *et al.*, 1978). However, the conditions that induce phosphorylation in *Tetrahymena* are virtually opposite to those leading to increased phosphorylation of S6 or its putative homolog in other species (Cuny *et al.*, 1985; Bratholm, 1986).

Now we have determined the primary structure of *Tetrahymena thermophila* S7, and surprisingly, it turned out to be homologous with mammalian ribosomal protein S4. In addition we have also identified the true *Tetrahymena* S6 homolog, named S3.

RESULTS AND DISCUSSION

Starvation and sodium induced phosphorylation of T. thermophila *ribosomal protein S7*

Figure 1 shows the location of *T. thermophila* ribosomal proteins S7 and S3 in a 2-D gel. The insert illustrates the appearance of the mono- and diphosphorylated forms observed upon starvation in the presence of sodium ions. Phosphorylation proceeds in a strictly sequential manner. The first site to be phosphorylated is a serine residue and the second a threonine (Bratholm, 1986). Although these residues reside in two separate tryptic peptides they are probably located rather close in the sequence in the C-terminal region of *T. thermophila* ribosomal protein S7 (unpublished results).

NATO ASI Series, Vol. H 76
Tyrosine Phosphorylation/Dephosphorylation and Downstream Signalling
Edited by L. M. G. Heilmeyer Jr.

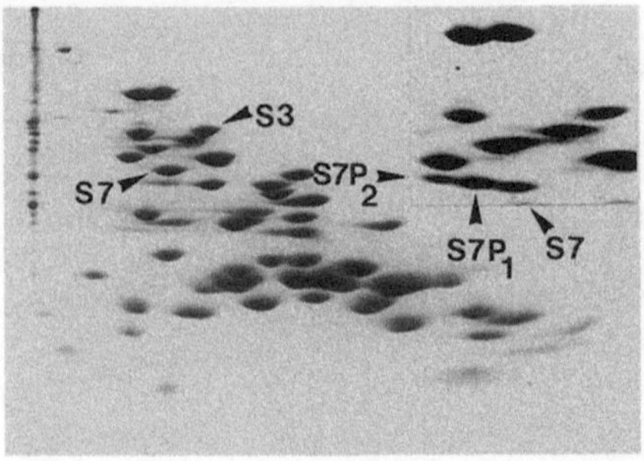

Figure 1. Separation of *T. thermophila* ribosomal proteins by 2-D PAGE using the pH 8.6/SDS system. Ribosomal protein S7 which becomes phosphorylated when cells are starved in the presence of sodium or lithium ions as well as ribosomal protein S3 which is homologous to mammalian S6 are indicated. Insert illustrates the appearance of the mono- and diphosphorylated forms of S7 in starved cells.

Isolation and sequencing of T. thermophila *ribosomal protein S7*

The primary structure of *Tetrahymena* ribosomal protein S7 was determined by a combination of different approaches. In one S7 separated by 2-D PAGE was digested in the gel with trypsin, *Staphylococcus aureus* V8 protease or chymotrypsin, or cleaved with cyanogen bromide. The generated peptides were eluted by diffusion and separated by narrowbore reversed-phase HPLC (Applied Biosystems model 140B). In the other approach ribosomal protein S7 was purified by three successive rounds of reversed-phase HPLC. The crude mixture was separated on a Dynosphere column using an acetic acid gradient, then the partly purified protein was separated on a Nucleosil C_{18} column using a gradient of aqueous 0.1% trifluoroacetic acid in acetonitrile and the final purification was carried out on a Nucleosil C_8 column using a gradient of aqueous 0.4% trifluoroacetic acid in acetonitrile. The purification was monitored by 2-D PAGE and electrospray mass spectrometry (Vestec). The electrospray mass spectrum of the purified S7 revealed the mass to be 29456 +/- 5 Da. The purified protein was digested and the resulting peptides separated as above. For each peptide the mass was determined by plasma desorption mass spectrometry (BioIon model 20), amino acid composition was determined following acid hydrolysis, and finally, the amino acid sequence was determined using a Knauer model 810 pulsed liquid sequencer.

```
       1                                                50
RatS4  ARGPKKHLKR VAAPKHWMLD KLTGVFAPRP STGPHKLREC LPLIIFLRNR
TthS7  ARGPKKHLKR INAPKSWMLN KLGGIWATRP SQGPHKLRES LPLSVLLKER

RatS4  LKYALTGDEV KKICMQRFIK IDGKVRTDIT YPAGFMDVIS IDKTGENFRL
TthS7  LNYALNGRDV TLILNDKEGN VFVDQKXDKG YPTGLMDVVR IEKTDQSFRI

RatS4  IYDTKGRFAV HRITPEEAKY KLCKVRKIFV GTKGIPHLVT HDARTIRYPD
TthS7  LYDTKGRFVL KSLSKEEAKY KLLKVTAKEI GPNQIPYKPT YDSRTIRFPN

RatS4  PLIKVNDTIQ IDLETGKITD FIKFDTGNLC MVTGGANLGR IGVITNRERH
TthS7  PEIKIGDTLK YDLVNNKIEN FAHLESENVL YIQQGNNIGR VGILQHIEKH

RatS4  PGSFDVVHV KDANGNSFAT RLSNIFVIGK GNKPWISLPR GKGIRLTIAEE
TthS7  QGSFDIGHV KXAKGNAFAT RLGNIFVLGQ GKKLYIELPS GDGVRETILEE

RatS4  RDKRLAAKQS SG
TthS7  RKRKFSY... ..
```

Figure 2. Alignment of *T. thermophila* ribosomal protein S7 with rat ribosomal protein S4.

The combined set of data allowed deduction of 99 % of the amino acid sequence of *T. thermophila* S7. A search against the PIR database clearly showed that *T. thermophila* S7 is homologous with mammalian S4 (Fig. 2). To the best of our knowledge phosphorylation of this ribosomal protein has never been observed in other non-infected eukaryotic cells, although it is subject to weak phosphorylation in vesicular stomatitis virus infected L cells (Marvaldi and Lucas-Lenard, 1977).

This posed the question as to whether *Tetrahymena* was lacking a ribosomal protein homologous with mammalian S6. To answer this question 40S ribosomal proteins with a M_r above 25,000 were cut out from 2-D gels, concentrated by electrophoresis through ProSieve agarose gels, blotted onto polyvinylidene difluoride membranes, and subjected to Edman degradation. The sequence of the 42 N-terminal amino acid residues of *T. thermophila* ribosomal protein S3 was determined using this approach and shown to exhibit a convincing similarity to rat S6 (Fig. 3).

```
T.therm. S3    MKFNISYPLTGAQKGIEIDDDKKGNIFMDKKMGQEVEGDXLG
Rat S6         MKLNISFPATGCQKLIEVDDERKLRTFYEKRWATEVAADALG..
```

Figure 3. Alignment of the N-terminal region *T. thermophila* ribosomal protein S3 with that of rat ribosomal protein S6.

In *Tetrahymena* ribosomal protein S3 has never been found in a phosphorylated state. However, when isolated *Tetrahymena* ribosomes were phosphorylated *in vitro* the two major phosphate acceptors were ribosomal proteins S3 and S7 (Fig. 4). Thus, both ribosomal proteins are clearly accessible for phosphorylation by a protein kinase present in exponentially growing cells. Sodium induced S7 phosphorylation in short time starved cells is not prevented by prior addition of inhibitors of transcription or translation (Bratholm, 1986) suggesting that S7 phosphorylation results from activation of a kinase or is triggered by a conformational change of the ribosome.

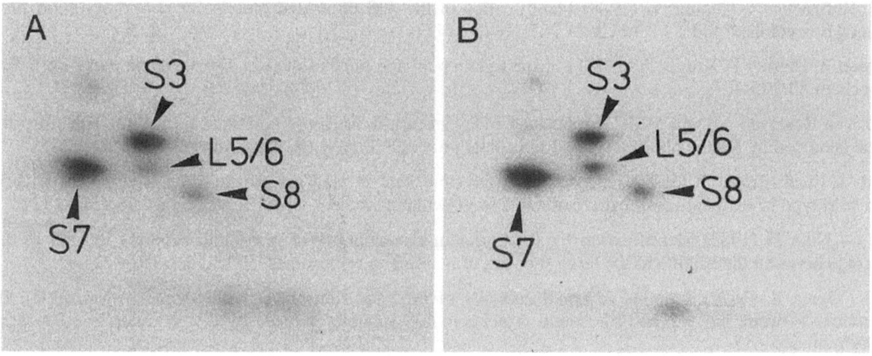

Figure 4. Phosphorylation *in vitro* of *Tetrahymena* ribosomes. Log phase ribosomes were incubated with 80 μM [γ-^{32}P]ATP and 10 μM cAMP in the presence (A) or absence (B) of a dialyzed S100 supernatant from starved cells. Ribosomal proteins were extracted, separated by two-dimensional gel electrophoresis and the gels exposed to an X-ray film with an intensifying screen.

UV cross-linking experiments have suggested that mammalian S6 is involved in mRNA binding (Terao and Ogata, 1979) and in keeping with this immunoelectron microscopy has indicated that S6 is located in the cleft where mRNA binds (Bommer *et al.*, 1980). We have shown that *Tetrahymena* S7 is located at the interface between the two ribosomal subunits

(Kristiansen *et al.*, 1978), and the same localization has been reported for mammalian S4 (Nygård and Nika, 1982; Uchiumi *et al.*, 1986). Thus both ribosomal proteins occupy a position in the 40S ribosomal subunit which is likely to be important for mRNA binding.

The lack of phosphorylation *in vivo* of the *Tetrahymena* S6 homolog is intriguing. Yeast S10 lacks the last three serine residues phoshorylated in mammalian S6 (Leer *et al.*, 1982; Krieg *et al.*, 1988). It is possible that *Tetrahymena* S3 is further truncated or that the serine residues phosphorylated in mammalian S6 have been replaced with other amino acid residues in *Tetrahymena* S3. Apparently such a replacement is responsible for the lack of P-protein phosphorylation in *T. thermophila* (Hansen *et al.*, 1991).

Finally, our finding that the only ribosomal protein subject to phosphorylation *in vivo* in *Tetrahymena* is homologous with mammalian S4 means that some caution is warranted concerning the general assumption that the major phosphorylated ribosomal protein in lower eukaryotes is homologous with mammalian S6.

REFERENCES

Bommer U-A, Noll F, Lutsch G, Bielka H (1980) Immunochemical detection of proteins in the small subunit of rat liver ribosomes involved in binding of the ternary initiation complex. FEBS Lett 111:171-174.

Bratholm PS (1986) Phosphorylation of ribosomal protein S7 in *Tetrahymena*. Thesis, University of Odense.

Cuny M (1985) Contribution a l'etude du ribosome de *Tetrahymena*. Thesis, University of Paris VII.

Cuny M, Sripati CE, Hayes DH (1985) The specific phosphorylation of a 40S ribosomal protein in growth-arrested *Tetrahymena* is induced by sodium. J Cell Phys 124:349-357.

Hansen TS, Andreasen PH, Dreisig H, Højrup P, Nielsen H, Engberg J, Kristiansen K (1991) *Tetrahymena thermophila* acidic ribosomal protein L37 contains an archaebacterial type of C-terminus. Gene 105:143-150.

Johnson SP, Warner JR (1987) Phosphorylation of Saccharomyces cerevisiae equivalent of ribosomal protein S6 has no detectable effect on growth. Mol Cell Biol 7:1338-1345.

Krieg J, Hofsteenge J, Thomas G (1988) Identification of the 40S ribosomal protein S6 phosphorylation sites induced by cycloheximide. J Biol Chem 263:11473-11477.

Kristiansen K, Plesner P, Krüger A (1978) Phosphorylation *in vivo* of ribosomes in *Tetrahymena pyriformis*. Eur J Biochem 83:395-403.

Leer RJ, van Raamsdonk-Duin MMC, Molenaar CMT, Cohen LH, Mager WH, Planta RJ (1982) The structure of the gene coding for the phosphorylated ribosomal protein S10 in yeast. Nucl Acids Res 10:5869-5878.

Marvaldi J, Lucas-Lenard J (1977) Differences in the ribosomal protein gel profile after infection of L cells with wild type or temperature-sensitive mutants of vesicular stomatitis virus. Biochemistry 16:4320-4327.

Nygård O, Nika H (1982) Identification by RNA-protein cross-linking of ribosomal proteins located at the interface between the small and the large subunits of mammalian ribosomes. EMBO J 1:357-362.

Terao K, Ogata K (1979) Proteins of small subunits of rat liver ribosomes that interact with poly(U). II. Crosslinks between poly(U) and ribosomal proteins in 40S subunits induced by UV irradiation. J Biochem (Tokyo) 86:605-617.

Thomas G, Siegmann M, Kubler A-M, Gordon J, Jimenez de Asua L (1980) Regulation of 40S ribosomal protein S6 phosphorylation in swiss mouse 3T3 cells. Cell 19:1015-1023.

Thomas G, Martin-Pérez J, Siegmann M, Otto AM (1982) The effect of serum, EGF, $PGF_{2\alpha}$ and insulin on S6 phosphorylation and the initiation of protein and DNA synthesis. Cell 30:235-242.

Uchiumi T, Kikuchi M, Ogata K (1986) Cross-linking study on protein neighborhoods at the subunit interface of rat liver ribosomes with 2-iminothiolane. J Biol Chem 261:9663-9667.

Wool IG (1979) The structure and function of eukaryotic ribosomes. Ann Rev Biochem 48:719-754.

NUCLEOCYTOPLASMIC TRANSPORT OF RIBOSOMAL PROTEIN S6

J. Kruppa, C. Schmidt
Department of Molecular Biology
University of Hamburg
Grindelallee 117
2000 Hamburg 13
Germany

Ribosomal protein S6 in the 40S subunit of higher eukaryotic cells is remarkable for its ability to exist in a variety of states of phosphorylation (Gressner, Wool, 1974; Thomas et al, 1979; Kruppa et al, 1983). Up to 5 mol of phosphate/mol of S6 protein are incorporated upon stimulation of cells by cycloheximide (Krieg et al, 1988), growth factors and oncogene products (Erikson, 1991; Sturgill, Wu, 1991). Results in vivo and in vitro indicate that S6 phosphorylation leads to activation of translation and cell growth by facilitating the initiation process of protein synthesis (Duncan, McConkey, 1982; Thomas et al, 1982). The phosphate group is transferred to specific serine residues at the C-terminal domain of S6 (Krieg et al, 1988) by different kinases, e.g. cAMP dependent protein kinase and S6 kinase. The mitogen-activated S6 kinase is able to phosphorylate four of the five sites in vivo (Krieg et al, 1988).

In order to study the biological functions of S6 protein and the importance of its phosphorylation by recombinant techniques we have isolated and sequenced a cDNA of human S6 protein (Heinze et al, 1988). In this report we are going to describe one specific functional aspect of S6 protein: its transport from the cytoplasm into the nucleus.

Gene expression of the components of the ribosomal subunits and their processing, transport and final assembly is a carefully controlled process that has not yet been fully elucidated. The regulation of the synthesis of the numerous ribosomal proteins must be balanced in concert with

NATO ASI Series, Vol. H 76
Tyrosine Phosphorylation/Dephosphorylation
and Downstream Signalling
Edited by L. M. G. Heilmeyer Jr.

transcription and processing of rRNA. Individual ribosomal proteins are translated in the cytosol and have to be imported into the nucleus where they must accumulate in the nucleolus at the site of subunit assembly. The mature ribosomal subunits must be exported from the nucleus to their site of function, the cytoplasm.

Theoretically, there exist several possibilities for the import of S6 protein into the nucleus:

1. S6 protein should be able to diffuse through the nuclear pores into the nucleus because it has a calculated molecular weight of 28,661 d (Heinze et al, 1988).
2. S6 protein associates with several other ribosomal proteins into a complex which cannot freely diffuse through the nuclear pores. This complex is actively transported into the nucleus because not S6 but one of the other components carries a nuclear location signal (NLS).
3. S6 protein contains at least one nuclear location signal permitting an efficient nuclear import with a higher rate than the diffusion rate.

In our search for the nuclear location signals we fused portions of the S6 sequence to the cytoplasmic enzyme ß-galactosidase. Specific restriction fragments were inserted into the expression vector pPbxnlslacZ (Hilberg et al, 1987) at the 5'end of the lacZ gene. Mouse L cells were transfected with the constructs using the DEAE-Dextran procedure. Fusion proteins with ß-galactosidase activity were localized in the different cell compartments after incubation with X-Gal. The fusion protein FP2 is targeted into the nucleus and nucleolus and FP3 only into the nucleus. The fusion protein FP1 containing roughly the amino terminal half of the S6 sequence is not transported into the nucleus at all but instead remains in the cytoplasm (Table 1).

Specifically, we looked in FP2 and FP3 for short sequences of basic amino acids which were located in hydrophilic domains of the S6 protein and had a high surface probability. The corresponding oligonucleotides to these selected sequences were synthesized and also cloned into the expression vector at the 5'end of the ß-galactosidase gene. Expressed fusion proteins containing the underlined sequences in figure 1 accumulated in the nucleus.

Table 1: Cellular localization of fusion proteins.

Fusion protein	S6 sequence	Cytoplasma	Nucleus	Nucleolus
FP1	^{33}A...R^{137}	+	-	-
FP2	^{138}A...L^{213}	-	+	+
FP3	^{214}A...R^{238}	-	+	-
FP4	^{3}L...R^{238}	-	+	+
FP5	^{3}L...R^{238} without NLSs	+	-	-

Deletion mutagenesis showed that removal of all three NLS in FP5 gave rise to a cytoplasmic staining of the transfected cells by the ß-galactosidase fusion protein whereas the control FP4 containing the original signals appeared in the nucleus and nucleolus (Table 1). Taken together, three nuclear targeting signals which are underlined in figure 1 were defined in S6 by these experiments.

1 **MKLNISFPAT GCQKLIEVDD ERKLRTFYEK RMATEVAADA LGEEWKGYVV**

51 **RISGGNDKQG FPMKQGVLTH GRVRLLLSKG HSCYRPRRTG ERKRKSVRGC**

101 **IVDANLSVLN LVIVKKGEKD IPGLTDTTVP RRLGPKRASR IRKLFNLSKE**

151 **DDVRQYVVRK PLNKEG<u>KKPR</u> TKAPKIQRLV TPRVLQH<u>KRR R</u>IALKKQRTK**

201 **KNKEEAAEYA KLLAKRMKEA KEKRQEQIA<u>K RRR</u>LSSLRAS TSKSESSQK**

Fig. 1: Amino acid sequence of the S6 protein.

Our studies indicate that the multiple nuclear targeting sequences result in an efficient translocation of S6-ß-galactosidase fusion proteins into the nucleus. Additional experiments will clarify whether S6 phosphorylation has any influence on the rate of nuclear uptake. (Supported by DFG)

References:

Duncan R, McConkey EH (1982) Preferential utilization of phosphorylated 40S ribosomal subunits during initiation complex formation. Eur J Biochem 123:535-538

Erikson RL (1991) Structure, expression, and regulation of protein kinases involved in the phosphorylation of ribosomal protein S6. J Biol Chem 266:6007-6010

Gressner AM, Wool IG (1974) The phosphorylation of liver ribosomal proteins in vivo. J Biol Chem 249:6917-6925

Heinze H, Arnold HH, Fischer D, Kruppa J (1988) The primary structure of the human ribosomal protein S6 derived from a cloned cDNA. J Biol Chem 263:4139-4144

Hilberg F, Stocking C, Ostertag W, Grez M (1987) Functional analysis of a retroviral host-range mutant: Altered long terminal repeat sequences allow expression in embryonal carcinoma cells. Proc Natl Acad Sci USA 84:5232-5236

Krieg J, Hofsteenge J, Thomas G (1988) Identification of the 40S ribosomal protein S6 phosphorylation sites induced by cycloheximide. J Biol Chem 263:11473-11477

Kruppa J, Darmer D, Kalthoff H, Richter D (1983) The phosphorylation of ribosomal protein S6 from progesterone-stimulated <u>Xenopus laevis</u> oocytes. Eur J Biochem 129:537-542

Sturgill TW, Wu J (1991) Recent progress in characterization ofprotein kinase cascades for phosphorylation of ribosomal protein S6. Biochim Biophys Acta 1092:350-357

Thomas G, Siegmann M, Gordon J (1979) Multiple phosphorylation of ribosomal protein S6 during transition of quiescent 3T3 cells into early G1, and cellular compartmentalization of the phosphate donor. Proc Natl Acad Sci USA 76:3952-3956

Thomas G, Martin-Perez J, Siegmann M, Otto AM (1982) The effect of serum, EGF, PGF2a and insulin on S6 phosphorylation and the initiation of protein and DNA synthesis. Cell 30:235-242

VI. Cell-Cycle-Control

REGULATING THE G_2 CHECKPOINT IN THE CELL CYCLE

J.L. Maller, B.G. Gabrielli, L.M. Roy, D.H. Walker, and T. Izumi

Howard Hughes Medical Institute and
Department of Pharmacology
University of Colorado School of Medicine
Denver, Colorado 80262, U.S.A.

1. INTRODUCTION

Recent progress in the field of cell cycle control has clearly identified the key enzymes involved in regulating the onset of mitosis in eucaryotic cells (1,2). The seminal finding that led to this advance was the purification of an enzyme known as maturation-promoting factor (MPF) (3). Elucidation of its structure revealed a heterodimeric complex of two proteins that had previously been identified genetically in the fission yeast, *Schizosaccharomyces pombe*, as key components regulating the $G_2 \Rightarrow M$ transition. These components were comprised of the 34 kilodalton serine/threonine protein kinase known as cdc2 and a 45-50 kilodalton protein known as cdc13 in yeast, and later identified as a protein known as cyclin B in marine invertebrate egg systems (4-7). The finding that two genetically-defined control elements in the initiation of mitosis were, in fact, the essential components of the enzyme that catalyzes the initiation of mitosis, MPF, united the genetic and biochemical approaches to cell cycle control and led to rapid progress in unravelling the mechanism of mitotic initiation.

Initial biochemical studies in *Xenopus* demonstrated that cdc2 activity was low under conditions in interphase where cdc2 was highly phosphorylated, and upon activation during entry into mitosis, cdc2 became dephosphorylated (8). Subsequent studies identified threonine 14 and tyrosine 15 in the cdc2-ATP binding site as the site of these inhibitory phosphorylations (9,10), and

NATO ASI Series, Vol. H 76
Tyrosine Phosphorylation/Dephosphorylation
and Downstream Signalling
Edited by L. M. G. Heilmeyer Jr.

block suggested that there was a mechanism involving exclusively phosphatase 1 in this pathway, although it is possible that phosphatase 2A is also involved in the effects of okadaic acid. Subsequent studies by Dr. Walker showed that periodic oscillations in the activity of protein phosphatase 1 occurred in a normal cell cycle. In particular, protein phosphatase 1 activity was high in mid-mitosis, declined upon exit from mitosis, and became elevated again during S-phase. At the end of S-phase protein phosphatase 1 declined, and increased again after mitosis had been initiated, becoming maximal again at mid-metaphase.

These two biochemically-defined peaks of protein phosphatase 1 were of particular interest because of previous genetic studies in several genetically tractable organisms which had identified two phenotypes derived from mutations in the highly conserved protein phosphatase 1 gene (31-33). One type of mutation best exemplified by the *dis2+* gene in *S. pombe* involved defects in sister chromatid exchange, in spindles, and in the extent of chromosome condensation. The existence of this class of mutants in protein phosphatase 1 was consistent with the peak of protein phosphatase 1 in mid-mitosis observed biochemically in the *Xenopus* egg extracts. The other phenotype associated with protein phosphatase 1 is known as bws1+ in *S. pombe* (34). This suppressor mutation involving protein phosphatase 1 suggests an action at a time in the cell cycle associated with the normal oscillation of protein phosphatase 1 at the end of S-phase. Interestingly, Dr. Walker was able to demonstrate that in extracts in which DNA synthesis was inhibited, the activity of protein phosphatase 1 was elevated due to the activation of a feedback control block. This suggested the hypothesis that the activity of protein phosphatase 1 was at least part of the mechanism regulating the activation of cdc25 at the G_2 restriction point. In order to get at this question more specifically, Dr. Walker cloned the *Xenopus* homolog of protein

phosphatase 1 and expressed this protein in bacteria carrying a histidine tag, allowing easy purification and recovery of the enzyme from bacterial or egg extracts. Impressively, Dr. Walker found that addition of the full-length protein phosphatase 1 to egg extracts resulted in the added phosphatase activity undergoing the same periodic oscillations in activity as for endogenous activity described above. This suggested the possibility that the exogenously added phosphatase could be used to study the mechanisms regulating protein phosphatase 1 activity during the cell cycle. Dr. Walker observed that the C-terminus of protein phosphatase 1 in *Xenopus* was unique compared to other protein phosphatase 1 family members identified in other species. This unique sequence also contained a cdc2 consensus site defined as T P X K. Moreover, Dr. Walker found that full-length protein phosphatase 1 produced in bacteria was a very good substrate *in vitro* for MPF, whereas a C-terminally-truncated mutant that lacked this phosphorylation site was not phosphorylated to any significant extent by MPF *in vitro*. This raises the possibility that phosphorylation of the cdc2-consensus site could regulate the activity of protein phosphatase 1 in the cell cycle.

Further studies are necessary to see whether the phosphorylation of protein phosphatase 1 *in vivo* changes its activity in a manner that is regulated by the activation of the G_2 restriction point. However, it is clear at this point that the general strategy of reconstructing one step at a time additional enzymatic steps that regulate the activation of *cdc2* is a productive one in terms of ultimately understanding how the failure to fully complete replication can create a G_2 block in the cell cycle and prevent the activation of mitosis. A fundamental understanding of this process will have important consequences in understanding the nature of oscillations in cellular activities during the cell cycle and the molecular composition of cellular clocks.

3. REQUIREMENT OF CDK2 FOR CSF ACTIVITY IN *XENOPUS*

A second type of experimental system that has been developed in this laboratory concerns the nature of cytostatic factor. As mentioned above, CSF is the name given to the activity which maintains unfertilized eggs in metaphase of meiosis II (15). The traditional assay for CSF was microinjection into one blastomere of a two-cell embryo. Upon microinjection, the injected blastomere arrests at the next metaphase, whereas the uninjected or buffer-injected control blastomere continues to divide, resulting ultimately in a hemi-blastula (15). Nearly four years ago Vande Woude and associates demonstrated that microinjection of the c-*mos*^{xe} proto-oncogene mRNA into oocytes also caused a metaphase arrest and moreover, immunodepletion of c-*mos*^{xe} from unfertilized egg cytoplasm caused significant reduction in CSF activity (35). These results led to the suggestion that c-*mos*^{xe} could be a component of CSF, although the data did not actually distinguish whether mos was upstream or downstream of CSF, or possibly even on a parallel pathway.

It is known from studies from Vande Woude and others that the synthesis of mos is increased during oocyte maturation (35,36) and that this increased synthesis is necessary for both meiosis I and meiosis II. Since cyclin pre-exists in oocytes, it was thought that the protein synthesis requirement from meiosis in *Xenopus* was accounted for by the *mos* proto-oncogene and not by cyclin. This hypothesis was recently validated for meiosis I by Vande Woude and co-workers (37), who succeeded in producing a soluble, bacterially-expressed mos-MBP fusion protein that they reported became activated upon microinjection into oocytes. They suggested that microinjection of this construct into oocytes led to its eventual activation as judged by *in vitro* autophosphorylation of immunoprecipitates from the injected cells, and moreover that the injected MBP-*mos*^{xe} was able to cause germinal vesicle breakdown (GVBD) in the presence of cycloheximide, suggesting that indeed *mos* was the sole protein synthesis being required for meiosis I. In contrast, microinjection of *mos* into oocytes in meiosis II was not sufficient to cause meiosis, suggesting that an additional newly-synthesized component was required for meiosis II and probably for CSF arrest. This situation became of great interest to this laboratory because of on-going work that was studying the synthesis and regulation of the *cdk2* gene in *Xenopus*. Like the mRNA for c-*mos*, the mRNA for *cdk2* becomes newly polyadenylated following progesterone stimulation of meiosis and the polypeptide begins to accumulate in meiosis I (38). Studies showed however that activity of *cdk2* was low at meiosis I and high at meiosis II. This raised the possibility that *cdk2* could be a candidate for the additional component required along with c-*mos*$^{\underline{xe}}$ to enter meiosis II and to cause a CSF-mediated metaphase II arrest.

To evaluate the requirement for *cdk2* in CSF arrest, Dr. Brian Gabrielli in the laboratory designed antisense oligodeoxynucleotides that ablated *cdk2* synthesis and increased the *cdk2* H1 kinase activity normally observed during maturation. He then monitored the kinase activity of *cdc2*, or MPF, during progesterone-induced meiosis I or meiosis II. Dr. Gabrielli observed that both meiosis I and meiosis II occurred with essentially unchanged kinetics or amplitude of *cdc2* kinase activation. However, sense-injected oligodeoxynucleotides oocytes arrested at meiosis II for several hours, whereas the antisense-injected oocytes rapidly exited meiosis II into the first embryonic cell cycle (39). This result suggested that, indeed, the presence of *cdk2* was required for metaphase II arrest. Important controls in these experiments included the demonstration that re-introduction of purified monomeric *cdk2* along with the antisense-injected oligodeoxynucleotides led to the activation of the injected *cdk2* during maturation, and the restoration of a metaphase II arrest in the cell cycle (39). The metaphase II arrest was characterized by high levels of *cdc2* kinase activity, high levels of cyclin B2, and high levels of c-mos protein kinase activity and c-mos polypeptide. In contrast, in oocytes that had exited meiosis II due to *cdk2* ablation, cyclin was degraded suggesting the exit was by the normal mechanism. Moreover *mos* activity was lower after exit, and several hours later decreased amounts of the c-*mos* polypeptide could be observed in the oocytes, as is thought to occur at fertilization.

These results then made an interesting prediction, namely, that any M-phase in which both c-mos^{xe} and *cdk2* were expressed should lead to metaphase arrest. In this regard, it seemed notable that whereas both kinases are present in meiosis II, in mitosis I *cdk2* is present but c-mos^{xe} is not, whereas in meiosis I *cdk2* is low whereas c-*mos* is high. Therefore we predicted that if *cdk2* could be expressed at metaphase in meiosis I it could cause a metaphase arrest. To carry out this experiment we purified from metaphase II-arrested eggs the kinase-active, high molecular weight complex of *cdk2* and microinjected it into oocytes at meiosis I under conditions where exit from meiosis I rapidly occurred. Dr. Gabrielli observed that exit from meiosis I was delayed from 1 to 2 hours following microinjection of *cdk2* at this stage consistent with the hypothesis that a metaphase arrest, at least transiently, could be induced by bringing *cdk2* activity up at meiosis I. In these particular experiments, because of the inability to get highly concentrated *cdk2*, the concentration of *cdk2* in the microinjected oocytes is only about 60% of that present normally at metaphase II. Moreover, studies by Vande Woude and co-workers has suggested that the level of c-mos^{xe} is greater at meiosis II than at meiosis I, although this has not been directly substantiated in any mass measurement to date. Nevertheless, these possibilities suggest that the shorter duration of meiosis arrest in the metaphase I experiments could be due to the lower concentrations of either or both of the required components. These experiments do not establish, necessarily, that *cdk2* is a component of CSF any more than the initial experiments with *mos* established that it was a component of CSF. Either or both of these kinases could be upstream of CSF, be a target downstream of CSF, or be on a parallel but required pathway with CSF to promote metaphase arrest. Only the actual purification of CSF will allow conclusions to be made about what comprise the actual components of CSF. Nevertheless, these results provide a potential explanation for why a metaphase arrest is unique in biology to meiosis II in vertebrates and illustrate the interaction of a cell cycle control element with a proto-oncogene product that is involved in the regulation of cell growth. It is intriguing to speculate that the oncogenic effects of *mos* in cells transformed with the Moloney sarcoma virus, which express the v-*mos* oncogene, are in fact mediated by the ability of v-*mos* to participate in processes occurring at metaphase.

4. ACKNOWLEDGEMENT

This work was supported by a grant from the National Institutes of Health (GM26743). L.M.R. is a recipient of a National Research Service Award (F32 CA08914), and J.L.M. is an Investigator of the Howard Hughes Medical Institute.

5. REFERENCES

1) MALLER, J. (1991). Mitotic control. Curr. Opin. Cell Biol., 3, 269.

2) NURSE, P. (1990). Universal control mechanism regulating onset of M-phase. Nature, 344, 503.

3) LOHKA, M.J., HAYES, M. and MALLER, J.L. (1988). Purification of maturation-promoting-factor, an intracellular regulator of early mitotic events. Proc. Natl. Acad. Sci. U.S.A., 85, 3009.

4) GAUTIER, J., NORBURY, C., LOHKA, M., NURSE, P. and MALLER, J.L. (1988). Purified maturation-promoting factor contains the product of a *Xenopus* homolog of the fission yeast cell cycle control gene *cdc* 2^+. Cell, 54, 433.

5) GAUTIER, J., MINSHULL, J., LOHKA, M., GLOTZER, M., HUNT, T. and MALLER, J.L. (1990) Cyclin is a component of MPF from *Xenopus*. Cell 60, 487.

6) EVANS, T., ROSENTHAL, E.T., YOUNGBLOM, J., DISTAL, D., and HUNT, T. (1983). Cyclin: a protein specified by maternal mRNA in sea urchin eggs that is destroyed at each cleavage division. Cell, 33, 389.

7) DUNPHY, W.G., BRIZUELA, L., BEACH, D., and NEWPORT, J. (1988). The *Xenopus* cdc2 protein is a component of MPF, a cytoplasmic regulator of mitosis. Cell, 54, 423.

8) GAUTIER, J., MATSUKAWA, T., NURSE, P. and MALLER, J.L. (1989). Dephosphorylation and activation of *Xenopus* $p34^{cdc2}$ protein kinase during the cell cycle. Nature, 339, 626.

9) PARKER, L. ATHERTON-FESSLER, S., LEE, M.S., OGG, S., FALK, J.L., SWENSON, K.I., and PIWNICA-WORMS, H. (1991). Cyclin promotes the tyrosine phosphorylation of $p34^{cdc2}$ in a *wee1*+ dependent manner. EMBO J., 10, 1255.

10) GOULD, K.L., and NURSE, P. (1989). Tyrosine phosphorylation of the fission yeast cdc2+ protein kinase regulates entry into mitosis. Nature, 342, 39.

11) KUMAGAI, A., and DUNPHY, W.G. (1991). The cdc25 protein controls tyrosine dephosphorylation of the cdc2 protein in a cell free system. Cell, 64, 903.

12) DASSO, M., and NEWPORT, J.W. (1990). Completion of DNA replication is monitored by a feedback system that controls the initiation of mitosis *in vitro*. Cell, 61, 811.

13) LOHKA, M.J. and MALLER, J.L. (1985). Induction of nuclear envelope breakdown, chromosome condensation, and spindle formation in cell-free extracts. J. Cell Biol., 101, 518.

14) MURRAY, A.W., SOLOMON, M.J., and KIRSCHNER, M.W. (1989). The role of cyclin synthesis and degradation in the control of maturation-promoting factor activity. Nature, 339, 280.

15) MASUI, Y. and MARKERT, C.L. (1971). Cytoplasmic control of nuclear behaviour during meiotic maturation of frog oocytes. J. Exp. Zool., 177, 129.

16) MURRAY, A.W., and KIRSCHNER, M. (1989a). Cyclin synthesis drives the early embryonic cell cycle. Nature, 339, 275.

17) WATANABE, N., VANDE WOUDE, G.F., IKAWA, Y., and SAGATA, N. (1989). Specific proteolysis of the *c-mos* proto-oncogene product by calpain on fertilization of *Xenopus* eggs. Nature, 342, 505.

18) SMYTHE, C. and NEWPORT, J.W. (1992). Coupling of mitosis to the completion of S-phase in *Xenopus* occurs via modulation of the tyrosine kinase that phosphorylates $p34^{cdc2}$. Cell, 68, 787.

19) WALKER, D.W. and MALLER, J.L. (1991). Role for cyclin A in the dependence of mitosis on completion of DNA replication. Nature, 354, 314.

20) STRAUSFELD, U., LABBÉ, J.C., FESQUET, D., CAVADORE, J.C., PICARD, A., SADHU, K., RUSSELL, P. and DORÉE, M. (1991). Dephosphorylation and activation of a $p34^{cdc2}$/cyclin B complex *in vitro* by human cdc25 protein. Nature, 351, 242.

21) GAUTIER J., SOLOMON, M.J., BOOHER, R.N., BAZER, J.F., and KIRSCHNER, M.W. (1991). Cdc25 is a specific tyrosine phosphatase that directly

activates p34cdc2. Cell, 68, 787.

22) ENOCH, T. and NURSE, P. (1990). Mutation of fission yeast cell cycle control genes abolishes dependence of mitosis on DNA replication. Cell, 60, 665.

23) PARKER, L.L. and PIWNICA-WORMS, H. (1992) Inactivation of the p34cdc2-cyclin B complex by the human wee1 tyrosine kinase. Science, 257, 1955.

24) LEE, M.S., OGG, S., XU, M., PARKER, L.L., DONOGHUE, D.J., MALLER, J.L. and PIWNICA-WORMS, H. (1992). cdc25+ Encodes a Protein Phosphatase that Dephosphorylates p34cdc2. Mol. Biol. Cell 3, 73.

25) ENOCH, T., GOULD, K.L., and NURSE, P. (1991). Mitotic checkpoint control in fission yeast. Cold Spring Harbor Symp. Quant. Biol., 56, 409.

26) MILLAR, J.B.A., MCGOWAN, C.H., LENAERS, G., JONES, R., and RUSSELL, P. (1991). p80cdc25 mitotic inducer is the tyrosine phosphatase that activates p34cdc2 kinase in fission yeast. EMBO J., 10, 4301.

27) KUMAGAI, A., and DUNPHY, W.G. (1991). Molecular mechanism of the final steps in the activation of MPF. Cold Spring Harbor Symp. Quant. Biol., 56, 585.

28) IZUMI, T., WALKER, D.H. and MALLER, J.L. (1992). Periodic changes in phosphorylation of the *Xenopus* cdc25 phosphatase regulate its activity. Mol. Biol. Cell, 3, 927.

29) COHEN, P. (1989). The structure and regulation of protein phosphatases. Annu. Rev. Biochem., 58, 453.

30) WALKER, D.H., DEPAOLI-ROACH, A.A. and MALLER, J.L. (1992). Multiple roles for protein phosphatase 1 in regulating the *Xenopus* early embryonic cell cycle. Mol. Biol. Cell, 3, 687.

31) OHKURA, H. KINOSHITA, N., MIYATANI, S., TODA, T., and YANAGIDA, M. (1989). The fission yeast *dis2+* gene required for chromosome disjoining encodes one of two putative type 1 protein phosphatases. Cell, 57, 997.

32) DOONAN, J.H. and MORRIS, N.R. (1989). The *bimG* gene of *Aspergillus nidulans*, required for completion of anaphase, encodes a homolog of mammalian phosphoprotein phosphatase 1. Cell, 57, 987.

33) AXTON, J.M., DOMBRÁDI, V., COHEN, P.T.W. and GLOVER, D.M. (1990). One of the protein phosphatase 1 isoenzymes in *Drosophila* is essential for mitosis. Cell, 63, 33.

34) BOOHER, R., and BEACH, D. (1989). Involvement of a type 1 protein phosphatase encoded by *bws1+* in fission yeast mitotic control. Cell, 57, 1009.

35) SAGATA, N., WATANABE, N., VANDE WOUDE, G.F., and IKAWA, Y. (1989). The c-*mos* proto-oncogene product is a cytostatic factor responsible for meiotic arrest in vertebrate eggs. Nature, 342, 512.

36) SAGATA, N., OSKARSSON, M., COPELAND, T., BRUMBAUGH, J. and VANDE WOUDE, G.F. (1988). Function of c-*mos* proto-oncogene product in meiotic maturation in *Xenopus* oocytes. Nature, 355, 519.

37) YEW, N., MELLINI, M.L., and VANDE WOUDE, G.F. (1992). Meiotic initiation of the *mos* protein in *Xenopus*. Nature, 355, 649.

38) PARIS, J., LEGUELLEC, R., COUTURIER, A., LEGUELLEC, K., OMILI, F., CAMORIS, J., MACNEIL, S. and PHILIPPE, M. (1991). Cloning by differential screening of a *Xenopus* cDNA coding for a protein highly homologous to cdc2. Proc. Natl. Acad. Sci. U.S.A., 88, 1039.

39) GABRIELLI, B.G., ROY, L.M. and MALLER, J.L. (1992). Requirement for cyclin-dependent kinase 2 in metaphase II arrest in unfertilized *Xenopus* eggs, submitted for publication.

CD2-Dependence on the CD3-Zeta-Chain in T Cell Signalling - Shown by Inhibitory and Modifying Effects of Cholera Toxin

G. F. Böl, B. M. Haack, K. Resch
Institute of Molecular Pharmacology, Hannover Medical School, Konstanty-Gutschow-Str. 8, D- 3000 Hannover 61, Germany, Fax ++49 511 532 2794

Introduction

T cells can be activated through the TCR/CD3 complex or, alternatively (*Meuer '87*), through the CD2 molecule. The TCR/CD3 complex consists of the α/β heterodimer (TCR) and the subunits γ, δ, ε and the ζ-ζ-homodimer (together CD3), ζ being the main signalling element with a long intracellular part (*Irving '91*). CD2 is a single chain, surface-associated glycoprotein with a molecular mass of 50 kD, expressed on all human thymocytes and mature T cells.

Triggering of the cells with monoclonal anti TCR/CD3- or anti CD2- antibodies leads to hydrolysis of phosphatidylinositol-4,5-bisphosphate (PIP_2) by phospholipase C (PLC) into inositol-1,4,5-triphosphate (IP_3) and diacylglycerol (DAG) as well as to an increase in intracellular free calcium, which together with DAG activates the proteinkinase C (PKC).

This signalling machinery can be blocked by cholera toxin (CTX). Recently we reported (*Haack*), that CTX covalently modifies the CD3 zeta chain in a stepwise manner. Here we demonstrate, that this modification correlates well with the time courses of the CTX-inhibited, activation dependent generation of IP_3 and increase of cytosolic Ca^{2+}. Triggering through CD2 leads to similar time courses of inhibition as through CD3, although CD2 itself is not modified by CTX. These experiments provide further evidence for the dependence of CD2 - signal transduction on CD3-ζ in T cells.

Materials and Methods

Jurkat cells were loaded with CTX (0,5 μg/ml) for the times indicated.

Ca^{2+}: For the measurement of Ca^{2+} FURA 2-acetoxymethylester (5 μM) was added for the last 30 min of incubation, then cells were washed and resuspended in Hank's solution. Cells were stimulated with 10 μg/ml OKT3 (anti CD3 ε) or a combination of T11.2 and T11.3 (1:425, from ascites, anti CD2). Fluorescence signals were detected on a spectrofluorometer as described elsewhere (*Haack, Böl, in prep.*).

IP_3: IP_3 was measured by loading the cells with ^{3}H-myo-inositol (2 μCi/ml) in inositol free medium over night. Loaded cells were stimulated with anti receptor antibodies 10 min before

NATO ASI Series, Vol. H 76
Tyrosine Phosphorylation/Dephosphorylation and Downstream Signalling
Edited by L. M. G. Heilmeyer Jr.

the reaction was stopped with chloroform/methanol. IP_n were extracted and separated using an ammoniumformiate-gradient (*Berridge '84*) and were determined by liquid scintillation counting.

Immunoprecipitation: Cells were lysed in 1% digitonin and postnuclear supernatants immunoprecipitated with protein A-sepharose-beads coupled to monoclonal anti CD3 ε- or anti CD2- antibodies. After boiling these complexes for 5 min in Laemmli buffer, supernatants were run on a 10 or 12,5 % SDS gel and blotted onto nitrocellulose. Proteins were detected with specific anti ζ- or anti CD2- antiserum, alkaline phosphatase conjugated second antibody and the nitroblue-tetrazolium/5-bromo-4-chloro-3-indoylphosphate- system.

Results

IP_3: Inhibition of IP_3-release after stimulation was dependent on the time of pretreatment with CTX (fig. 1a). The anti receptor antibody induced increase was standardized to 100 % in order to demonstrate the inhibitory effects of CTX. No remarkable differences in the time courses of inhibition were detected comparing the different triggering mechanisms via CD3 or via CD2.

Ca^{2+}: The time courses of CTX inhibited, activation dependent increases in intracellular free calcium were nearly identical for the two different triggering pathways (fig. 1b). Cholera toxin ADP-ribosylates the stimulating G-protein of the adenylate cyclase, which results in an increased intracellular cAMP-level. Control experiments employing dibutyric-cAMP (data not shown) showed that the inhibitory CTX effects were not mediated by the adenylate cyclase/cAMP system.

Immunoprecipitation: We reported (*Haack*), that cholera toxin covalently modified the ζ-

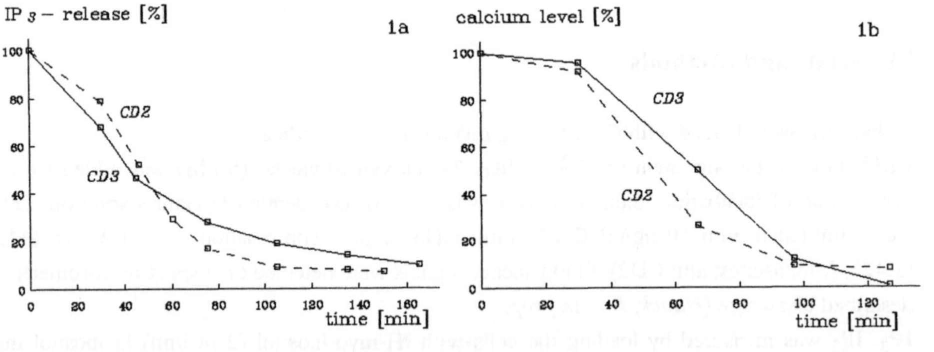

Fig.1: CTX-inhibition of IP_3- (a) and Ca^{2+}- (b) release in anti receptor antibody stimulated cells
Cells were incubated with CTX (0,5 μg/ml) for the times indicated and then stimulated with OKT3 (anti CD3, solid line) or T11.2 and T11.3 (anti CD2, dashed line). The resulting releases of IP_3 and Ca^{2+} are % of control (without CTX).

chain of the CD3 complex resulting in a decrease in electrophoretic mobility as shown in fig. 2a. The time course of this modification correlated well with the inhibition in Ca^{2+}- and IP_3-release. Fig. 2b shows the CD2 molecule in different glycosylated forms (50-55 kD) precipitated from cells pretreated with CTX. There was no modification visible, demonstrating that CTX did not attack the CD2 molecule.

Fig.2a: Modification of CD3-ζ by CTX in Jurkat cells
Cells were incubated with CTX (0,5 μg/ml) for the times indicated [min]. After lysis in 1% digitonin proteins were immunoprecipitated by incubating postnuclear supernatants with OKT3 - protein A - sepharose beads. Shown is a Western blot of a 12,5 % gel, specifically incubated with anti-ζ-antiserum. Standards are indicated in kD on the left. Zeta appears as a homodimer at 32 kD.

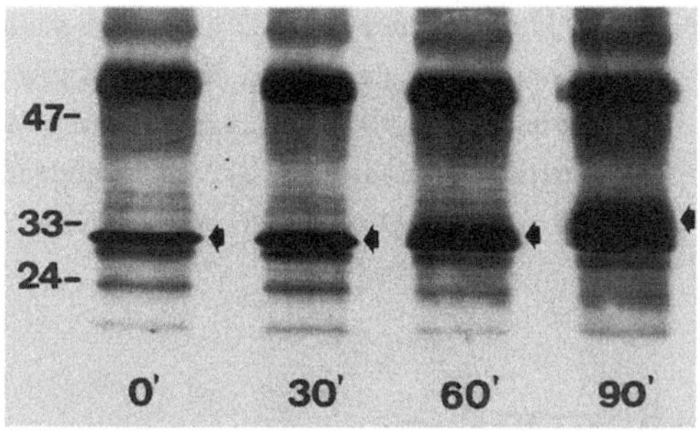

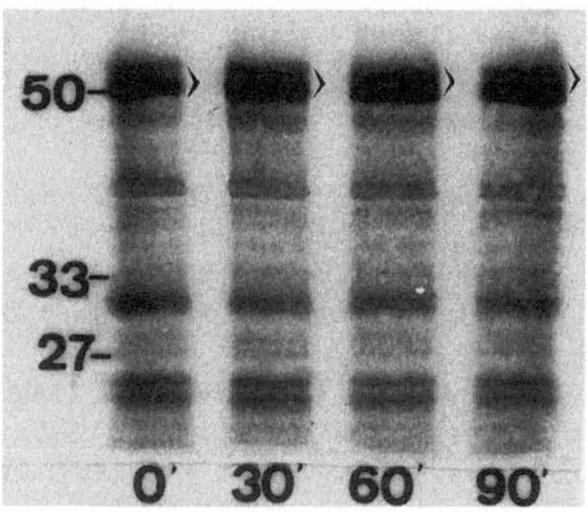

Fig.2b: CD2 is not modified by CTX in Jurkat cells
Cells were incubated with CTX (0,5 μg/ml) for the times indicated [min]. After lysis in 1% triton X-100 proteins were immunoprecipitated by incubating postnuclear supernatants with OKT11 - protein A - sepha-rose beads. Shown is a Western blot of a 10 % gel, specifically incubated with anti-CD2-antiserum. Standards are indicated in kD on the left. CD2 appears in different glycosylated forms at 50-55 kD.

Discussion

Basing on the findings of *Imboden* (*'86*), that cholera toxin inhibits TCR/CD3-triggered IP_3- and Ca^{2+}-releases, we measured the kinetics for this inhibitory effects and checked the same for the alternative way of activation through CD2. We demonstrated, that the kinetics of the inhibitory effects of CTX on IP_3- and Ca^{2+}-releases in Ab-stimulated Jurkat cells were nearly the same for TCR/CD3 and CD2 mediated signal transduction. As IP_3 opens intracellular calcium stores and - together with IP_4 - Ca^{2+}-channels in the plasma membrane, it was likely that the Ca^{2+}-release would follow the course of IP_3-release inhibited by CTX.

We showed in Western blotting experiments, that CTX modified the CD3 ζ-chain of the T cell receptor complex (*Haack*). For the function of CD2 in T cell signalling it was therefore interesting to look at possible analogous modifications caused by CTX. No modification of CD2 by CTX was detectable. If CD2 transduces signals through CD3 ζ (*Bockenstedt '88*) and a modification of ζ by CTX inhibits signalling, there is no need for a special, additional modification of CD2 by CTX. Therefore these experiments with CTX give further evidence for the CD3 ζ-dependency of CD2 in T cell signalling.

Altogether these data suggest, that CTX can interfere with signal transduction in T lymphocytes by modifying the CD3 ζ subunit and provide further evidence, that CD3 ζ is required for signal transduction through CD2 - independently of wether a protein tyrosine kinase (*Samelson '90*) or a common G-protein (*Sommermeyer '90*) mediates the signals.

References

Berridge MJ (1984) Inositol triphosphat, a novel second messenger in cellular signal transduction. Nature 312:315

Bockenstedt LK, Goldsmith MA, Dustin M, Olive D, Springer TA, Weiss A (1988) The CD2 ligand LFA-3 activates T cells but depends on the expression and function of the antigen receptor. J Immunol 141:1904

Haack BM, Emmrich F, Resch K Cholera toxin inhibits T cell receptor signalling by covalent modification of the CD3-zeta subunit. Accepted for publication in J Immunol

Imboden JB, Shoback DM, Pattison G, Stobo JD (1986) Cholera toxin inhibits the T cell antigen receptor-mediated increases in inositol triphosphate and cytoplasmatic free calcium. Proc Natl Acad Sci USA 83:5673

Irving BA, Weiss A (1991) The cytoplasmatic domain of the T cell receptor ζ chain is sufficient to couple to receptor-associated signal transduction pathways. Cell 64:891

Meuer SC, Hauer M, Deusch K, Moebius U, Meyer zum Büschenfelde KH (1987) Two pathways of T cell activation. Boehring Inst Mitt 81:15

Samelson LE, Phillips AF, Luong ET, Klausner RD (1990) Association of the fyn protein-tyrosine kinase with the T-cell antigen receptor. Proc Natl Acad Sci USA 87:4358

Sommermeyer H, Schwinzer R, Kaever V, Behl B, Resch K (1990) The G protein coupling T cell antigen receptor/CD3-complex and phospholipase C in the human T cell lymphoma Jurkat is not a target for cholera toxin. Eur J Immunol 20:1881

cdc25 protein phosphatase in mammalian fibroblasts: cell cycle expression and intracellular localization

Franck Girard, Anne Fernandez and Ned Lamb
CRBM-CNRS-INSERM
BP5051
34033 Montpellier cedex, France

INTRODUCTION

The serine-threonine protein kinase $p34^{cdc2}$ is a crucial component of the regulatory mechanisms that bring about coordinate division in eukaryotic cells (1). Whilst $p34^{cdc2}$ protein level is constant throughout the cell division cycle, its kinase activity reaches a maximum at the G2/M transition, coincident with mitotic entry. The complete activation of $p34^{cdc2}$ kinase at mitosis involves two distinct mechanims. First, $p34^{cdc2}$ binds a regulatory subunit, named cyclin B, whose expression oscillates during the cell cycle. A number of different cyclins have been cloned, but their precise function, in particular the nature of the cdk protein they bind or the time of the cell cycle when these cyclin-dependent kinases are activated, is still not fully elucidated (2). Secondly, $p34^{cdc2}$ kinase activation involves changes in the phosphorylation state of p34, in particular the dephosphorylation of the residues thr14 and tyr15, located in the amino terminus of the molecule and within the ATP-binding site of $p34^{cdc2}$ (3).These specific dephosphorylations are under the control of the cdc25 protein phosphatase, which is the first member of a new phosphatase class showing a dual specificity for both tyrosine and threonine residues (4,5). Homologs of the fission yeast cdc25 gene have now been identified in a variety of organisms, including human (5). Interestingly, three cdc25 genes named A, B and C have been cloned in human cells (6,7), but their substrate specificity towards the different cylin-dependent kinases and their function during the cell cycle is still unknown.
In the present study, we have examined the cell cycle expression and intracellular distribution of the cdc25 protein phosphatase in non transformed mammalian fibroblasts, in order to determine if changes in the protein levels or cellular localization may play a role in the regulation of the $p34^{cdc2}$ kinase activity.

RESULTS AND DISCUSSION

Constant expression of cdc25 protein during the mammalian cell cycle:

We have developed different affinity-purified anti cdc25 antibodies, raised against either the entire human cdc25-C protein overexpressed in bacteria, or a synthetic peptide corresponding to the catalytic site of the phosphatase. This sequence motif, CEFSSKRGPDLLR, is conserved

NATO ASI Series, Vol. H 76
Tyrosine Phosphorylation/Dephosphorylation
and Downstream Signalling
Edited by L. M. G. Heilmeyer Jr.

throughout evolution. Two particular residues (cysteine and arginine in position 1 and 7 respectively) have been shown neccessary for the biological function and tyrosine phosphatase activity of cdc25 (8,9,10). Since the results we obtained were the same with these different antibodies, we will refer in the text as anti cdc25C. By western blot, anti cdc25C detected a major 67 kD protein species in different cell lines (rat embryonic fibroblasts REF52, human fibroblasts HS68 and Hela cells), and an additional protein of relative molecular weight 75 kD. Preincubation of the antibodies with the immune peptide or the human cdc25-C protein completely abolished the immunoreactivity of both bands, confirming the specificity of the antibodies. When we performed two dimensional immunoblots in which the cell extracts were treated with urea, the 67 kD protein resolved at 55 kD, which is in agreement with the expected size of the human cdc25-C protein (7). The 75 K is most probably an other cdc25 form, since a protein of similar size has been detected in Hela cells (6). Finally, these anti cdc25C were also able to inhibit cdc25-dependent p34cdc2 dephosphorylation in an in vitro assay. The expression profile of cdc25 during the mammalian cell cycle was assessed by immunoblot using serum-synchronized HS68 cells and is shown in figure 1.A. The major 67 kD protein is expressed at a constant level throughout the cell cycle, as also observed with the 75 kD form (not shown).

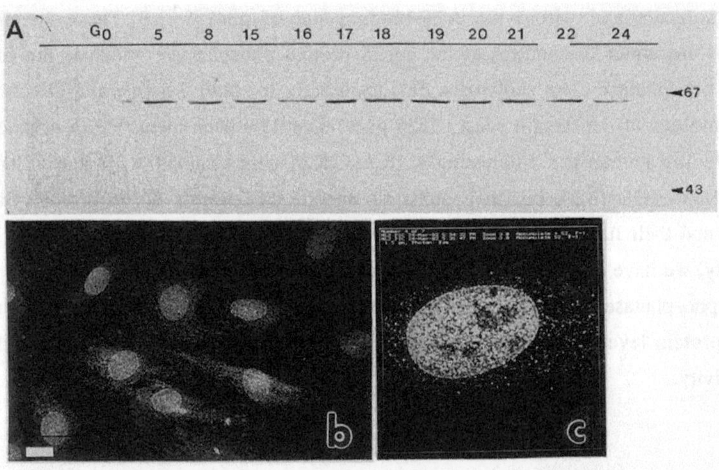

Figure 1: <u>cdc25 protein is constantly expressed and localized into the nucleus during the cell cycle of non transformed mammalian fibroblasts.</u>
A: Western blotting of serum synchronized cell extracts (numbers are given as hours after serum stimulation) with polyclonal anti cdc25C antibodies showing the constant expression of cdc25 protein during the cell cycle of HS68 fibroblasts. G1-phase covers the period 0-16 h, S-phase 16-19 h, G2-phase the period 19-24 h when cells enter mitosis. Molecular weight are in kilodaltons.
B-C: cdc25 protein is essentially localized into the nucleus and at the nuclear envelope, as revealed by indirect immunofluorescence (panel B) and confocal microscopy (panel C). Shown are G2 phase HS68 cells stained with anti cdc25C, and biotinylated anti rabbit/streptavidin Texas red. Bar represents 5 µm.

Results were the same in REF52 and Hela cells. We conclude that, as previously observed in Hela cells (10) and as also observed for p34^{cdc2}, cdc25 phosphatase is constantly expressed during the cell cycle of non transformed mammalian cells, which appears to be fundamentally different with the observation made in fission yeast that the cdc25 mitotic inducer expression increased during G2 to peak at mitosis (11).

Nuclear localization of cdc25 protein in mammalian cells:

Since we observed no changes in cdc25 protein levels during the cell cycle, we conclude that the control of cdc25 phosphatase activity clearly does not involve translational regulation. We then examined the intracellular distribution of cdc25 by indirect immunofluorescence using serum synchronized cells growing onto glass coverslips (see legend to figure 1). Throughout interphase, cdc25 localizes into the nucleus (fig1B shows G2-phase HS68 cells), with a more intense staining of the nuclear envelope as revealed by confocal microscopy (fig1C). Preincubation of anti cdc25C with immune peptide completely eliminated the nuclear staining, including that associated with the nuclear envelope.

We also have performed double immunofluorescence staining for both cdc25 and nuclear lamins A and C (see legend to figure 2). During early prophase, as judged by the low level of chromatin condensation, both cdc25 (fig2A) and lamin stainings (fig2B) are restricted to the nucleus, as during interphase. During late prophase, as chromatin is heavily condensed and some lamin staining begins to diffuse to the cytoplasm (fig2D), we observed a substantial decrease of the cdc25 nuclear staining concommitant with an increase in the overall cdc25 cytoplasmic staining (fig2C). During metaphase, cdc25 is distributed throughout the cell, without association with the condensed chromatine, and is completely relocated into the two reforming nuclei during telophase (not shown).

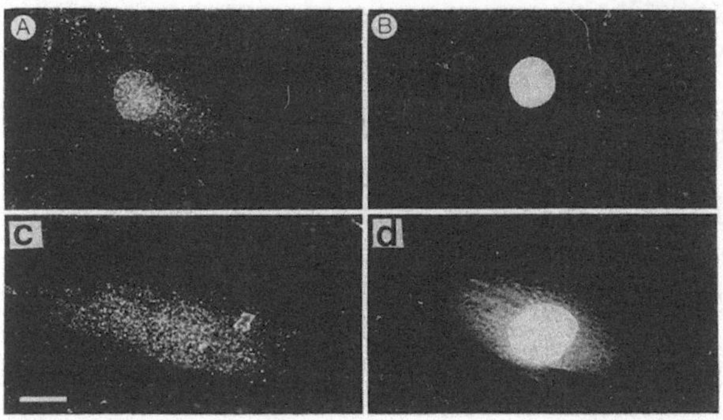

Figure 2: <u>distribution of cdc25 during prophase</u>
HS68 cells were fixed 22 hours after serum stimulation, and stained for cdc25 with anti cdc25C (panels A and C) and nuclear lamins A and C (panels B and D) with monoclonal anti-lamins. Incubation was carried out with biotinylated anti rabbit/streptavidin Texas red and fluorescein conjugated anti mouse in order to visualize cdc25 and lamins respectively. Bar represents 5 µm.

The nuclear distribution of cdc25 is of importance with respect to the time and locale at which the $p34^{cdc2}$-cyclin B kinase is activated through cdc25-dependent dephosphorylation. In mammalian cells, cyclin B is present into the cytoplasm during G2-phase until prophase when the protein translocates to the nucleus (12). The localization of $p34^{cdc2}$ is more complexe, due to the increasing number of cdc2-related proteins identified. Together with our results on cdc25, one can suggest two hypotheses for $p34^{cdc2}$ activation. The first one is that $p34^{cdc2}$-cyclin B accumulates during G2 into the cytoplasm as an inactive complex, which is fully activated into the cytoplasm by cdc25 during early prophase as a result of cdc25 cytoplasmic translocation. In the second model, cdc25-dependent dephosphorylation would occur into the nucleus, the decrease in cdc25 staining we observed during prophase would in that case result from the physical association between cyclin B and cdc25 thus masking the cdc25 epitope. In this later case, the mechanism regulating the nuclear translocation of the cdc2-cyclin B complex would be independent of cdc25-dephosphorylation. cdc25 phosphatase activity appears to be tightly regulated by phosphorylation. Indeed it was recently shown in Xenopus that cdc25 is active at a low level during interphase. Activation of the phosphatase at mitosis is coincident with hyperphosphorylation of cdc25 (13,14). The kinase responsible for cdc25 phosphorylation is still unknown. Nevertheless, we have evidence that $p34^{cdc2}$ itself phosphorylates cdc25 in vitro (U.Strausfeld, F.G and N.L, manuscript in preparation). Furthermore, microinjection of $p34^{cdc2}$ phosphorylated cdc25 into interphase cells induces phenotypic changes similar to those observed upon microinjection of active $p34^{cdc2}$ kinase (15), thus suggesting that $p34^{cdc2}$ might participate in the regulation of its own phosphatase.

REFERENCES

1- Nurse, P. (1990) Nature 344: 503-508
2- Hunt, T. (1991) Nature 350: 462-463
3- Krek, W. and Nigg, E.A. (1991) EMBO J.10: 3331-3341
4- Strausfeld, U. et al, (1991), Nature 351: 242-245
5- Millar, J.B. and Russell, P. (1992) Cell 68: 407-410
6- Galaktionov, K. and Beach, D. (1991) Cell 67: 1181-1194
7- Sadhu, K. et al, (1990) Proc.Natl.Acad.Sci.USA 87: 5139-5143
8- Dunphy, W.G. and Kumagai, A. (1991) Cell 67: 189-196
9- Gautier, J. et al, (1991) Cell 67: 197-211
10- Millar, J.B. et al, (1991) EMBO J. 10: 4301-4309
11- Moreno, S. et al, (1990) Nature 344: 4-549-552
12- Pines, J. and Hunter, T. (1991) J.Cell.biol. 115: 1-17
13- Kumagai, A. and Dunphy, W.G. (1992) Cell 70: 139-151
14- Izumi, T. et al. (1992) Mol.Biol.Cell 3: 927-939
15- Lamb, N.J.C. et al, (1990) Cell 60: 151-165

VII. Control of Intracellular Ca^{2+}

THE ROLE OF SMALL MOLECULAR WEIGHT GTP-BINDING PROTEINS IN STIMULUS SECRETION COUPLING OF PANCREATIC ACINAR CELLS

Irene Schulz, Stefan Zeuzem* and Petra Zimmermann
Physiology II
University of the Saarland
6650 Homburg, FRG

INTRODUCTION

In the exocrine pancreas, intracellular transport of proteins and enzyme secretion is stimulated in response to a variety of hormones and neurotransmitters such as cholecystokinin (CCK), vasoactive intestinal peptide (VIP) and secretin (3,23). Receptor-mediated activation of phospholipase C (PLC) by cholecystokinin-octapeptide (CCK-OP) and direct activation of PLC-coupling G-proteins by weakly hydrolyzable analogues of GTP, such as guanosine 5'-0-(3-thiotriphosphate) (GTPγS), lead to phosphatidylinositol 4,5-biphosphate breakdown and formation of inositol 1,4,5-trisphosphate (IP_3; see Ref. 24) and diacylglycerol (DG; see Ref. 19) in both intact and permeabilized pancreatic acinar cells. Whereas IP_3 releases Ca^{2+} from intracellular Ca^{2+} stores (26), DG or its synthetic analogue 12-0-tetradecanoylphorbol 13-acetate (TPA) activates protein kinase (PK) C (7) and subsequently stimulates enzyme secretion (15). Both secretin and VIP stimulate enzyme secretion (11) by an increase in intracellular adenosine 3',5'-cyclic monophosphate (cAMP) levels, which leads to activation of PKA (5).

Although these mechanisms involved in signal transduction and in the generation of intracellular messengers in secretory cells are relatively well understood, little is known about intracellular effectors that control regulated exocytosis.

Recent evidence suggests that a variety of small molecular weight guanosine 5'-triphosphate (GTP)-binding proteins (smg-proteins) have a role in many steps of intracellular membrane traffic and exocytosis (4,10,12).

*J.W.v.Goethe Universität, Gastroentereologie, Theodor Stern-Kai, 6000 Frankfurt 70, FRG

NATO ASI Series, Vol. H 76
Tyrosine Phosphorylation/Dephosphorylation and Downstream Signalling
Edited by L. M. G. Heilmeyer Jr.

In addition to the family of heterotrimeric GTP-binding proteins (G-proteins) which serve as transducers for various transmembrane signalling systems (6) there exists a large family of small molecular weight GTP-binding proteins (smg-proteins) with molecular masses between 18 and 30 kDa, which have been suggested to play a role in intracellular membrane traffic (4,12) and protein transport through the secretory pathway (10,25). We have detected seven different smg-proteins with molecular masses between 18-27 kDa in subfractions of pancreatic acinar cells using SDS-polyacrylamid gel electrophoresis (SDS/PAGE) with subsequent protein transfer on nitrocellulose and [α-^{32}P]GTP-binding (32). Two-dimensional separation of proteins revealed the presence of up to 30 differently charged smg-proteins (21,22).

We report here that receptor-mediated stimulation of intracellular pathways with activation of subsequent steps such as PLC-DG-PKC leads to decreased [α-^{32}P]GTP-binding to yet unidentified 21 kDa smg-protein(s) and increased [α-^{32}P]GTP-binding to 23 kDa smg-protein(s) in microsomal membranes (MM). Furthermore, these or one of these 21 kDa smg-protein(s) of the microsomal membrane fraction seems to be involved in the CCK-OP-stimulated protein-secretion pathway of pancreatic acinar cells. Our data show that Brefeldin A (BFA) a fungal antibiotic which induces rapid disassembly of the Golgi complex and blocks protein transport within the secretory pathway beyond the Golgi complex (9,18) also blocks CCK-OP-induced enzyme secretion from isolated pancreatic acinar cells. Using permeabilized acinar cells protein secretion could be stimulated by GTP (10^{-3}M). BFA had no inhibitory effect on protein secretion in the presence of GTP, indicating that BFA might act on a GTP-binding protein. Investigating the effect of BFA on smg-proteins we observed that [α-^{32}P]GTP-binding to 21 kDa smg-protein(s) in the microsomal fraction was increased in the presence of BFA. We conclude that these 21 kDa smg-protein(s) may be a molecular target of BFA in pancreatic acinar cells and that these 21 kDa smg-protein(s) might play a key role in the intracellular protein transport within the secretory pathway (30). If both

CCK-OP and BFA act on the same or on different 21 kDa smg-proteins remains to be investigated.

METHODS

Cell preparation

Acinar cells were isolated from the pancreas of male Wistar rats (200-250 g) by collagenase digestion as previously described (27).

Preparation of the microsomal membrane fraction from rat pancreas

Isolated cells were incubated in the absence (control) or presence of the secretagogue CCK-OP for 30 min at 37°C in a standard Krebs- Ringer N-2-hydroxyethylpiperazine-N'-2-ethanesulfonic acid (HEPES) (KRH) buffer and gassed continuously with O_2 as described (27). At the end of the incubation period, cells were centrifuged at 100xg for 4 min at 4°C. The pellet was washed once in homogenization buffer containing 28×10^{-2}M mannitol, 5×10^{-3}M HEPES, 10^{-2}M KCl, 10^{-3}M $MgCl_2$, 10^{-3}M benzamide, 25×10^{-5}M phenylmethylsulfonyl fluoride (PMSF), 10^{-6}M leupeptin and 20 μg/ml trypsin inhibitor (pH 7.0). Subsequently, cells were homogenized, the homogenate was centrifugated at 1,000xg for 12 min and the resulting supernatant was subsequently centrifuged at 11,000xg for 15 min in a Beckman JA-20 rotor (2,29). The fluffy layer on the top of the 11,000xg pellet, which contains microsomal membranes (MM) of the endoplasmic reticulum and the Golgi complex (29), was removed and resuspended in the same buffer at a concentration of 30 mg/ml protein.

[α-32P]GTP-binding assay

Proteins were resolved by SDS-PAGE gels followed by an electrophoretical transfer to nitrocellulose paper as described (21,22,32). Binding of [α-^{32}P]GTP to proteins transferred to nitrocellulose blots was carried out as previously described (32). Briefly, nitrocellulose blots were incubated for 60 min in a buffer containing $5x10^{-2}M$ tris(hydroxymethyl)aminomethane hydrochloride (Tris·HCl), pH 7.5, $10^{-3}M$ ethylene glycol-bis(ß-aminoethyl ether)-N,N,N',N'-tetraacetic acid (EGTA), $5x10^{-3}M$ $MgCl_2$, 0.3% Tween 20, and 1 μCi [α-^{32}P]GTP/ml. To decrease non-specific binding, the incubation buffer was supplemented with $10^{-3}M$ K_2HPO_4/KH_2PO_4. Blots were washed six times, dried, and autoradiographed for 2-7 days at -70°C using Kodak X-OMAT AR films. The autographs were scanned by a computerized laser densitometer (Ultrascan LKB).

Pulse-chase experiments

Isolated intact cells were preincubated with or without Brefeldin A at the indicated concentration in Eagles "minimum essential medium" (MEM), pulse-labelled for 10 min with [^{3}H]leucine (20 μCi/ml) in Eagles MEM lacking unlabelled leucine and subsequently chased for 120 min at 37°C in complete Eagles MEM supplemented with HEPES 20 mM, glucose $5x10^{-3}M$, BSA 2g/l, and trypsin inhibitor 0.1g/l(13). Isolated cells were permeabilized as indicated with digitonin (5μg/ml) (15,30) and permeabilized cells were chased in KCl-buffer (KCl $12x10^{-3}M$, KH_2PO_4 $1.2x10^{-3}M$, $MgCl_2$ $2x10^{-3}M$, $CaCl_2$ $10^{-4}M$, HEPES $20x10^{-3}M$, glucose $10^{-2}M$, BSA 2g/l, trypsin inhibitor 0.1 g/l, pH 7.0 adjusted with Tris) which was supplemented with MEM amino acids, ATP $10^{-3}M$ and an ATP-regenerating system (8). BFA was present in the medium throughout the pulse-chase period at concentrations as indicated. Total protein synthesis was determined by measuring the tricholoroacetic acid insoluble radioctivity incorporated into the cells during a 10 min pulse.

RESULTS

Effect of CCK-OP on GTP-binding to smg-protein(s)

After stimulation of pancreatic acini with the secretagogue CCK-OP (10^{-11}M) for 30 min at 37°C and subsequent subcellular fractionation, we found decreased [α-^{32}P]GTP-binding to 21 kDa smg-protein(s) in microsomal membranes (MM) as compared with MM from unstimulated cells whereas the [α-^{32}P]GTP-binding to 23 kDa smg-protein(s) were enhanced (Fig.1A).

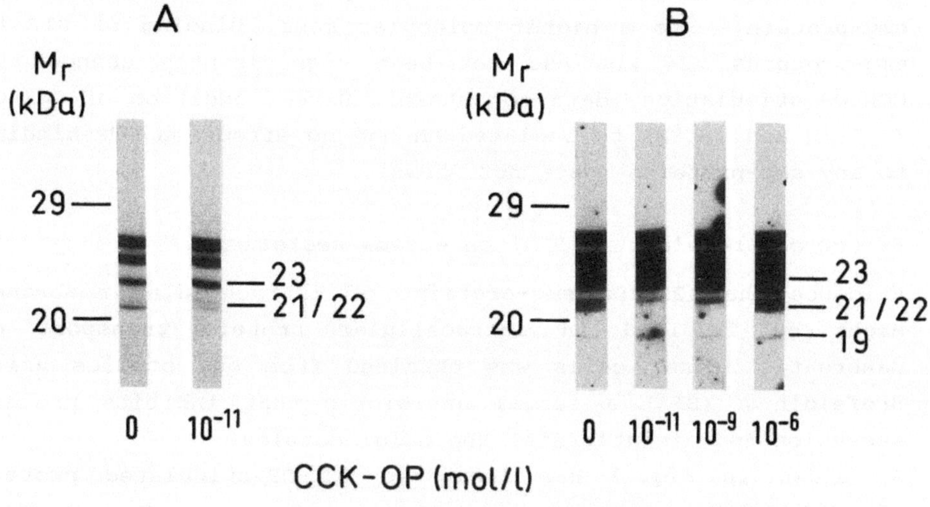

Fig. 1. Effect of different concentrations of CCK-OP on [α-^{32}P]GTP-binding to smg-proteins in MM from rat pancreatic acinar cells. Pancreatic acini were incubated for 30 min at 37° C with or without CCK-OP at different concentrations as indicated and subsequently used for preparation of MM. Equal amounts of membrane protein (100 μg/lane) were separated by SDS-PAGE and transferred to nitrocellulose; smg-proteins were detected by [α-^{32}P]GTP-binding and autoradiography for 3 (A) and 6 (B) days. Experiment shown is representative of 3 separate experiments (from 32).

CCK-OP had no effect on [α-^{32}P]GTP-binding to smg-proteins from membranes enriched in mitochondria or the cytosol (data not shown). The effect of CCK-OP stimulation on [α-^{32}P]GTP-binding to the 21 kDa protein(s) in the MM was dependent on the CCK-OP concentration (Fig. 1B), showing a decrease in [α-^{32}P]GTP-binding with 10^{-11}M CCK-OP and 10^{-9}M CCK-OP as compared to the control. With a high concentration of 10^{-6}M CCK-OP, [α-^{32}P]GTP-binding was again increased as compared to effects of 10^{-11}M and 10^{-9}M CCK-OP. At a CCK-OP concentration of 10^{-14}M, no effect of CCK-OP on [α-^{32}P]GTP-binding to the 21 kDa smg-protein(s) could be observed (data not shown). The decrease in [α-^{32}P]GTP-binding to the 21 kDa smg-protein(s) after stimulation of cells with 10^{-11}M and 10^{-9}M CCK-OP was accompanied by a concomitant increase in [α-^{32}P]GTP-binding to the 23 kDa smg-protein(s) indicating a shift of the 21 kDa smg-protein(s) to a higher molecular mass. Binding of GTP to smg-proteins >24 kDa had not been significantly changed by CCK-OP stimulation (data not shown). Direct addition of CCK-OP (10^{-11}M and 10^{-6}M) to isolated MM had no effect on GTP-binding to any smg-proteins (data not shown).

Effect of Brefeldin A (BFA) on enzyme secretion

Evidence that 21 kDa smg-proteins of intracellular membranes might be involved in intracellular protein transport of pancreatic acinar cells was obtained from our studies using Brefeldin A (BFA) a fungal antibiotic that inhibits protein secretion and disintegrates the Golgi complex.

As shown in Fig. 2 BFA inhibited CCK-OP-stimulated protein secretion from intact acinar cells as a function of chase time after pulse labelling with [^{3}H]leucine for 10 min. After the chase period of 120 min both basal and CCK-OP-stimulated protein release was inhibited by 67±2% and 86±2%, respectively. No significant difference was observed in the protein secretion rate of cells treated with BFA (1μg/ml) alone or cells treated with CCK-OP (10^{-9}M) plus BFA (1 μg/ml)(Fig.2)

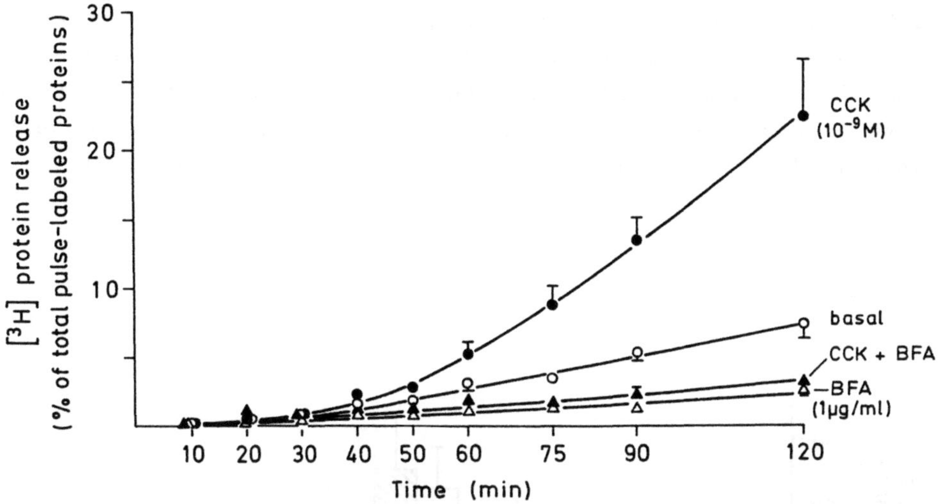

Fig. 2. Effect of BFA (1 µg/ml) on total protein secretion in isolated intact cells as a function of chase time after pulse-labelling with [^{3}H]leucine for 10 min. At the indicated times of chase aliquots were taken, the trichloroacetic acid-insoluble radioativity released into the medium was measured and expressed as percent of total incorporated radioactivity (n=3) (from 30).

To investigate the effects of GTP and GTPγS on intracellular protein transport permeabilized pancreatic acinar cells were used. In the presence of ATP and an ATP-regenerating system CCK-OP (10^{-9}M) and GTP (10^{-3}M) stimulated protein secretion during a 120 min chase period by 72 ± 9% (P<0.001, n=7) and 112 ± 28% (P<0.0001, n=7), respectively. GTPγS (10^{-4}M) had no significant effect on basal protein secretion (P=n.s., n=6)(Fig. 3). However, incubation of CCK-OP-stimulated cells with GTPγS resulted in a significant inhibition of protein secretion by 68 ±6% (P<0.01, n=3)(data not shown). When permeabilized acinar cells were incubated in an ATP-depletion system (no ATP, no ATP-regenerating system, no glucose, oligomycin 10^{-5}M, sodiumazide 10^{-2}M) GTP did not stimulate protein release, in fact basal protein release in the presence of GTP was drastically reduced (Fig. 3). Similar results were obtained in cells incubated with CCK-OP in an ATP-depletion system (data not shown). BFA (30 µg/ml) significantly inhibited basal and CCK-OP-stimulated protein secretion (P<0.05 and P<0.01, respectively), but had no effect in cells incubated with GTP (10^{-3}M) or with GTPγS (10^{-4}M) (Fig. 3).

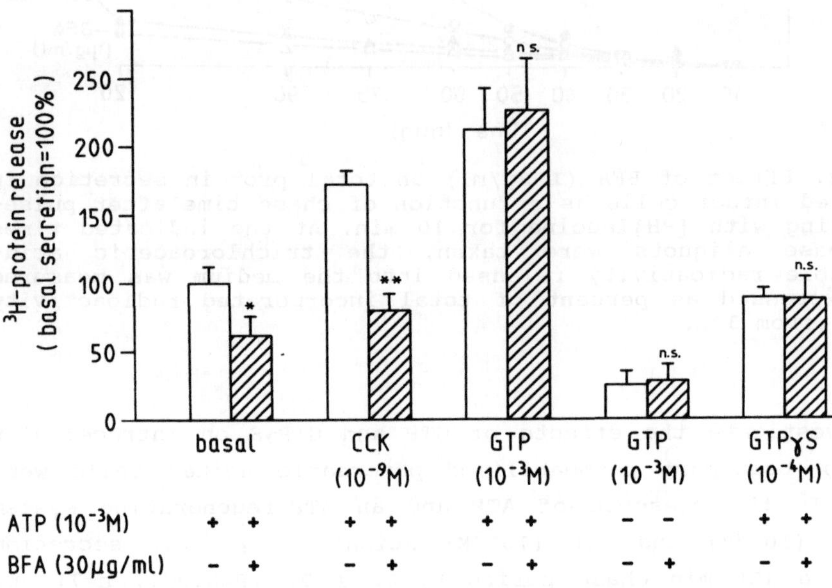

Fig. 3. Pulse-labelled permeabilized acinar cells were incubated with substances as indicated. Total protein release was measured during a 120 min chase. Results are expressed as percentage of basal secretion (= 100 %). Columns represent mean values ± SE of 3-7 experiments. *P < 0.05, **P < 0.01 (from 30).

These results imply that BFA might act on GTP-binding proteins. We therefore investigated the effect of BFA on [α-^{32}P]GTP-binding to the group of smg-proteins, which are suggested to regulate intracellular vesicular transport. Isolated intact acinar cells were incubated for 20 min at 37°C

with and without BFA (1 μg/ml) and subsequently fractionated. As shown in Fig. 4 (lane 1 and 2) no effect of BFA was observed in the total cell homogenate. However, in MM [α-^{32}P]GTP-binding to the 21 kDa smg-protein(s) was enhanced in the presence of BFA (Fig. 4, lane 3 and 4). Previous studies of our group (32) have shown that [α-^{32}P]GTP-binding to these 21 kDa smg-proteins is decreased when cells were preincubated with CCK-OP (10^{-9}M) (Fig. 4, lane 6). This CCK-OP-effect was not observed in isolated acinar cells incubated with both CCK-OP (10^{-9}M) and BFA (1 μg/ml). Indeed [α-^{32}P]GTP-binding to the 21 kDa smg-protein(s) was increased similar as in cells with BFA alone (Fig. 4, lane 5). In permeabilized acinar cells BFA (30 μg/ml) had no effect on [α-^{32}P]GTP-binding to the 21 kDa smg-protein(s) when GTP (10^{-3}M) or GTPγS (10^{-4}M) were present (data not shown).
Furthermore, we previously observed that vesicular membrane binding of the 21 kDa smg-protein(s) is regulated by intravesicular pH established by a vacuolar-type H^+-ATPase (31). However, since BFA has no effect on the vacuolar-type H^+-ATPase of pancreatic acinar cells (data not shown), the effect of BFA on the 21 kDa smg-protein(s) can not be explained by a change in intravesicular acidification.

DISCUSSION

Effect of CCK-OP on the 21 kDa smg-protein(s)

The decrease in GTP-binding to the 21 kDa smg-protein(s) and the increase in GTP-binding to 23 kDa smg-protein(s) in microsomal membranes could indicate that stimulation with CCK-OP results in phosphorylation of 21 kDa smg-protein(s) mediated by CCK-OP-induced activation of phospholipase C followed by an increase in diacylglycerol and consequent stimulation of proteinkinase C. This would shift the electrophoretic mobility of the 21 kDa smg-protein(s) to a higher (23 kDa) molecular mass area. Phosphorylation of smg-proteins (ras, rap 1B) by PKC and PKA has already been shown (14,16,17). Furthermore, it has been reported that phosphorylation of a 22 kDa smg-protein (termed rap 1/smg p21/krev 1) by PKA was associated with the translocation from membrane to cytosolic fractions in human platelets and with a simultaneous shift of its mobility to the higher molecular mass of 24 kDa in SDS-PAGE (16).

The bell-shaped concentration dependence of CCK-OP on [α-^{32}P]GTP-binding to the 21 kDa smg-protein(s) (see Fig. 1) is reminiscent to a similar bell-shaped CCK-OP-response curve in enzyme secretion (11) and could indicate different activation of secretory pathways by high- and low-affinity CCK-receptors, respectively (11).

Effect of Brefeldin A on the 21 kDa smg-protein(s) and intracellular protein transport

We have demonstrated in this study that BFA inhibits basal and CCK-OP-stimulated protein secretion in rat pancreatic acinar cells. About 10-fold higher concentrations of BFA were needed to inhibit protein release from isolated permeabilized cells as compared to intact cells, a phenomenon already described previously (20). These results suggest that BFA might be concentrated in isolated intact cells by transport processes.

In the secretory pathway GTP-hydrolysis is required for vesicular transport between the ER and the Golgi complex, between different Golgi compartments and between the Golgi complex and the plasma membrane. The weakly hydrolyzable GTP-analog GTPγS inhibits vesicular transport between ER and Golgi complex as well as between Golgi compartments (for review see 1). In contrast, GTPγS increases secretion (exocytosis) of proteins which are already stored in secretory granules by stimulation of heterotrimeric G-proteins. GTP has little or no effect on secretion (exocytosis) of proteins which are already stored in secretory granules (28).

Using the pulse chase technique which monitors vesicular transport from ER to final secretion (exocytosis included), we could show that GTP (10^{-3}M) stimulates this event in permeabilized cells (Fig. 3). In our experiments in which permeabilized cells were metabolically inhibited, no effect of GTP was observed. This implies an ATP-dependent GTP effect on intracellular protein transport. Furthermore GTPγS did not stimulate protein transport in our pulse chase experiments, supporting the importance of GTP hydrolysis for protein transport from ER to the site of exocytosis. The stimulation of intracellular protein transport by GTP seems to be due to activation of smg-proteins. According to a model proposed by Bourne (4) smg-proteins direct vesicles to specific target organelles in a GTP-bound active state. GTP hydrolysis would be required to trigger a conformational change and the inactive GDP form of the protein would then return to the cytosol to target another vesicle in a cyclic process (4). BFA blocks vesicular transport from the ER to the Golgi complex and enhances [α-^{32}P]GTP-binding to the 21 kDa smg-protein(s) (Fig. 4). This suggests that one or more 21 kDa smg-protein(s) might be involved in the vesicular transport between these two compartments and that BFA might act on one or more of these 21 kDa smg-proteins which subsequently leads to inhibition of vesicular transport and accumulation of these proteins in the ER membrane.

BFA failed to inhibit GTP-stimulated protein transport indicating that in the GTP-bound state the protein might have taken a conformation which was not attackable by BFA. Similarily in the presence of GTPγS basal protein release was not inhibited by BFA. Furthermore, no effect of BFA on [α-^{32}P]GTP-binding to the 21 kDa smg-protein(s) was observed when permeabilized cells were incubated with GTP (10^{-3}M) or GTPγS (10^{-4}M).

Taken together evidence is accumulating that BFA acts on the 21 kDa smg-protein(s) in the ER, as a molecular target. If this are the same 21 kDa smg-proteins, which are also modified by stimulation of pancreatic acinar cells with CCK-OP, remains to be investigated.

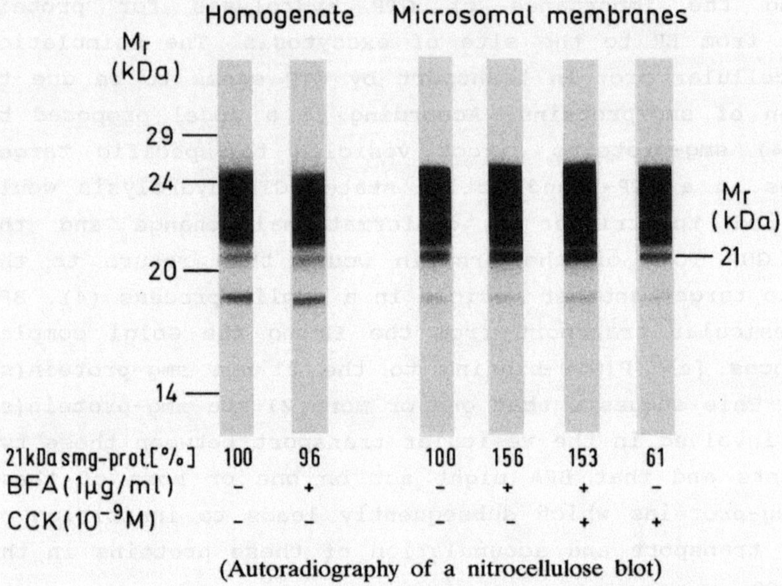

Fig.4. Effect of BFA (1μg/ml) and CCK-OP (10^{-9}M) on [α-^{32}P]GTP-binding to smg-proteins. Isolated cells were incubated for 20 min with and without BFA and CCK-OP, respectively. Homogenization, subcellular fractionation and the [α-^{32}P]GTP-binding assay were performed as described in the "Methods". [α-^{32}P]GTP-binding to smg-proteins of the total cell homogenate and the microsomal membranes are shown. The protein bands of the 21 kDa smg-proteins were scanned by laser densitometry and the respective density expressed as percent of control (= 100 %). The results are representative for five separate experiments (from 30).

REFERENCES

1. Balch WE (1989) Biochemistry of interorganelle transport. J Biol Chem. 264: 16965-16968

2. Bayerdörffer E, Eckhardt L, Haase W, Schulz I (1984) Characterization of calcium uptake into rough endoplasmic reticulum of rat pancreas. J Membr Biol 81:69-82

3. Bieger W, Martin-Achard A, Bassler M, Kern HF (1976) Studies on intracellular transport of secretory proteins in the rat exocrine pancreas. Cell Tissue Res 165:435-453

4. Bourne HR (1988) Do GTPases direct membrane traffic in secretion? Cell 53:669-671

5. Burnham DB, Williams JA (1984) Activation of protein kinase activity in pancreatic acini by calcium and cAMP. Am J Physiol 246 (Gastrointest Liver Physiol 9) G500-G508

6. Casey PJ, Gilman AG (1988) G protein involvement in receptor-effector coupling. J Biol Chem 263:2577-2580

7. Castagna M, Takai M, Kaibushi K, Sano K, Kikkawa U, and Nishisuka Y (1982) Direct activation of calcium-activated, phospholipid-dependent protein kinase by tumor-promoting phorbol esters. J Biol Chem 257:7847-7851

8. Dehlinger-Kremer M, Zeuzem S, Schulz I (1991) Interaction of caffeine-, IP_3- and vanadate-sensitive Ca^{2+} pools in acinar cells of the exocrine pancreas. J Membrane Biol 119:85-100

9. Doms RW, Russ G, Yewdell JW (1989) Brefeldin A redistributes resident and itinerante Golgi proteins to the endoplasmic reticulum. J Cell Biol 109:61-72

10. Fischer von Mullard G, Südhoff TC, Jahn R (1991) A small GTP-binding protein dissociates from synaptic vesicles during exocytosis. Nature Lond 349:79-81

11. Gardner JP, Jensen RT (1987) Regulation of pancreatic enzyme secretion in vitro. In Physiology of the Gastrointestinal Tract. Johnson LR (ed) New York Raven 1109-1127

12. Hall A (1990) The cellular functions of small GTP-binding proteins. Sience Wash DC 249:635-640

13. Jamieson JD, Palade GE (1971) Synthesis, intracellular transport, and discharge of secretory proteins in stimulated pancreatic exocrine cells. J Cell Biol 50:135-158

14. Jeng AY, Srivastava SK, Lacal JC, Blumberg PM (1987) Phosphorylation of RAS oncogene product by protein kinase C. Biochem Biophys Res Commun 145:782-788

15. Kimura T, Imamura K, Eckhardt L, Schulz I (1986) Ca^{2+}-, phorbol ester-, and cAMP-stimulated enzyme secretion from permeabilized rat pancreatic acini. Am J Physiol 250 (Gastrointest Liver Physiol 13):G698-G708

16. Lapetina EG, Lacal JC, Reep BR, Vedina LMY (1989) A ras-related protein is phosphorylated and translocated by agonists that increase cAMP levels in human platelets. Proc Natl Acad Sci USA 86:3131-3134

17. Lazarowski E, Lacal JC, Lapetina E (1989) Agonist-induced phosphorylation of an immunologically ras-related protein in human erythroleukemia cells. Biochem Biophys Res Commun 161:972-978

18. Lippincott-Schwartz J, Yuan LC, Bonifacino JS, Klausner RD (1989) Rapid redistribution of Golgi proteins into the ER in cells treated with brefeldin A: evidence for membrane cycling from the Golgi to ER. Cell 56:801-813

19. Matozaki T, Wiliams JA (1989) Multiple sources of 1,2-diacylglycerol in isolated rat pancreatic acini stimulated by cholecystokinin. J Biol Chem 264:14729-14734

20. Orci L, Tagaya M, Amherdt M, Perrelet A, Donaldson JG, Lippincott-Schwartz J, Klausner RD, Rothman JE (1991) Brefeldin A, a drug that blocks secretion, prevents the assembly of non-clathrin-coated beeds on Golgi cisternae. Cell 64:1183-1195

21. Pröfrock A, Zimmermann P, Schulz I (1992) Bombesin receptors interact with G_i and $p21^{ras}$ proteins in plasma membranes from rat pancreatic acinar cells. Am J Physiol 263 (Gastrintest Liver Physiol):G240-G247

22. Schnefel S, Zimmermann P, Pröfrock A, Jahn R, Aktories K, Zeuzem S, Haase W, Schulz I (1992) High- and small-molecular weight GTP-binding proteins in zymogen granule membranes of rat pancreatic acinar cells. Cell Physiol Biochem 2:77-89

23. Schulz I (1986) Signaling transduction in hormone- and neurotransmitter-induced enzyme secretion from the exocrine pancreas. In Handbook of Physiology, The Gastrointestinal System. Salivary, Gastric, Pancreatic, and Hepatobilary Secretion. Bethesda, MD: Am. Physiol. Soc. Vol III Chapt 22:443-463

24. Schnefel S, Banfic H, Eckhardt L, Schultz G, Schulz I (1988) Acetylcholine and cholecystokinin receptors functionally couple by different G Proteins to phospholipase C in pancreatic acinar cells. FEBS Lett 230:125-130

25. Stearns T, Willingham MC, Bottstein D, Kahn RA (1990) ADP-ribosylation factor is functionally and physically associated with the Golgi complex. Proc Natl Acad Sci USA 87:1238-1242

26. Streb H, Irvine RF, Berridge MJ, Schulz I (1983) Release from Ca^{2+} from a nonmitrochondrial intracellular store in pancreatic acinar cells by inositol 1,4,5-triphosphate. Nature Lond 306:67-69

27. Streb H, Schulz I (1983) Regulation of cytosolic free Ca^{2+} concentration in acinar cells of rat pancreas. Am J Physiol 245:G347-G357

28. Stutchfield J, Cockcroft S (1988) Guanine nucleotides stimulate polyphosphoinositide phosphodiesterase and exocytic secretion from HL 60 cells permeabilized with streptomycin. Biochem J 250:375-382

29. Thèvenod F, Kemmer TP, Christian AL, Schulz I (1989) Characterization of MgATP-driven H^+ uptake into microsomal vesicle fraction from pancreatic acinar cells. J Membr Biol 107:263-275

30. Zeuzem S, Zimmermann P, Schulz I (1991) Guanosine nucleotides modulate the inhibitory effect of Brefeldin A on protein secretion. FEBS Lett 288: 143-146

31. Zeuzem S, Feick P, Zimmermann P, Haase W, Kahn RA, Schulz I (1992) Intravesicular acidification correlates with binding of ADP-ribosylation factor to microsomal membranes. Proc Natl Acad Sci (Cell Biology) 89:6619-6623

32. Zimmermann P, Schnefel S, Zeuzem S, Pröfrock A, Haase W, Schulz I (1992) Effects of agonists on $p21^{ras}$ and ras-related proteins in rat pancreatic acinar cells. Am J Physiol 263 (Gastrointest Liver Physiol): G396-G406

CHARACTERIZATION OF CALCIUM TRANSPORT ATPase ISOZYMES IN DIFFERENT BLOOD CELLS

Magócsi M, Penniston J.T.*, and Gárdos G.
National Institute of Haematology
and Blood Transfusion,
Budapest, Daróczi u. 24., H-1113, Hungary
* Mayo Clinic, Rochester, MN, USA

INTRODUCTION

To maintain very low intracellular free Ca^{2+} concentration is essential for all eukaryotic cells. The P-type ATPase, responsible for the active transport of Ca^{2+} through the plasma membrane (PMCA) has been studied and characterized in red blood cells (see Carafoli et al., 1987). By screening cDNA libraries of various species at least five different mammalian isozymes encoded by separate genes have been discovered. The different gene products within a given species have only 76 to 86% identity in their primary sequences. There are five particularly high sequence divergences between the isozymes (Strehler, 1991), including the putative acidic phospholipid sensitive region and the C terminal regulatory domain that contains the calmodulin binding site and the cAMP dependent phosphorylation site. This C terminal part of the molecule contains two of the four domains where different alternative splicings are also indicated (see Carafoli, 1992). The co-expression of different PMCAs in the same tissue make detailed studies of the molecular properties of these enzymes particularly difficult. Based on PCR technology we worked out a method that makes possible to follow the expression of multiple PMCA mRNAs in different cell types. Any cell line which expresses only one type of the three different human isozymes may serve as a suitable model to study the functional properties in the proper membrane environment. The present study is concerned with the physiological significance of the PMCA isozyme diversity in different blood cells and the characteristic change of the isozyme distribution during the cell maturation and differentiation.

MATERIALS AND METHODS

Preparation of total RNA: DNA free total RNA was prepared from different cell lines according to the method of Chirgwin et al. 1979.
Reverse transcription (RT) was performed by using a Gene Amp RNA PCR kit (Perkin Elmer Cetus) with random hexamer primers from 1 μg of total RNA.
Polimerase chain reaction (PCR) was performed with 100 μg of each isozyme-specific primer. The exact location of the pairs of primers in the mRNA sequences were as follows: 2143-2167 and 2860-2879 for isozyme hPMCA1; 2394-2415 and 2855-2882 for isozyme hPMCA2; and 2331-2357 and 3260-3281 for isozyme hPMCA4. Each primer had <63% identity with any part of the known mRNA sequence of the other two isozymes. The amplification, the analysis of PCR product and Southern blotting were performed as described by Magócsi et al., 1992. For high-stringency hybridization 10 pmol ^{32}P-labelled

NATO ASI Series, Vol. H 76
Tyrosine Phosphorylation/Dephosphorylation and Downstream Signalling
Edited by L. M. G. Heilmeyer Jr.

oligonucleotide probes were used. The exact location for the probes in mRNA sequences of the three hPMCA isozymes were as follows: 2191-2290 for isozyme hPMCA1; 2394-2493 for isozyme hPMCA2; and 2863-3962 for isozyme hPMCA4. The exposure time of the autoradiography was 24 hours.

RESULTS AND DISCUSSION

Different human leukemic cell lines were used for studying the isozyme distribution of PMCA. These cell lines represented T and B lymphoblast, myeloblast, megakaryoblast and erythroid cells. In Fig.1 we show a representative pattern of the analysis of RT-PCR amplified DNA products for mRNAs of human PMCA isozymes. The RT-PCR products were transblotted and probed by hybridization of oligonucleotides selective for the three different PMCA isozyme's PCR products with high stringency. The 736-bp RT-PCR product in the autoradiogram of Fig 1/a confirmed the identity of this DNA sequence as a part of isozyme hPMCA1, whereas the 488-bp DNA product of RT-PCR is evidently an integral part of isozyme hPMCA2 (Fig 1/b). The 950-bp DNA sequence belongs to the gene coding isozyme hPMCA4 (Fig 1/c). Each experiment was performed at least three times and representative results are presented.

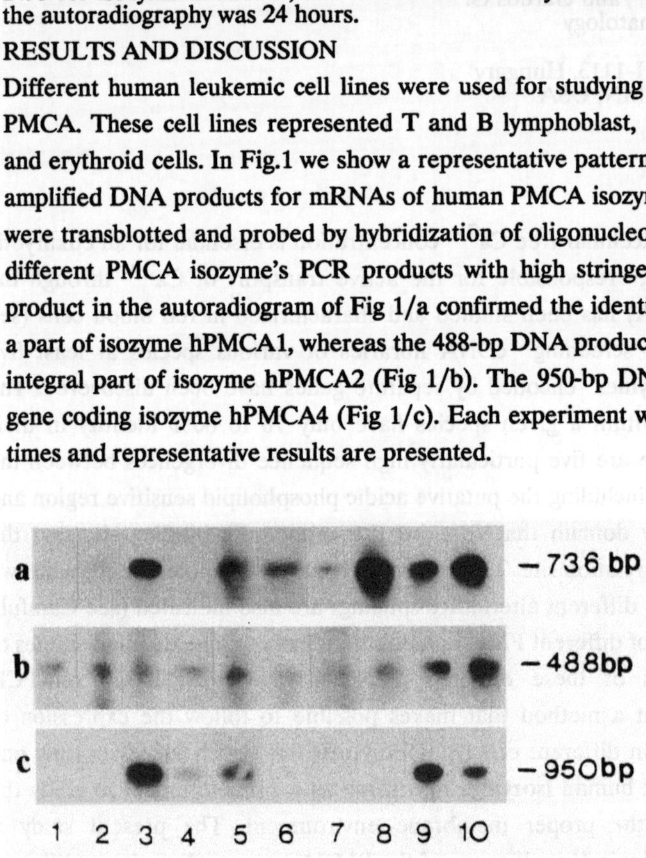

Fig.1 Autoradiogram of the Southern blot of RT-PCR products of mRNAs coding for hPMCA1 (Panel A), hPMCA2 (Panel B) and hPMCA4 (Panel C). The order of the cell lines in each panel is the following Molt4 (1), Yure (2), Jurkat (3), Raji (4), HL60 (5), K-11 (6), lymphocytes from AML patient (7), Dami (8), Meg (9) and HEL (10).

Since the RT-PCR analysis was not quantitated, even a semiquantitative comparison of relative mRNA amounts between different cells should be viewed with caution. The results show that mRNA coding for at least one of the hPMCA isozymes is present virtually in all leukemic blood cell lines examined. Shull and Greeb (1988) suggested that PMCA1 has a general housekeeping role because this isozyme was found in nearly the same amount of all studied tissues by Northern blot. By using homogeneous cells, we found two cell lines, Molt4 and Raji, where no mRNA for the hPMCA1 isozyme were detectable but, interestingly, hPMCA2 seems to be the most typical isozyme of those cells. A major role of the hPMCA2 isozyme was found in tubular epithelium cells too, when microdissected nephron segments were studied in rat kidney (Magócsi et al., 1992). In blood cells the diversity of hPMCA4 isozyme, in which the cAMP dependent phosphorylation site is missing, is even more pronounced than that of hPMCA1 (Fig 1/c). Its presence in Jurkat lymphoblast or Meg, HEL megakaryoblast is evident, but other lymphoblastic cells as Molt4 and Yure, or Dami megakaryoblast have no detectable mRNA coding for hPMCA4. Presently, we can conclude that leukemic blood cells significantly differ in their mRNA expression coding for different hPMCA isozymes, even though the exact physiological role of the different isozymes is still unknown. Molt4 cells, in which only the mRNA coding for hPMCA2 could be identified seem to be suitable models for these kind of studies.

Extensive peptide sequencing efforts made on purified human erythrocyte Ca^{2+} pump protein suggest that two different isozymes of this enzyme are present in human red blood cells. (Verma et al., 1988, Strehler et al., 1990), namely the hPMCA1 and hPMCA4. These studies could not identify any peptide sequence corresponding to hPMCA2. Due to the absence of mRNA in mature red blood cells, the study of the Ca pump isoforms in erythrocytes was undertaken in an erythroleukemic cell line (K562) that can be differentiated to red blood cell. We treated the K562 cells with 20uM tiazofurin for 24 hours and the increase of haemoglobin synthesis was used as differentiation marker. Fig.2 shows the autoradiogram of RT-PCR products after Southern hybridization for detecting mRNAs coding for PMCA isozymes in undifferentiated and differentiated K562 cells (one from 3 independent experiments).

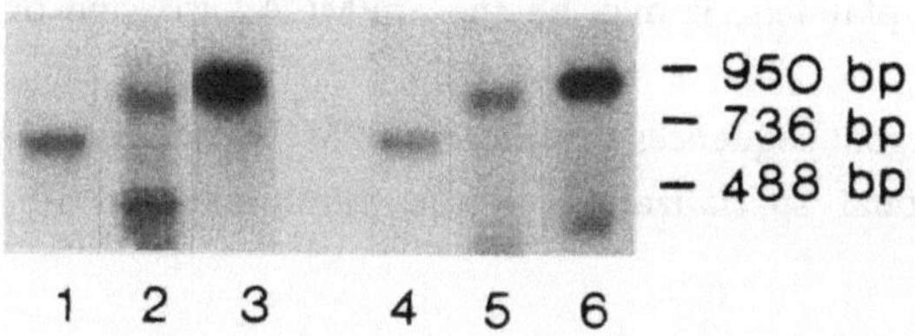

Fig.2 Change in PMCA isozyme distribution as a result of differentiation of K562 (lanes 4, 5, and 6) compared to the undifferentiated cells (lanes 1, 2, and 3). Autoradiogram of the Southern blot RT-PCR reaction with specific primers and a probe for hPMCA1 (lanes 1 and 4), for hPMCA2 (lanes 2 and 5) and for hPMCA4 (lanes 3 and 6).

The 488-bp band, representing the mRNA coding for hPMCA2, disappeared from the differentiated cells, while hPMCA1 and hPMCA4 remained intact in accordance with the protein analysis data.

Membrane Ca^{2+} transporting proteins play a major role in the regulation of the signalling function of Ca^{2+}. In platelet, the endoplasmic reticulum Ca^{2+} pumps (SERCA) refilling the intracellular Ca^{2+} storages, are extensively studied. However, the presence and physiological importance of a plasma membrane type Ca^{2+} pump in platelets is controversial and unproved (Papp et al., 1991, Enouf et al., 1987, Enyedi et al., 1986).

In order to look for PMCA(s) in platelets, we studied the expression of mRNAs coding for the three previously described Ca^{2+} pump isoforms in the platelet precursor megakaryoblasts. Fig. 3 shows the hPMCA isozyme distribution in megakaryoblasts as compared to the distribution of the isozymes in K562 cells and in human platelets.

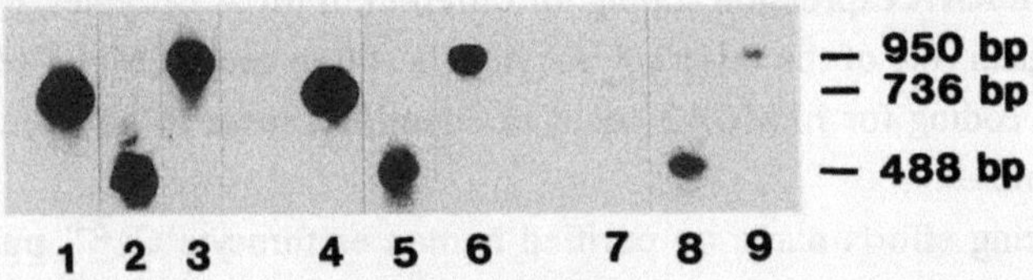

Fig.3 Identification of hPMCA mRNAs in K562 cells (lanes 1, 2 and 3), megakaryoblasts (lanes 4, 5 and 6) and platelets (lanes 7, 8 and 9). Autoradiogram of the Southern blot after RT-PCR reaction with specific primers and a probe for hPMCA1 (lanes 1, 4 and 7), for hPMCA2 (lanes 2, 5 and 8) and for hPMCA4 (lanes 3, 6 and 9).

Because of the lack of a significant amount of mRNA in mature platelets, and since the RT-PCR method can pick up mRNAs originated from less than 1% contaminating blood cells e.g. lymphocytes or monocytes, these latter results must be viewed with caution.

Our present conclusion is that the presence of a plasma membrane type Ca pump in platelet can not be ruled out, because the PMCA coding mRNAs expressed in the precursor cells are similar to those expressed in the erythroid cells. Furthermore, if there is such a pump protein in mature platelets, it may be the hPMCA2 isozyme or its alternatively spliced isoform.

All these data suggest that the divergent sequences of different hPMCA isozymes might serve as a regulatory and functional specialization of the pump adapted to the physiological needs of the cell.

REFERENCES

Carafoli E (1987) Ann. Rev. Biochem 56:395-433
Carafoli E (1992) J.B.C. 267:2115-2118
Chirgwin JM, Przybula AE, McDonald RY, Rutter W (1979) Biochemistry 18:5294-5299
Enouf J, Bredoux B, Bourdeau N, Lévy-Toledano S (1987) J. Biol. Chem. 262:9293-9297
Enyedi Å, Sarkadi B, Földes-Papp Z, Monostory S., Gárdos G (1986) J. Biol. Chem. 261:9558-9564
Magócsi M, Yamaki M, Penniston JT, Dousa TP (1992) Am. J. Physiol. 263:F7-F14
Papp B, Enyedi Å, Kovács T, Sarkadi B, Wuytack F, Thastrup O, Gárdos G, Bredoux R, Lévy-Toledano S, Enouf J (1991) J. Biol. Chem. 266:14593-14596
Shull GE, Greeb J (1988) J. Biol. Chem. 263:8646
Strehler EE, James P, Fischer R, Heim R, Vorherr T, Filoteo AG, Penniston JT, Carafoli E (1990) J. Biol. Chem. 265:2835-2842
Strehler EE (1991) J. Membrane Biol. 120:1-15
Verma AK, Filoteo AG, Stanford DR, Wieben ED, Penniston JT, Strehler EE, Fischer R, Heim R, Vogel G, Matthews S, Strehler-Page MA, James P, Vorherr T, Krebs J, Carafoli E (1988) J. Biol. Chem. 263:14152-14159

INDEX

NATO ASI Series H

Vol. 21: **Cellular and Molecular Basis of Synaptic Transmission.**
Edited by H. Zimmermann. 547 pages. 1988.

Vol. 22: **Neural Development and Regeneration. Cellular and Molecular Aspects.**
Edited by A. Gorio, J. R. Perez-Polo, J. de Vellis, and B. Haber. 711 pages. 1988.

Vol. 23: **The Semiotics of Cellular Communication in the Immune System.**
Edited by E.E. Sercarz, F. Celada, N.A. Mitchison, and T. Tada. 326 pages. 1988.

Vol. 24: **Bacteria, Complement and the Phagocytic Cell.**
Edited by F. C. Cabello und C. Pruzzo. 372 pages. 1988.

Vol. 25: **Nicotinic Acetylcholine Receptors in the Nervous System.**
Edited by F. Clementi, C. Gotti, and E. Sher. 424 pages. 1988.

Vol. 26: **Cell to Cell Signals in Mammalian Development.**
Edited by S.W de Laat, J.G. Bluemink, and C.L. Mummery. 322 pages. 1989.

Vol. 27: **Phytotoxins and Plant Pathogenesis.**
Edited by A. Graniti, R. D. Durbin, and A. Ballio. 508 pages. 1989.

Vol. 28: **Vascular Wilt Diseases of Plants. Basic Studies and Control.**
Edited by E. C. Tjamos and C. H. Beckman. 590 pages. 1989.

Vol. 29: **Receptors, Membrane Transport and Signal Transduction.**
Edited by A. E. Evangelopoulos, J. P. Changeux, L. Packer, T. G. Sotiroudis, and K.W.A. Wirtz. 387 pages. 1989.

Vol. 30: **Effects of Mineral Dusts on Cells.**
Edited by B.T. Mossman and R.O. Begin. 470 pages. 1989.

Vol. 31: **Neurobiology of the Inner Retina.**
Edited by R. Weiler and N.N. Osborne. 529 pages. 1989.

Vol. 32: **Molecular Biology of Neuroreceptors and Ion Channels.**
Edited by A. Maelicke. 675 pages. 1989.

Vol. 33: **Regulatory Mechanisms of Neuron to Vessel Communication in Brain.**
Edited by F. Battaini, S. Govoni, M.S. Magnoni, and M. Trabucchi. 416 pages. 1989.

Vol. 34: **Vectors asTools for the Study of Normal and Abnormal Growth and Differentiation.**
Edited by H. Lother, R. Dernick, and W. Ostertag. 477 pages. 1989.

Vol. 35: **Cell Separation in Plants: Physiology, Biochemistry and Molecular Biology.** Edited by D. J. Osborne and M. B. Jackson. 449 pages. 1989.

Vol. 36: **Signal Molecules in Plants and Plant-Microbe Interactions.**
Edited by B.J.J. Lugtenberg. 425 pages. 1989.

Vol. 37: **Tin-Based Antitumour Drugs.** Edited by M. Gielen. 226 pages. 1990.

NATO ASI Series H

Vol. 38: **The Molecular Biology of Autoimmune Disease.** Edited by A.G. Demaine, J-P. Banga, and A.M. McGregor. 404 pages. 1990.

Vol. 39: **Chemosensory Information Processing.** Edited by D. Schild. 403 pages. 1990.

Vol. 40: **Dynamics and Biogenesis of Membranes.** Edited by J. A. F. Op den Kamp. 367 pages. 1990.

Vol. 41: **Recognition and Response in Plant-Virus Interactions.** Edited by R. S. S. Fraser. 467 pages. 1990.

Vol. 42: **Biomechanics of Active Movement and Deformation of Cells.** Edited by N. Akkas. 524 pages. 1990.

Vol. 43: **Cellular and Molecular Biology of Myelination.** Edited by G. Jeserich, H. H. Althaus, and T. V. Waehneldt. 565 pages. 1990.

Vol. 44: **Activation and Desensitization of Transducing Pathways.** Edited by T. M. Konijn, M. D. Houslay, and P. J. M. Van Haastert. 336 pages. 1990.

Vol. 45: **Mechanism of Fertilization: Plants to Humans.** Edited by B. Dale. 710 pages. 1990.

Vol .46: **Parallels in Cell to Cell Junctions in Plants and Animals.** Edited by A. W Robards, W. J . Lucas, J. D. Pitts, H . J . Jongsma, and D. C. Spray. 296 pages. 1990.

Vol. 47: **Signal Perception and Transduction in Higher Plants.** Edited by R. Ranjeva and A. M. Boudet. 357 pages. 1990.

Vol. 48: **Calcium Transport and Intracellular Calcium Homeostasis.** Edited by D. Pansu and F. Bronner. 456 pages. 1990.

Vol. 49: **Post-Transcriptional Control of Gene Expression.** Edited by J. E. G. McCarthy and M. F. Tuite. 671 pages. 1990.

Vol. 50: **Phytochrome Properties and Biological Action.** Edited by B. Thomas and C. B. Johnson. 337 pages. 1991.

Vol. 51: **Cell to Cell Signals in Plants and Animals.** Edited by V. Neuhoff and J. Friend. 404 pages. 1991.

Vol. 52: **Biological Signal Transduction.** Edited by E. M . Ross and K . W. A. Wirtz. 560 pages. 1991.

Vol. 53: **Fungal Cell Wall and Immune Response.** Edited by J. P. Latge and D. Boucias. 472 pages. 1991.

Vol. 54: **The Early Effects of Radiation on DNA.** Edited by E. M. Fielden and P. O'Neill. 448 pages. 1991.

Vol. 55: **The Translational Apparatus of Photosynthetic Organelles.** Edited by R. Mache, E. Stutz, and A. R. Subramanian. 260 pages. 1991.

NATO ASI Series H

Vol. 56: **Cellular Regulation by Protein Phosphorylation.**
Edited by L. M. G. Heilmeyer, Jr. 520 pages. 1991.

Vol. 57: **Molecular Techniques in Taxonomy.**
Edited by G . M . Hewitt, A. W. B. Johnston, and J. P. W. Young .
420 pages. 1991.

Vol. 58: **Neurocytochemical Methods.**
Edited by A. Calas and D. Eugene. 352 pages. 1991.

Vol. 59: **Molecular Evolution of the Major Histocompatibility Complex.**
Edited by J. Klein and D. Klein. 522 pages. 1991.

Vol. 60: **Intracellular Regulation of Ion Channels.**
Edited by M. Morad and Z. Agus. 261 pages. 1992.

Vol. 61: **Prader-Willi Syndrome and Other Chromosome 15q Deletion Disorders.**
Edited by S. B. Cassidy. 277 pages. 1992.

Vol. 62: **Endocytosis. From Cell Biology to Health, Disease and Therapie.**
Edited by P. J. Courtoy. 547 pages. 1992.

Vol. 63: **Dynamics of Membrane Assembly.**
Edited by J. A. F. Op den Kamp. 402 pages. 1992.

Vol. 64: **Mechanics of Swelling. From Clays to Living Cells and Tissues.**
Edited by T. K. Karalis. 802 pages. 1992.

Vol. 65: **Bacteriocins, Microcins and Lantibiotics.**
Edited by R. James, C. Lazdunski, and F. Pattus. 530 pages. 1992.

Vol. 66: **Theoretical and Experimental Insights into Immunology.**
Edited by A. S. Perelson and G. Weisbuch. 497 pages. 1992.

Vol. 67: **Flow Cytometry. New Developments.**
Edited by A. Jacquemin-Sablon. 1993.

Vol. 68: **Biomarkers. Research and Application in the Assessment of Environmental Health.** Edited by D. B. Peakall and L. R. Shugart.
138 pages. 1993.

Vol. 69: **Molecular Biology and its Application to Medical Mycology.**
Edited by B. Maresca, G. S. Kobayashi, and H. Yamaguchi. 271 pages.
1993.

Vol. 70: **Phospholipids and Signal Transmission.**
Edited by R. Massarelli, L. A. Horrocks, J. N. Kanfer, and K. Löffelholz.
448 pages. 1993.

Vol. 71: **Protein Synthesis and Targeting in Yeast.**
Edited by A. J. P. Brown, M. F. Tuite, and J. E. G. McCarthy. 425 pages.
1993.

Vol. 72: **Chromosome Segregation and Aneuploidy.**
Edited by B. K. Vig. 425 pages. 1993.

NATO ASI Series H

Vol. 73: **Human Apolipoprotein Mutants III. In Diagnosis and Treatment.** Edited by C. R. Sirtori, G. Franceschini, B. H. Brewer Jr. 302 pages. 1993.

Vol. 74: **Molecular Mechanisms of Membrane Traffic.** Edited by D. J. Morré, K. E. Howell, and J. J. M. Bergeron. 429 pages. 1993.

Vol. 75: **Cancer Therapy. Differentiation, Immunomodulation and Angiogenesis.** Edited by N. D'Alessandro, E. Mihich, L. Rausa, H. Tapiero, and T. R.Tritton. 299 pages. 1993.

Vol. 76: **Tyrosine Phosphorylation/Dephosphorylation and Downstream Signalling.** Edited by L. M. G. Heilmeyer Jr. 388 pages. 1993.

Vol. 77: **Ataxia-Telangiectasia.** Edited by R. A. Gatti, R. B. Painter. 306 pages. 1993.

MIX
Papier aus verantwortungsvollen Quellen
Paper from responsible sources
FSC® C105338

If you have any concerns about our products,
you can contact us on
ProductSafety@springernature.com

In case Publisher is established outside the EU,
the EU authorized representative is:
Springer Nature Customer Service Center GmbH
Europaplatz 3, 69115 Heidelberg, Germany

Printed by Libri Plureos GmbH
in Hamburg, Germany